GLOBAL STUDIES

CHINA

FOURTEENTH EDITION

Dr. Zhiqun Zhu
Bucknell University

Christopher J. Sutton
Western Illinois University

OTHER BOOKS IN THE GLOBAL STUDIES SERIES

- Africa
- Europe
- Islam and the Muslim World
- India and South Asia
- Japan and the Pacific Rim
- Latin America
- The Middle East
- Russia, the Baltics and Eurasian Republics, and Central/Eastern Europe

The McGraw·Hill Companies

GLOBAL STUDIES: CHINA, FOURTEENTH EDITION

Published by McGraw-Hill, a business unit of The McGraw-Hill Companies, Inc., 1221 Avenue of the Americas, New York, NY 10020.

Global Studies is published by the **Contemporary Learning Series** group within the McGraw-Hill Higher Education division.

1 2 3 4 5 6 7 8 9 0 QDB/QDB 1 0 9 8 7 6 5 4 3 2 1

ISBN 978-0-07-802619-5
MHID 0-07-802619-9
ISSN 1050-2025

Managing Editor: *Larry Loeppke*
Senior Developmental Editor: *Jill Meloy*
Senior Permissions Coordinator: *Lenny J. Behnke*
Senior Marketing Communications Specialist: *Mary Klein*
Senior Project Manager: *Jane Mohr*
Design Coordinator: *Brenda A. Rolwes*
Cover Graphics: *Rick D. Noel*

Compositor: Laserwords Private Limited
Cover Image: © Getty Images RF

www.mhhe.com

Academic Advisory Board

Member of the Academic Advisory Board are instrumental in the final selection of articles for each edition of *Global Studies.* Their review of articles, content, level, and appropriateness provides critical direction to the editors and staff. We think that you will find their careful consideration well reflected in this volume.

ACADEMIC ADVISORY BOARD MEMBERS

CHINA

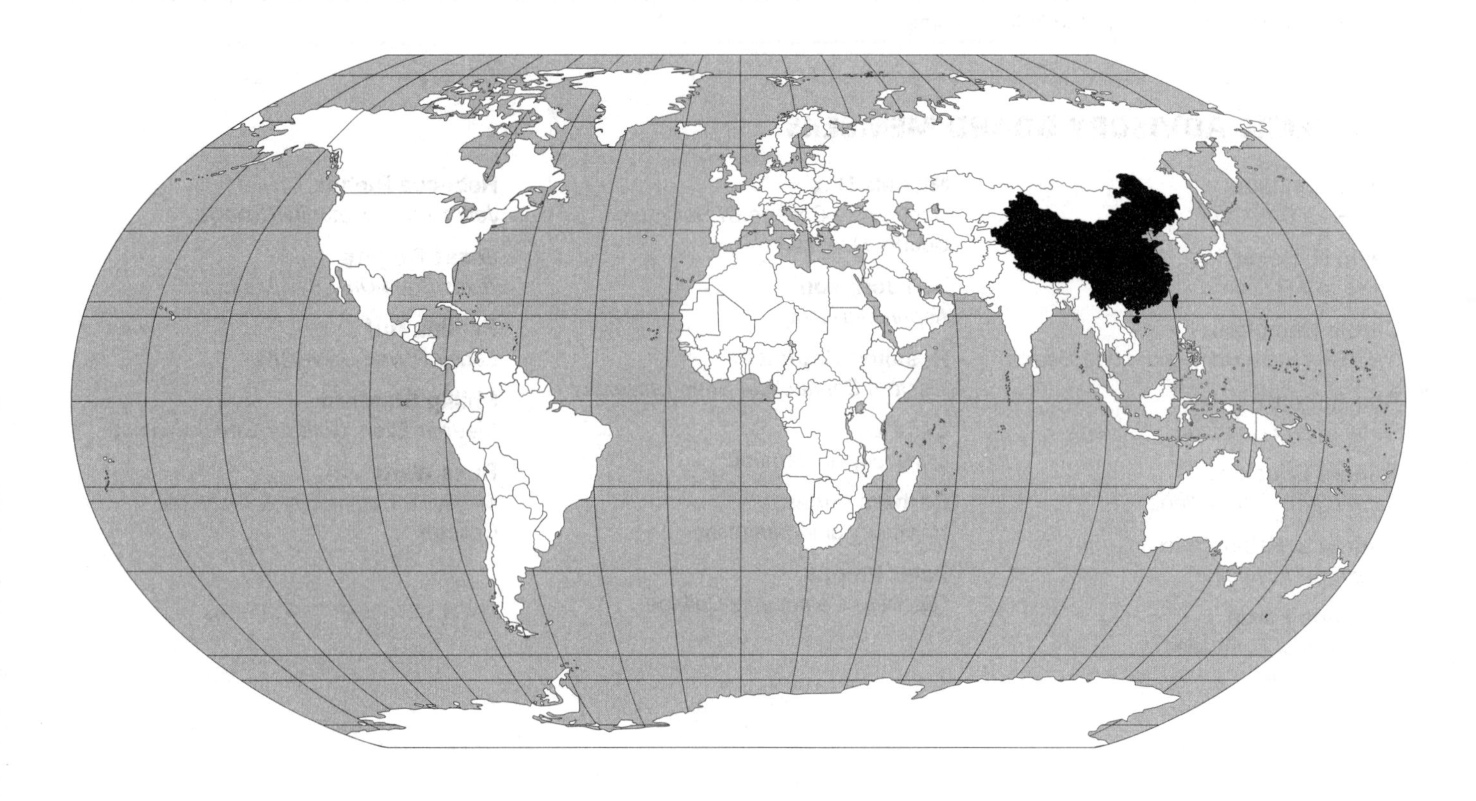

AUTHOR/EDITOR

Dr. Zhiqun Zhu

Dr. Zhiqun Zhu is John D. and Katherine T. MacArthur China in East Asian Politics and Associate Professor of Political Science and International Relations at Bucknell University in Pennsylvania. His teaching and research interests focus on Chinese politics, Chinese foreign policy, East Asian political economy, and U.S.-China relations.

His recent publications include *China's New Diplomacy: Rationale, Strategies and Significance* (Ashgate, 2010); *The People's Republic of China Today: Internal and External Challenges* [ed.] (World Scientific Publishing, 2010); *Understanding East Asia's Economic "Miracles"* (Association for Asian Studies, 2009); and *U.S.-China Relations in the 21st Century: Power Transition and Peace* (Routledge, 2006).

Contents

Global Studies: China

Using Global Studies: China viii
Selected World Wide Web Sites x
U.S. Statistics and Map xii
Canada Statistics and Map xiii
World Map xiv

China Map 2

China (People's Republic of China) 4

Hong Kong Map 49

Hong Kong Special Administrative Region 50

Taiwan Map 71

Taiwan 72

Articles from the World Press

1. **Think Again: China,** Harry Harding, *Foreign Policy,* March/April 2007. It's often said that China is walking a tightrope: Its economy depends on foreign money, its leadership is set in its ways, and its military expansion threatens the world. But the Middle Kingdom's immediate dangers run deeper than you realize. 95
2. **Think Again: China's Military,** Drew Thompson, *Foreign Policy,* March/April 2010. Is China's military a growing threat? Will China's "one child" generation weaken its military? Does China's military have global aspirations? A closer look at the world's largest armed forces. 99
3. **The China Model,** Rowan Callick, *The American,* November/December 2007. Economic freedom plus tight political control—this Chinese model seems to be displacing the "Washington Consensus" and winning fans from regimes across Asia, Africa, the Middle East, and Latin America. One wonders, for how long? 103
4. **China: A Threat to or Threatened by Democracy?** Edward Friedman, *Dissent,* Winter 2009. How can one know whether China will or will not democratize? China has already evolved politically into a non-Stalinist authoritarianism. Joined by its authoritarian friends, China is well on the way to defeating the global forces of democracy. 112
5. **Is China Afraid of Its Own People?** Willy Lam, *Foreign Policy,* September 2010. The diplomatic tussle over the East China Sea has calmed down, but a bigger foreign-policy problem waits: China's newly empowered masses won't take "no" for an answer, and Beijing is right to be scared. 116
6. **Five Reasons Why China Will Rule Tech,** Patrick Thibodeau, *Computerworld,* July 2010. China's focus on science and technology is relentless, and it's occurring at all levels of its society. There are five reasons why China may yet succeed in its goal to achieve world dominance in technology. 118
7. **How Being Big Helps and Hinders China,** David Pilling, *Financial Times,* October 2010. China's size has been crucial in getting it this far. Its seemingly limitless supply of cheap labor was a magnet for foreign investment and technology. Now, its potentially endless queues of shoppers are having the same effect. 120
8. **China's Team of Rivals,** Cheng Li, *Foreign Policy,* March/April 2009. A financial meltdown promises to test the Communist Party's power in ways not seen since Tiananmen. But theirs is a house divided, as princelings take on populists and Pekinologists try to make sense of it all. 121
9. **China Offers Direct Line to Its Leaders,** Kathrin Hille, *Financial Times,* September 2010. The Communist Party has initiated a new PR strategy to invite Chinese citizens to email party leaders now as a way to improve the party's image and shape public opinions. 125
10. **Seven Notches on the Chinese Doorpost,** David Pilling, *Financial Times,* December 15, 2010. 2010 was a milestone year for China, not just because it surpassed Japan to become the second largest economy, but also because of developments in several other areas. These developments include frustrations in China's diplomacy, China's standoff with Google, its achievements in high-speed rail network, continued pressure on *Renminbi* despite its appreciation, Liu Xiaobo's Nobel Peace Prize, China's dominance in rare-earth production, and a spate of suicides at Taiwanese-invested Foxconn. 126

11. **China's Complicit Capitalists,** Kellee S. Tsai, *Far Eastern Economic Review,* January/February 2008. There are over 29 million private businesses that employ over 200 million people and generate two-thirds of China's industrial output. Will China's growing capitalist class overthrow the Communist Party and demand democracy based on the principle of "no taxation without representation"? **128**

12. **Bye Bye Cheap Labor: Guangdong Exodus,** Alexandra Harney, *Far Eastern Economic Review,* March 2008. Higher taxes, a new labor law and the growing demands of China's increasingly sophisticated workers are forcing manufacturers either up the value chain or toward the exits. **130**

13. **Liu Xiaobo and Illusions About China,** Fang Lizhi, *New York Times,* October 11, 2010. Human rights have not improved significantly despite a soaring economy. The Nobel Committee's decision to award the Peace Prize to the imprisoned Liu Xiaobo should be applauded. The committee challenged the West to re-examine a dangerous notion that has become prevalent since 1989: that economic development will inevitably lead to democracy in China. **133**

14. **China Winning Renewable Energy Race,** Steve Hargreaves, *CNN,* September 2010. China has already surpassed the U.S. in the amount of wind turbines and solar panels that it makes. China is also gaining on the U.S. when it comes to how much of their energy comes from renewable energy sources. **135**

15. **Mania on the Mainland,** Dexter Roberts, *Bloomberg Businessweek,* January 2010. Chinese who have the money are desperately snapping up apartments for fear prices will rise further. Think the U.S. real estate bubble was bad? China's could be worse. **137**

16. **China's Reform Era Legal Odyssey,** Jerome A. Cohen, *Far Eastern Economic Review,* December 2008. Thirty years ago, China was a legal shambles. China is still building a healthy legal system. It may take another 30 years to see significant improvements. **139**

17. **China's Final Frontier,** Sophie Elmhirst, *New Statesman,* February 2009. The Chinese are latecomers to space, and desperate to catch up. Like it or not, the space race with the U.S. has begun. **142**

18. **China's Land Reform: Speeding the Plough,** Tom Orlik and Scott Rozelle, *Far Eastern Economic Review,* November 2008. Neither farmers nor migrants are yet able to participate fully in the benefits of China's ongoing modernization. Nothing can benefit the 700 million strong rural population more than land reform. **145**

19. **Chinese Acquire Taste for French Wine,** Patti Waldmeir, *Financial Times,* September 2010. China overtook the UK and Germany to become the top export market for Bordeaux wine in 2010. Wine consumption habits are becoming more sophisticated in China. **148**

20. **China Won't Revalue the Yuan,** John Lee, *Foreign Policy,* September 2010. No amount of pressure or hectoring by the West is going to change the calculus of Chinese leaders. An undervalued currency may be critical to their very survival. **149**

21. **China's Unbalanced Growth Has Served It Well,** Yukon Huang, *Financial Times,* October 2010. With both GDP and consumption increasing rapidly, why should China give up its unbalanced growth approach? The major concern is rather whether its high levels of investment will continue to generate adequate returns or are sustainable in a broader sense. **151**

22. **China Will not Be the World's Deputy Sheriff,** David Pilling, *Financial Times,* January 2010. The world expects a lot from China these days, but China is not ready, or willing, to take up the leadership role. It prefers to keep a low profile and continue to build its economy. **153**

23. **China Extends Trade with Iran,** Najmeh Bozorgmehr, *Financial Times,* February 2010. China has overtaken the EU to become Iran's largest trading partner, which underlines China's reluctance to agree to any further economic sanctions on Iran. **155**

24. **Africa Builds as Beijing Scrambles to Invest,** David Pilling, *Financial Times,* December 2009. It would be wrong to be wide-eyed about China's investments. Whatever its side-effects, a scramble to invest in Africa has got to be better than the European precedent–a scramble to carve it up. **156**

25. **A New China Requires a New US Strategy,** David Shambaugh, *Current History,* September 2010 The United States needs to revise its China strategy to deal with a complex new China. The worst thing Washington could do is to operate on autopilot, to assume that past strategies and policies are ipso facto indefinitely useful. **158**

Taiwan and Hong Kong Articles

26. **China and Taiwan Sign Landmark Deal,** Robin Kwong, *Financial Times,* June 2010. The Economic Cooperation Framework Agreement (ECFA) is the centerpiece of Ma Ying-jeou's effort to mend relations with mainland China. Taiwan hopes the deal will also smooth the path to sign free trade agreements with other key trading partners. **165**

27. **Beijing and Taiwan Try Their Hand at Détente,** Sandra Schulz, *Spiegel Online,* July 2008. Both Beijing and Taipei seem to understand that "soft power" carries more influence than "hard" military power. Part of the reconciliation stems from strong cultural and business ties as well as Taiwan relaxing its stance on "national" labels. **166**

28. **Who's Listening to Taiwan's People?,** Julian Baum, *Far Eastern Economic Review,* November 2009. As President Ma Ying-jeou pursues an ambitious agenda that will require more accommodations with Beijing, he will need to deal with the popular affirmation of Taiwanese identity. **168**

29. Behind the Dalai Lama's Taiwan Visit, Julian Baum, *Far Eastern Economic Review,* September 2009. For Taiwan's separatist opposition, the Dalai Lama's visit was a political coup. Taiwan is open, Democratic and practices freedom of religion. In some aspects, Taiwan exemplifies what Tibetans would like China to be. **170**

30. Taiwan Caters to China's Giant Fish Appetite, Robin Kwong, *Financial Times,* September 2010. The China-Taiwan trade deal opens the door for Taiwanese grouper farmers, typically small family businesses, to develop into bigger companies by entering new markets such as supplying frozen fish to inland Chinese provinces. **173**

31. Wen Hints at Scrapping Taiwan-Facing Missiles, Robin Kwong, *Financial Times,* September 2010. Chinese Premier Wen Jiabao has for the first time raised the possibility of removing some of the thousand-plus missiles the PLA has deployed facing Taiwan. **175**

32. Hong Kong Closes in on Financial Top Spot, Brooke Masters, *Financial Times,* September 2010. London and New York are still the world's leading cities for banking and other financial services, but Hong Kong is breathing down their necks. **176**

Glossary of Terms and Abbreviations **177**

Bibliography **179**

Index **188**

Using *Global Studies*: *China*

THE GLOBAL STUDIES SERIES

The Global Studies series was created to help readers acquire a basic knowledge and understanding of the regions and countries in the world. Each volume provides a foundation of information—geographic, cultural, economic, political, historical, artistic, and religious—that will allow readers to better assess the current and future problems within these countries and regions and to comprehend how events there might affect their own well-being. In short, these volumes present the background information necessary to respond to the realities of our global age.

Each of the volumes in the Global Studies series is crafted under the careful direction of an author/editor—an expert in the area under study. The author/editors teach and conduct research and have traveled extensively through the regions about which they are writing.

MAJOR FEATURES OF THE GLOBAL STUDIES SERIES

The Global Studies volumes are organized to provide concise information on the regions and countries within those areas under study. The major sections and features of the books are described here.

Country Reports

Concise reports are written for each of the countries within the region under study. These reports are the heart of each Global Studies volume. *Global Studies: China, Fourteenth Edition,* contains three country reports: People's Republic of China, Hong Kong, and Taiwan.

The country reports are composed of five standard elements. Each report contains a detailed map that visually positions the country among its neighboring states; a summary of statistical information; a current essay providing important historical, geographical, political, cultural, and economic information; a historical timeline, offering a convenient visual survey of a few key historical events; and four "graphic indicators," with summary statements about the country in terms of development, freedom, health/welfare, and achievements.

A Note on the Statistical Reports

The statistical information provided for each country has been drawn from a wide range of sources. (The most frequently referenced are listed on page 184.) Every effort has been made to provide the most current and accurate information available. However, sometimes the information cited by these sources differs to some extent; and, all too often, the most current information available for some countries is somewhat dated. Aside from these occasional difficulties, the statistical summary of each country is generally quite complete and up to date. Care should be taken, however, in using these statistics (or, for that matter, any published statistics) in making hard comparisons among countries. We have also provided comparable statistics for the United States and Canada, which can be found on pages x and xi.

World Press Articles

Within each Global Studies volume is reprinted a number of articles carefully selected by our editorial staff and the author/ editor from a broad range of international periodicals and newspapers. The articles have been chosen for currency, interest, and their differing perspectives on the subject countries. There are 34 articles in *Global Studies: China, Fourteenth Edition.*

The articles section is preceded by an annotated table of contents. This resource offers a brief summary of each article.

WWW Sites

An extensive annotated list of selected World Wide Web sites can be found on pages viii–ix in this edition of *Global Studies: China.* In addition, the URL addresses for country-specific Web sites are provided on the statistics page of most countries. All of the Web site addresses were correct and operational at press time. Instructors and students alike are urged to refer to those sites often to enhance their understanding of the region and to keep up with current events.

Glossary, Bibliography, Index

At the back of each Global Studies volume, readers will find a glossary of terms and abbreviations, which provides a quick reference to the specialized vocabulary of the area under study and to the standard abbreviations used throughout the volume.

Following the glossary is a bibliography that lists general works, national histories, and current-events publications and periodicals that provide regular coverage on China.

The index at the end of the volume is an accurate reference to the contents of the volume. Readers seeking specific information and citations should consult this standard index.

Currency and Usefulness

Global Studies: China, like the other Global Studies volumes, is intended to provide the most current and useful information available necessary to understand the events that are shaping the cultures of the region today.

This volume is revised on a regular basis. The statistics are updated, regional essays and country reports revised, and world press articles replaced. In order to accomplish this task, we turn to our author/editor, our advisory boards, and—hopefully—to you, the users of this volume. Your comments are more than welcome. If you have an idea that you think will make the next edition more useful, an article or bit of information that will make it more current, or a general comment on its organization, content, or features that you would like to share with us, please send it in for serious consideration.

Selected World Wide Web Sites for *Global Studies: China*

Some websites continually change their structure and content, so the information listed here may not always be available.

GENERAL SITES

Access Asia
www.accessasia.org

Asia Intelligence Home Page
www.asiaint.com

Asia News
www.asianews.it

Asia Resources on the World Wide Web
www.aasianst.org/wwwchina.htm

Asia Source (Asia Society)
www.asiasource.org

Asia Times
www.atimes.com

BBC News
www.bbc.co.uk/news

Brookings Institution, Center for Northeast Asian Studies
www.brookings.edu/fp/cnaps/center_hp.htm

Center for Strategic & International Studies
www.csis.org/China

PEOPLE'S REPUBLIC OF CHINA

Beijing International
www.ebeijing.gov.cn/Tour/default.htm

Beijing Review
www.bjreview.com.cn

Carnegie Endowment for International Peace: China Program
www.ceip.org/files/events/events.asp?pr=16&EventID=674

Center for US-China Policy Studies
http://cuscps.sfsu.edu

China Business Information Center
www.cbiz.cn

The China Daily
www.chinadaily.com.cn

China Development Brief: Index of International NGOS in China
www.chinadevelopmentbrief.com/dingo/index.asp

China Digital News
www.ceip.org/files/events/events.asp?pr=16&EventID=674

China Elections and Governance
http://chinaelections.org/en/default.asp

Chinese Embassy in the U.S.
www.china-embassy.org/eng

China Law and Governance Review
http://chinareview.info

China Military Online
http://english.chinamil.com.cn

China's Ministry of Foreign Affairs
www.fmprc.gov.cn/eng

China Online
www.chinaonline.com

China Related Websites
http://orpheus.ucsd.edu/chinesehistory/othersites.htm

China's Official Gateway to News & Information
www.china.org.cn

Chinese Human Rights Web
www.chinesehumanrightsreader.org

Chinese Military Power Research Sites
www.comw.org/cmp/links.htm

Congressional-Executive Commission on China
www.cecc.gov

CSIS International Security Program
www.chinatopnews.com/MainNews/English

East Turkestan Information Center
www.uygur.org/enorg/h_rights/human_r.htm

Foreign Policy in Focus
www.fpif.org/index.html

Foreign Policy in Focus Policy Brief Missile Defense & China
www.fpif.org/briefs/vol6/v6n03taiwan.html
www.uschinaedu.org-Program.asp

Global Times
www.globaltimes.cn

Jamestown Foundation
www.jamestown.org

Mainland Affairs Council Malaysia News Center—China News
http://news.newmalaysia.com/world/china

Modern East-West Encounters
www.thescotties.pwp.blueyonder.co.uk/ew-asiapacific.htm

National Committee on U.S. China Relations
www.ncuscr.org

Needham Research Institute, Cambridge, England
www.nri.org.uk

People's Daily Online
http://english.peopledaily.com.cn

SCMP.com - Asia's leading English news channel
www.scmp.com

Sinologisches Seminar, Heidelberg University
www.sino.uni-heidelberg.de

Status of Population and Family Planning Programme in China by Province
www.unescap.org/esid/psis/population/database/chinadata/intro.htm

Tiananmen Square, 1989, The Declassified Story: A National Security Archive Briefing Book
www.gwu.edu/ nsarchiv/NSAEBB/NSAEBB16/documents/index.html

The Chairman Smiles - Chinese Posters 1966–1976
www.iisg.nl/exhibitions/chairman/chnintro2.html

The China Journal
http://rspas.anu.edu.au/ccc/journal.htm

The Chinese Military Power Page–The Commonwealth Institute
www.comw.org/cmp

U.S. China Education Programs
www.fpif.org/briefs/vol6/v6n03taiwan.html

U.S. Embassy
www.usembasy-china.org.cn

U.S. International Trade Commission
www.usitc.gov

UCSD Modern Chinese History Site
www.usitc.gov

United Nations: China's Millennium Goals, Progress
www.unchina.org/MDGConf/html/reporten.pdf

United Nations Human Development Reports, China
http://hdr.undp.org

US-China Education and Culture Exchange Center
www.uschinaedu.edu

Xinhua Net
http://news.xinhuanet.com/english

Yahoo! News and Media Newspapers by Region Countries China
http://dir.yahoo.com/News_and_Media/Newspapers/By_Region/Countries/China

HONG KONG

Chinese University of Hong Kong
www.usc.cuhk.edu.hk/uscen.asp

CIA
www.cia.gov/cia/publications/factbook/geos/hk.html

Civic Exchange, Christine Loh's Newsletter
www.civic-exchange.org/n_home.htm

Clean the Air
www.cleartheair.org.hk

Hong Kong Special Administrative Region Government Information
www.info.gov.hk/eindex.htm

Hong Kong Transition Project, 1982–2007
www.hkbu.edu.hk/ hktp

Shenzhen Government Online
http://english.sz.gov.cn

South China Morning Post
www.scmp.com

TAIWAN

The China Post
www.chinapost.com.tw

Mainland Affairs Council
www.mac.gov.tw

My Egov
http://english.www.gov.tw/e-Gov/index.jsp

Taipei Times
www.taipeitimes.com/News

Taipei Yearbook
http://english.taipei.gov.tw/yearbook/index.jsp?recordid=7345

Taiwan Economic and Cultural Representative Office in the U.S.
www.tecro.org

Taiwan Headlines
www.taiwanheadlines.gov.tw/mp.asp

Taiwan News
www.etaiwannews.com/Taiwan

Taiwan Security Research
www.taiwansecurity.org

Taiwan Review
http://taiwanreview.nat.gov.tw

Taiwan Today
www.taiwantoday.tw

See individual country report pages for additional Websites.

The United States (United States of America)

GEOGRAPHY

Area in Square Miles (Kilometers): 3,794,085 (9,826,630) (about 1/2 the size of Russia)
Capital (Population): Washington, DC (563,400)
Environmental Concerns: air and water pollution; limited freshwater resources, desertification; loss of habitat; waste disposal; acid rain
Geographical Features: vast central plain, mountains in the west, hills and low mountains in the east; rugged mountains and broad river valleys in Alaska; volcanic topography in Hawaii
Climate: mostly temperate, but ranging from tropical to arctic

PEOPLE

Population

Total: 301,139,947
Annual Growth Rate: 0.89%
Rural/Urban Population Ratio: 19/81
Major Languages: predominantly English; a sizable Spanish-speaking minority; many others
Ethnic Makeup: 82% white; 13% black; 4% Asian; 1% Amerindian and others
Religions: 52% Protestant; 24% Roman Catholic; 1% Jewish; 13% others; 10% none or unaffiliated

Health

Life Expectancy at Birth: 75 years (male); 81 years (female)
Infant Mortality: 6.37/1,000 live births
Physicians Available: 2.3/1000 people
HIV/AIDS Rate in Adults: 0.6%

Education

Adult Literacy Rate: 97% (official)
Compulsory (Ages): 7–16

COMMUNICATION

Telephones: 177,900,000 main lines
Daily Newspaper Circulation: 196.3/1,000 people
Televisions: 844/1,000 people
Internet Users: 208,000,000 (2006)

TRANSPORTATION

Highways in Miles (Kilometers): 3,986,827 (6,430,366)
Railroads in Miles (Kilometers): 140,499 (226,612)
Usable Airfields: 14,947
Motor Vehicles in Use: 229,620,000

GOVERNMENT

Type: federal republic
Independence Date: July 4, 1776
Head of State/Government: President Barak H. Obama is both head of state and head of government
Political Parties: Democratic Party; Republican Party; others of relatively minor political significance
Suffrage: universal at 18

MILITARY

Military Expenditures (% of GDP): 4.06%
Current Disputes: various boundary and territorial disputes; Iraq and Afghanistan; "war on terrorism"

ECONOMY

Per Capita Income/GDP: $43,800/$13.06 trillion
GDP Growth Rate: 2.9% (2006)
Inflation Rate: 3.2%
Unemployment Rate: 4.8%
Population Below Poverty Line: 12%
Natural Resources: many minerals and metals; petroleum; natural gas; timber; arable land
Agriculture: food grains; feed crops; fruits and vegetables; oil-bearing crops; livestock; dairy products
Industry: diversified in both capital and consumer-goods industries
Exports: $1.023 trillion (primary partners Canada, Mexico, Japan, China, U.K.)
Imports: $1.861 trillion (primary partners Canada, Mexico, Japan, China, Germany)

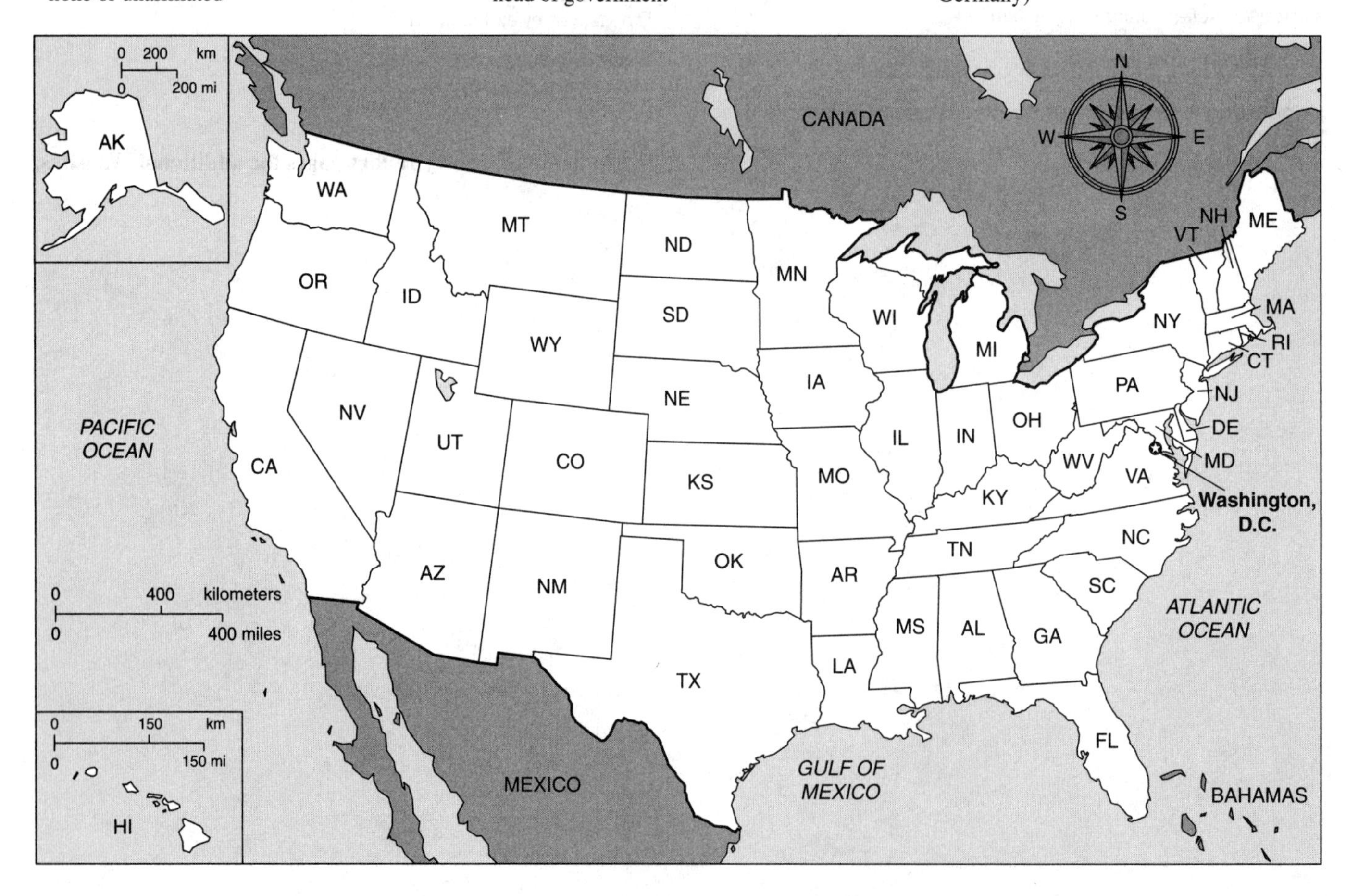

Canada

GEOGRAPHY

Area in Square Miles (Kilometers): 3,855,103 (9,984,670) (slightly larger than the United States)
Capital (Population): Ottawa (1,560,000)
Environmental Concerns: air and water pollution; acid rain; industrial damage to agriculture and forest productivity
Geographical Features: permafrost in the north; mountains in the west; central plains; lowlands in the southeast
Climate: varies from temperate to arctic

PEOPLE

Population

Total: 33,390,141(2007)
Annual Growth Rate: 0.87%
Rural/Urban Population Ratio: 20/80
Major Languages: both English and French are official
Ethnic Makeup: 28% British Isles origin; 23% French origin; 15% other European; 6% others; 2% indigenous; 26% mixed
Religions: 42.6% Roman Catholic; 27.7% Protestant; 12.7% others; 16% none.

Health

Life Expectancy at Birth: 77 years (male); 84 years (female)
Infant Mortality: 4.63/1,000 live births
Physicians Available: 2.1/1,000 people
HIV/AIDS Rate in Adults: 0

Education

Adult Literacy Rate: 97%
Compulsory (Ages): 6–16

COMMUNICATION

Telephones: 20,780,000 main lines
Daily Newspaper Circulation: 167.9/1,000 people
Televisions: 709/1,000 people
Internet Users: 22,000,000 (2006)

TRANSPORTATION

Highways in Miles (Kilometers): 646,226 (1,042,300)
Railroads in Miles (Kilometers): 29,802 (48,068)
Usable Airfields: 1,343
Motor Vehicles in Use: 18,360,000

GOVERNMENT

Type: federation with parliamentary democracy
Independence Date: July 1, 1867
Head of State/Government: Queen Elizabeth II; Prime Minister Stephen Harper
Political Parties: Conservative Party of Canada; Liberal Party; New Democratic Party; Bloc Québécois; Green Party
Suffrage: universal at 18

MILITARY

Military Expenditures (% of GDP): 1.1%
Current Disputes: maritime boundary disputes with the United States and Denmark (Greenland)

ECONOMY

Currency ($U.S. equivalent): 0.97 Canadian dollars = $1 (Oct. 2007)
Per Capita Income/GDP: $35,700/$1.181 trillion
GDP Growth Rate: 2.8%
Inflation Rate: 2%
Unemployment Rate: 6.4% (2006)
Labor Force by Occupation: 75% services; 14% manufacturing; 2% agriculture; and 8% others
Natural Resources: petroleum; natural gas; fish; minerals; cement; forestry products; wildlife; hydropower
Agriculture: grains; livestock; dairy products; potatoes; hogs; poultry and eggs; tobacco; fruits and vegetables
Industry: oil production and refining; natural-gas development; fish products; wood and paper products; chemicals; transportation equipment
Exports: $401.7 billion (primary partners United States, Japan, United Kingdom)
Imports: $356.5 billion (primary partners United States, China, Japan)

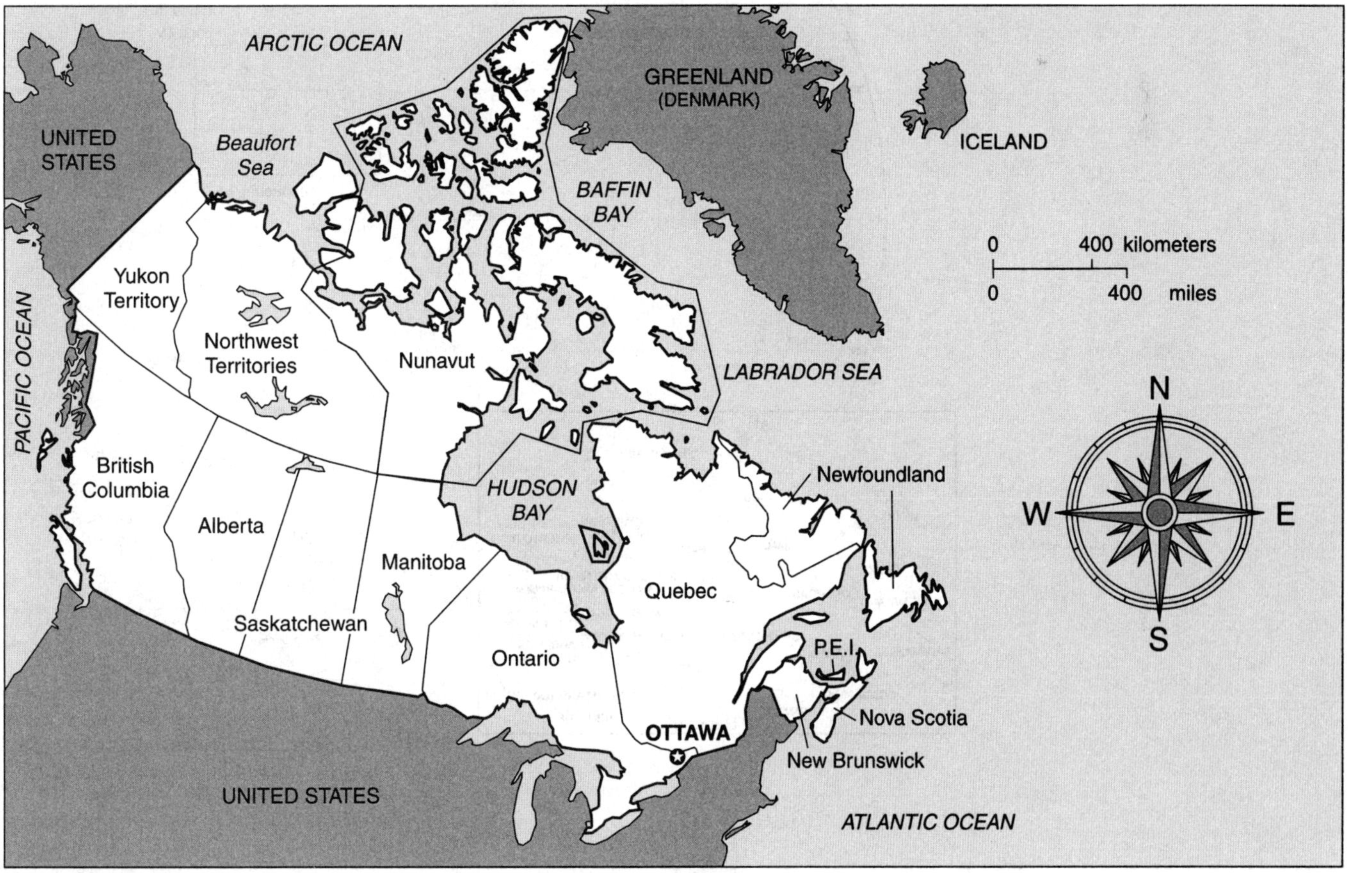

GLOBAL STUDIES

This map is provided to give you a graphic picture of where the countries of the world are located, the relationship they have with their region and neighbors, and their positions relative to major trade and power blocs. We have focused on certain areas to illustrate these crowded regions more clearly. China is shaded for emphasis.

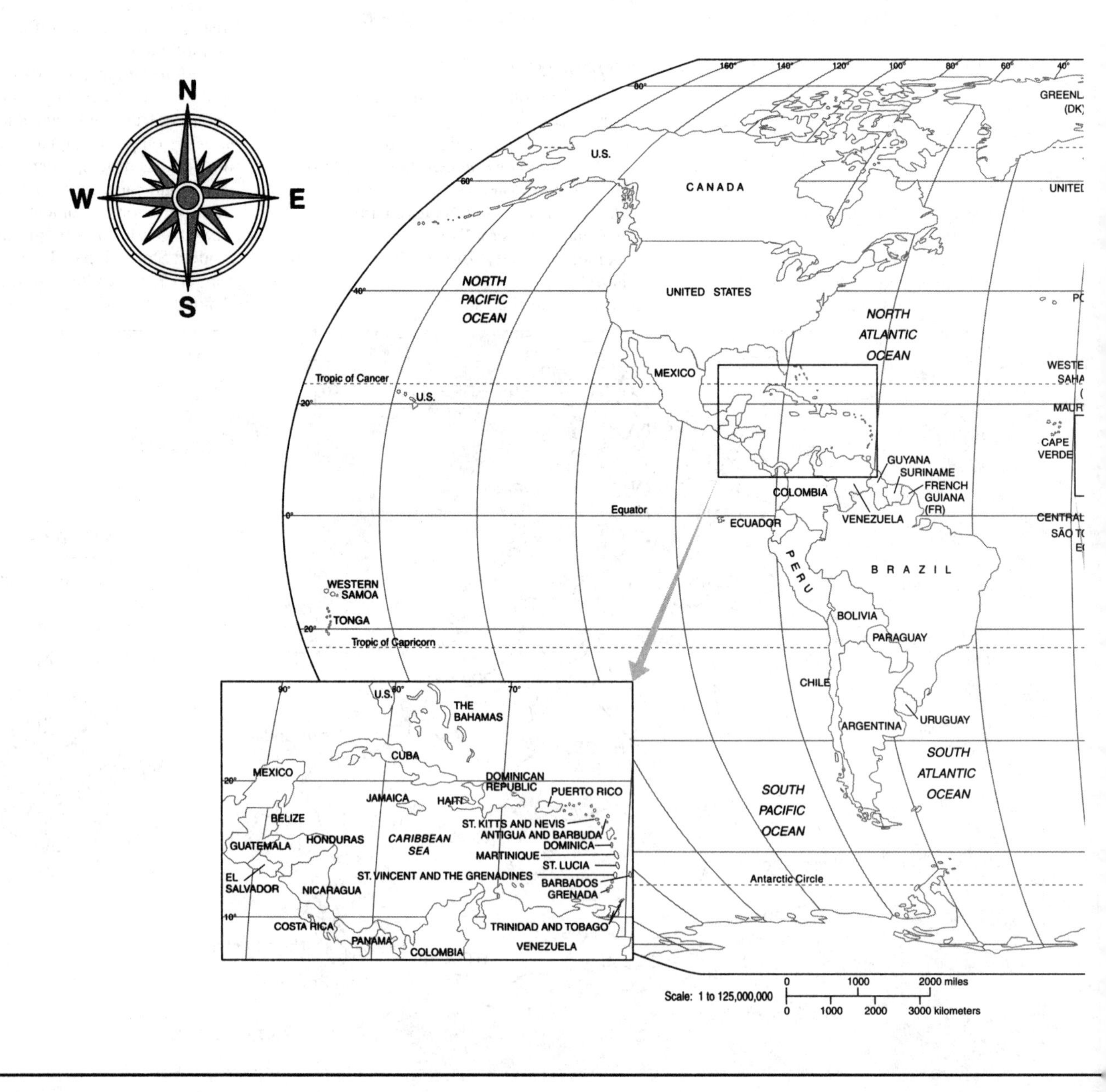

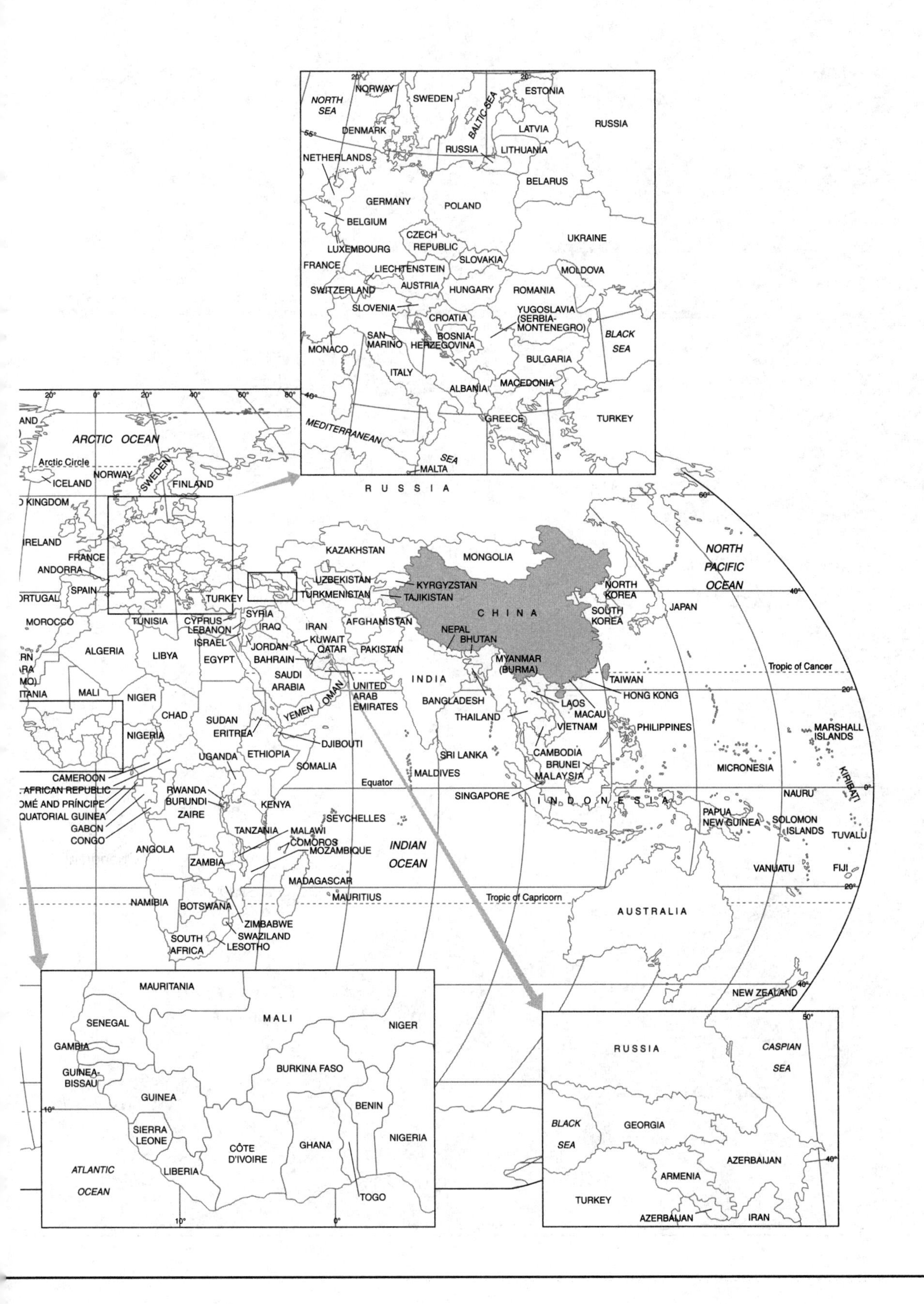

NORTH SEA
NORWAY
SWEDEN
BALTIC SEA
ESTONIA
RUSSIA
DENMARK
LATVIA
LITHUANIA
NETHERLANDS
BELARUS
GERMANY
POLAND
BELGIUM
CZECH REPUBLIC
UKRAINE
LUXEMBOURG
SLOVAKIA
FRANCE
LIECHTENSTEIN
MOLDOVA
SWITZERLAND
AUSTRIA
HUNGARY
ROMANIA
SLOVENIA
CROATIA
YUGOSLAVIA (SERBIA-MONTENEGRO)
BLACK SEA
SAN MARINO
BOSNIA-HERZEGOVINA
MONACO
BULGARIA
ITALY
ALBANIA
MACEDONIA
GREECE
TURKEY
MEDITERRANEAN SEA
MALTA
ARCTIC OCEAN
Arctic Circle
ICELAND
FINLAND
IRELAND
ANDORRA
SPAIN
MOROCCO
TUNISIA
CYPRUS
LEBANON
ISRAEL
SYRIA
IRAQ
IRAN
JORDAN
KUWAIT
QATAR
BAHRAIN
SAUDI ARABIA
UNITED ARAB EMIRATES
OMAN
YEMEN
ALGERIA
LIBYA
EGYPT
MALI
NIGER
CHAD
SUDAN
ERITREA
DJIBOUTI
NIGERIA
ETHIOPIA
SOMALIA
UGANDA
CAMEROON
RWANDA
BURUNDI
ZAIRE
KENYA
GABON
CONGO
EQUATORIAL GUINEA
TANZANIA
MALAWI
COMOROS
MOZAMBIQUE
ANGOLA
ZAMBIA
SEYCHELLES
INDIAN OCEAN
MADAGASCAR
MAURITIUS
NAMIBIA
BOTSWANA
ZIMBABWE
SWAZILAND
SOUTH AFRICA
LESOTHO
KAZAKHSTAN
MONGOLIA
UZBEKISTAN
TURKMENISTAN
KYRGYZSTAN
TAJIKISTAN
AFGHANISTAN
PAKISTAN
C H I N A
NEPAL
BHUTAN
I N D I A
BANGLADESH
MYANMAR (BURMA)
THAILAND
LAOS
VIETNAM
CAMBODIA
MACAU
HONG KONG
TAIWAN
NORTH KOREA
SOUTH KOREA
JAPAN
NORTH PACIFIC OCEAN
PHILIPPINES
SRI LANKA
MALDIVES
BRUNEI
MALAYSIA
SINGAPORE
I N D O N E S I A
Equator
Tropic of Cancer
Tropic of Capricorn
MARSHALL ISLANDS
MICRONESIA
KIRIBATI
NAURU
PAPUA NEW GUINEA
SOLOMON ISLANDS
TUVALU
VANUATU
FIJI
A U S T R A L I A
NEW ZEALAND
MAURITANIA
SENEGAL
GAMBIA
GUINEA-BISSAU
GUINEA
BURKINA FASO
BENIN
SIERRA LEONE
LIBERIA
CÔTE D'IVOIRE
GHANA
TOGO
ATLANTIC OCEAN
CASPIAN SEA
GEORGIA
AZERBAIJAN
ARMENIA

China Map

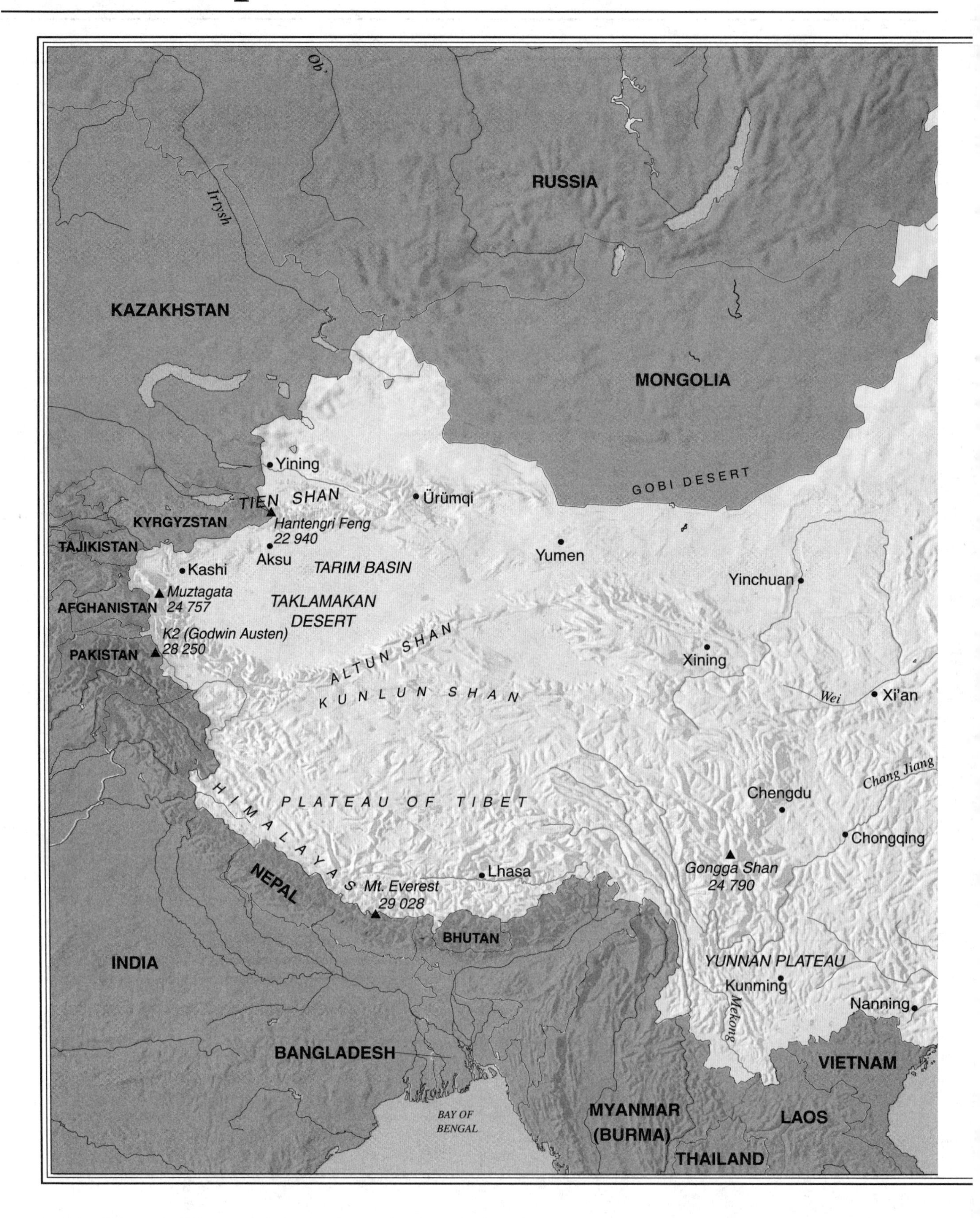

Ob'
RUSSIA
Irtysh
KAZAKHSTAN
MONGOLIA
Yining
GOBI DESERT
TIEN SHAN
Ürümqi
KYRGYZSTAN
Hantengri Feng
22 940
TAJIKISTAN
Aksu
Yumen
Kashi
TARIM BASIN
Yinchuan
Muztagata
24 757
AFGHANISTAN
TAKLAMAKAN
DESERT
K2 (Godwin Austen)
28 250
PAKISTAN
ALTUN SHAN
Xining
KUNLUN SHAN
Wei
Xi'an
Chang Jiang
HIMALAYAS
PLATEAU OF TIBET
Chengdu
Chongqing
Gongga Shan
24 790
Lhasa
NEPAL
Mt. Everest
29 028
BHUTAN
INDIA
YUNNAN PLATEAU
Kunming
Mekong
Nanning
BANGLADESH
VIETNAM
BAY OF
BENGAL
MYANMAR
(BURMA)
LAOS
THAILAND

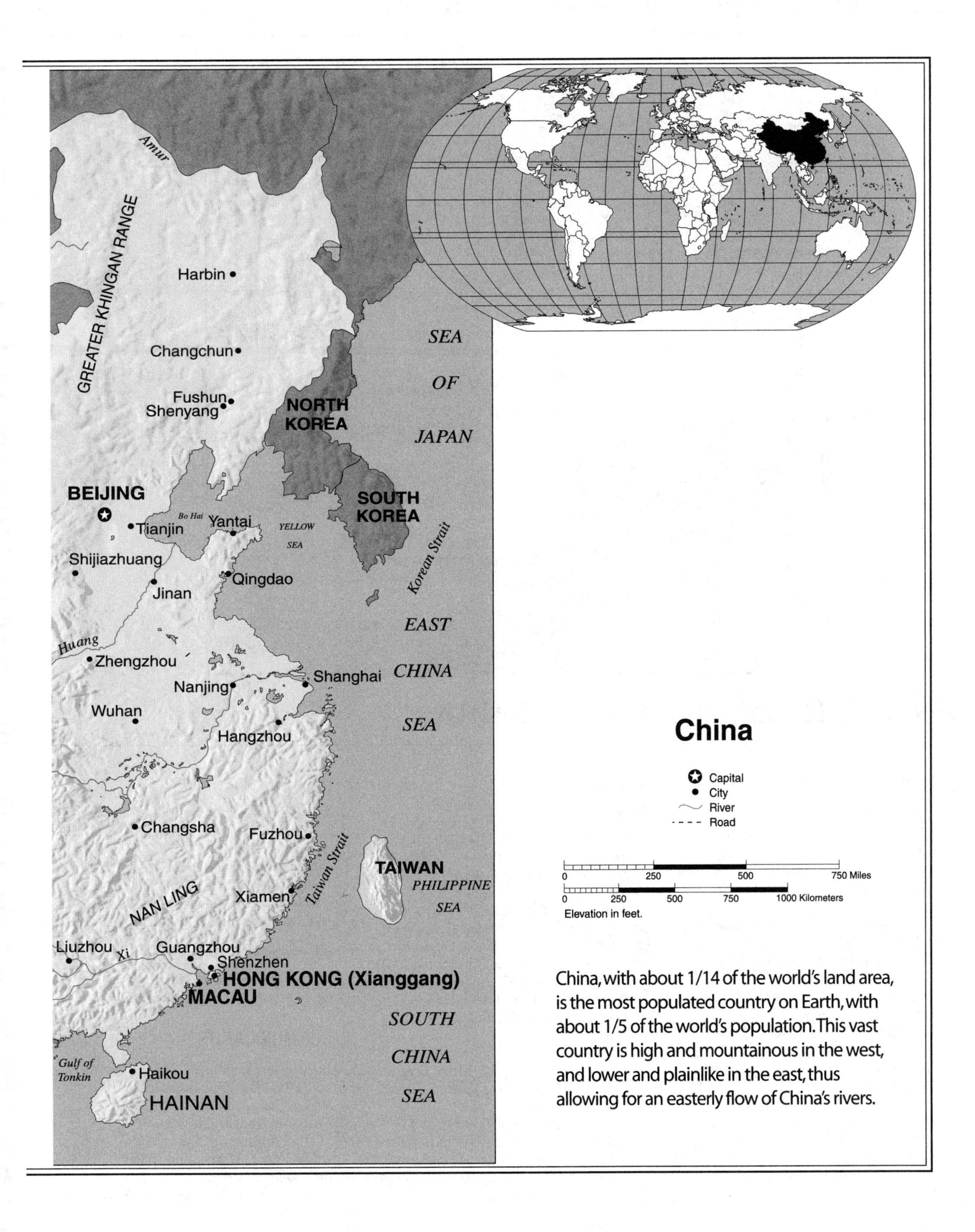

China, with about 1/14 of the world's land area, is the most populated country on Earth, with about 1/5 of the world's population. This vast country is high and mountainous in the west, and lower and plainlike in the east, thus allowing for an easterly flow of China's rivers.

China (People's Republic of China)

People's Republic of China Statistics

GEOGRAPHY

Area in Square Miles (Kilometers): 3,705,386 (9,596,960) (about the same size as the United States)

Capital (Population): Beijing, 8.7 million (city proper), 17.4 million (metro area)

Other Large Cities: Shanghai, Chongqing (Chungking), Tianjin, Wuhan, Shenyang (Mukden), Guangzhou, Harbin, Xi'an, Chengdu, and Nanjing (Nanking)

Environmental Concerns: air and water pollution; water shortages; desertification; trade in endangered species; acid rain; loss of agricultural land; deforestation

Geographical Features: mostly mountains, high plateaus, and deserts in the west and northwest; plains, deltas, and hills in the east

Climate: extremely diverse, from tropical to subarctic

PEOPLE

Population

Total: 1,330,141,295 (July 2010 est.)

Annual Growth Rate: 0.494% (2008 est.)

Rural/Urban Population Ratio: 60/40

Sex Ratio of Total Population: 1.06 male(s)/1.0 female (2010 est.)

Major Languages: Standard Chinese or Mandarin (Putonghua, based on the Beijing dialect), Yue (Cantonese), Wu (Shanghainese), Minbei (Fuzhou), Minnan (Hokkien-Taiwanese), Xiang, Gan, Hakka dialects, minority languages

Ethnic Makeup: Han Chinese 91.5%, Zhuang, Manchu, Hui, Miao, Uyghur, Tujia, Yi, Mongolian, Tibetan, Buyi, Dong, Yao, Korean, and other nationalities 8.5% (2000 census)

Religions: officially atheist; but popular religions include Daoism (Taoism), Buddhism, Islam, Christianity, ancestor worship, and animism

Health

Life Expectancy at Birth: 74.51 years (male: 71.37 years, female: 76.77 years) (2010 est.)

Infant Mortality Rate: 21.16 deaths/1,000 live births (2008 est.)

Number of Physicians Per 1000 People: 1.65

HIV/AIDS Rate in Adults: 0.1% (2003 est.)

Education

Adult Literacy Rate: 91.6%

Compulsory Years of Schooling: 9

COMMUNICATION

Telephones: Landlines: 314 million (2009); Cell phones: 800 million (2010)

Newspapers: 2,081

Daily Newspaper Circulation: 93.5 million

Journals and Magazines: 9,363

Publishing Houses: 600

Internet Users: 350 million (2010)

TRANSPORTATION

Airports: 502 (2010)
Major Ports: Shanghai, Dalian, Guangzhou, Ningbo, Qingdao, Qinhuangdao, Shenzhen, Tianjin
Railways: 77,834 km (24,433 km electrified) (2008)
Roadways: 3,583,715 km, 1,575,571 km paved (includes 53,913 km of expressways) (2007)
Waterways: 124,000 km navigable (2006)
Motor Vehicles in Use: 160 million, including over 30 million privately owned cars (2007)

GOVERNMENT

Type: one-party state ruled by the Communist Party
Independence Date: unification in 221 B.C. The People's Republic of China was established on October 1, 1949
Head of State/Government: President Hu Jintao; Premier Wen Jiabao
Political Parties: Chinese Communist Party; eight registered small parties under the leadership of the CCP
Suffrage: universal at 18 in village and urban district elections

MILITARY

Military Expenditures (% of GDP): 4.3% (2006)
People's Liberation Army (PLA): Ground Forces, Navy, Air Force, and Second Artillery Corps (strategic missile force); People's Armed Police (PAP)
Current Disputes: territorial disputes with a few countries, especially India, Japan, and Russia, and potentially serious disputes over Spratly and Paracel Islands with several countries

ECONOMY

Currency: yuan
Exchange Rate: US$1 = 6.671 yuan (December 2010)
GDP (purchasing power parity): $8.818 trillion (2009 est.)
GDP (official exchange rate): $4.985 trillion (2009 est.)
Per Capita Income (purchasing power parity): $6,700 (2009 est.)
Labor Force: 813.5 million (2009 est.)
Unemployment Rate: 4.3% unemployment in urban areas; substantial unemployment and underemployment in rural areas (2009 est.)
Distribution of *Family Income — Gini index:* 47 (2007)
Inflation Rate (consumer prices): –0.7% (2009 est.)
Investment (gross fixed): 46.3% of GDP (2009 est.)
Agricultural Products: rice, wheat, potatoes, corn, peanuts, tea, millet, barley, apples, cotton, oilseed, pork, fish
Industries: mining and ore processing; iron, steel, aluminum, and other metals, coal; machine building; armaments; textiles and apparel; petroleum; cement; chemicals; fertilizers; consumer products, including footwear, toys, and electronics; food processing; transportation equipment; telecommunications equipment, commercial space launch vehicles, satellites
Industrial Production Growth Rate: 9.9% (2009 est.)
Exports: $1.204 trillion f.o.b. (2009 est.)
Exports Commodities: machinery, electrical products, data processing equipment, apparel, textiles, steel, mobile phones
Exports Partners: US 19.4%, Hong Kong 15.2%, Japan 8.4%, South Korea 4.6%, Germany 4.1% (2009)
Imports: $954.3 billion f.o.b. (2009 est.)
Imports Commodities: machinery and equipment, oil and mineral fuels, plastics, LED screens, data processing equipment, optical and medical equipment, organic chemicals, steel, copper
Imports Partners: Japan 13.9%, South Korea 11%, Taiwan 10.6%, Hong Kong 10.06%, US 7.5%, Germany 4.7% (2009)

SUGGESTED WEB SITES

www.cia.gov/library/publications/the-world-factbook/geos/ch.html
www.DemographicsNowChina.com
www.infoplease.com/ipa/A0107411.html
www.chinatoday.com/data/data.htm

People's Republic of China Country Report

Chinese civilization originated in the Neolithic Period, which began around 5000 B.C., but scholars know little about it until the Shang Dynasty, which dates from about 2000 B.C. By that time, the Chinese had already developed the technology and art of bronze casting to a high standard; and they had a sophisticated system of writing with ideographs, in which words are portrayed by picturelike characters. From the fifth to the third centuries B.C., the level of literature and the arts was comparable to that of Greece in the Classical Period, which flourished at the same time. Stunning breakthroughs occurred in science, a civil service evolved, and the philosopher Confucius developed a highly sophisticated system of ethics for government and moral codes for society. Confucian values were dominant until the collapse of the Chinese imperial system in 1911, but even today they influence Chinese thought and behavior in China, and in Chinese communities throughout the world.

From several hundred years B.C. until the 15th century, China was the world's leader in technology, had the largest economy, and enjoyed the highest GDP per capita income in the world. By 1500, however, the government had closed China's doors to broad international trade, and Europe's GDP per capita surpassed China's. Still, it remained the world's largest economy, accounting for some 30 percent of the world's GDP in 1820. Over the next 130 years, war, revolution, and invasions ate away at China's productive capabilities. By the time the Chinese Communist Party came to power in October 1949, China's share of the world's GDP had dropped to 5 percent.[1] The historical baggage that China carried into the period of the People's Republic of China in 1949 was, then, substantial. China had fallen from being one of the world's greatest empires—not just from an economic perspective, but also from cultural and scientific perspectives—before 1500, to one of the poorest countries in the world. When the Chinese Communists interpreted that history through the lens of Marxism and Leninism, they saw feudalism, capitalism, and imperialism as the cause of China's problem. They saw China as a victim both of exploitation within their own society, and from abroad.

The Chinese Empire

By 221 B.C., the many feudal states ruled by independent princes had been conquered by Qin Shi Huang Di, the first ruler of a unified Chinese Empire. He established a system of governmental institutions and a concept of empire that continued in China until A.D. 1911. Although China was unified from the Qin Dynasty onward, it was far less concrete than the term *empire* might indicate.

China's borders really reached only as far as its cultural influence did. Thus China contracted and expanded according to whether or not other groups of people accepted the Chinese ruler and culture as their own.

Those peoples outside "China" who refused to acknowledge the Chinese ruler as the "Son of Heaven" or pay tribute to him were called "barbarians." In part, the Great Wall, which stretches more than 2,000 miles across north China and was built in stages between the third century B.C. and the seventeenth century A.D., was constructed in order to keep marauding "barbarians" out of China. Nevertheless, they frequently invaded China and occasionally even succeeded in subduing the Chinese—as in the Yuan (Mongol) Dynasty (1279–1368) and, later, the Qing (Manchu) Dynasty (1644–1911).

However, the customs and institutions of the invaders eventually yielded to the powerful cultural influence of the Chinese. Indeed, in the case of the Manchus, who seized control of the Chinese Empire in 1644 and ruled until 1911, their success in holding onto the throne for so long may be due in part to their willingness to assimilate Chinese ways and to rule through existing Chinese institutions, such as the Confucian-ordered bureaucracy. By the time of their overthrow, the Manchu were hardly distinguishable from the ethnic Han Chinese in their customs, habits, and beliefs. When considering today's policies toward the numerous minorities who inhabit such a large expanse of the People's Republic of China, it should be remembered that the central Chinese government's ability to absorb minorities was key to its success in maintaining a unified entity called China (*Zhongguo*—"the Central Kingdom") for more than 2,000 years.

The Imperial Bureaucracy

A distinguishing feature of the political system of imperial China was the civil service examinations through which government officials were chosen. These examinations tested knowledge of the moral principles embodied in the classical Confucian texts. Although the exams were, in theory, open to all males in the Chinese Empire, the lengthy and rigorous preparation required meant that, in practice, the sons of the wealthy and powerful with access to a good education had an enormous advantage. Only a small percentage of those who began the process actually passed the examinations and received an appointment in the imperial bureaucracy. Some of those who were successful resided in the capital to advise the emperor, while others were sent as the emperor's agents to govern throughout the far-flung realm.

The Decline of the Manchus

The vitality of Chinese institutions and their ability to respond creatively to new problems came to an end during the Manchu Dynasty (1644–1911). A stagnant agricultural system incapable of supporting the burgeoning population and the increasing exploitation of the peasantry who comprised the vast majority of Chinese people led to massive internal rebellions and the rise of warlords. As the imperial bureaucracy became increasingly corrupt and incompetent, the Manchu Dynasty gradually lost the ability to govern effectively.

China's decline in the nineteenth century was exacerbated by a social class structure that rewarded those who could pass the archaic, morality-based civil service examinations rather than those who had expertise in science and technology and could thereby contribute to China's development. An inward-looking culture contributed to the malaise by preventing the Chinese from understanding the dynamism of the Industrial Revolution then occurring in the West. Gradually, the barriers erected by the Manchu rulers to prevent Western culture and technology from polluting the ancient beauty of Chinese civilization crumbled, but too late to strengthen China to resist the West's military onslaughts.

The Opium War (1839–1842)

In the early nineteenth century, the British traded with China, but it was primarily a one-way trade. The British nearly drained their coffers buying Chinese silk, tea, and porcelain; China's self-satisfied rulers found little of interest to purchase from the rapidly industrializing British. The British were also frustrated by China's refusal to recognize the British Empire as an equal of the Chinese Empire, and to open up ports to trade with them along China's extensive coastline and rivers.

Opium, produced in the British Empire's colony India, proved to be the one product that the Chinese were willing to purchase, and it reversed the trade balance in favor of the British. Eventually they used the Chinese attack on British ships carrying opium as an excuse for declaring war on the decrepit Chinese Empire. The Opium War ended with defeat for the Chinese and the signing of the Treaty of Nanjing (sometimes called Nanking). This treaty ceded the island of Hong Kong to the British as a colony and allowed them to establish trading posts.

Subsequent wars with the British and other European powers brought further concessions—the most important of which was the Chinese granting of additional "treaty ports" to Europeans. The Chinese had hoped that they could contain and control Europeans within these ports. Although that was true to a degree, this penetration of China led to the spread of Western values that challenged the stagnant, and by then collapsing, Chinese Empire. As the West and, by the late nineteenth century, Japan, nibbled away at China, the Manchu rulers made a last-ditch effort at reform to strengthen and enrich China. But it was too late, and the combination of internal decay, warlordism, revolution, and foreign imperialism finally toppled the Manchu Dynasty. Thus ended more than 2,000 years of imperial rule in China.

(Public Domain (PD01-Confucius))

Confucius.

CONFUCIUS: CHINA'S FIRST "TEACHER"

Confucius (551–479 B.C.), whose efforts to teach China's rulers how to govern well were spurned, spent most of his life teaching his own disciples. Yet 300 years later, Confucianism, as taught by descendants of his disciples, was adopted as the official state philosophy. The basic principles of Confucianism include hierarchical principles of obedience and loyalty to one's superiors, respect for one's elders, and filial piety; and principles and practices for maintaining social order and harmony; and the responsibility of rulers to exercise their power benevolently.

REPUBLICAN CHINA

The 1911 Revolution, which led to the overthrow of Manchu rule and derived its greatest inspiration from Chinese nationalist Sun Yat-sen, resulted in the establishment of the Republic of China (R.O.C.).

OPIUM AS A PRETEXT FOR WAR

Although the opium poppy is native to China, large amounts of opium were shipped to China by the English-owned East India Company, from the British colony of India. Eventually India exported so much opium to China that 5 to 10 percent of its revenues derived from its sale.

By the late 1700s, the Chinese government had officially prohibited first the smoking and selling of opium, and later its importation or domestic production. But because the sale of opium was so profitable—and also because so many Chinese officials were addicted to it—the Chinese officials themselves illegally engaged in the opium trade. As the number of addicts grew and the Chinese government became more corrupted by its own unacknowledged participation in opium smuggling, so grew the interest of enterprising Englishmen in smuggling it into China for financial gain.

The British government was primarily interested in expanding trade with China. But it also wanted to establish a diplomatic relationship based on equality to supplant the existing one, in which the Chinese court demanded that the English kowtow to the Chinese emperor. In addition, it wanted to secure legal jurisdiction over its nationals residing in China to protect them against Chinese practices of torture of those suspected of having committed a crime.

China's efforts to curb the smuggling of opium and the Chinese refusal to recognize the British as equals reached a climax in 1839, when the Chinese destroyed thousands of chests of opium aboard a British ship. This served as an ideal pretext for the British to attack China with their sophisticated gunboats (pictured below destroying a junk in Canton's (Guangzhou's) harbor). Ultimately their superior firepower gave victory to the British.

Thus the so-called Opium War (1839–1842) ended with defeat for the Chinese Empire and the signing of the Treaty of Nanjing, which ceded the island of Hong Kong to the British and allowed them to establish trading posts on the Chinese Mainland.

(Library of Congress Prints and Photographs Division/LC-USZ62-86300)
Opium War (1839–1842). British attack on junks in Canton / Guangzhou Harbor.

It was, however, a "republic" only in name, for China was unable to successfully transfer Western forms of democratic governance to China. This was in no small part because of China's inability to remain united and to maintain law and order. China had been briefly united under the control of the dominant warlord of the time, Yuan Shikai; but with his death in 1916, China was again torn apart by the resurgence of contending warlords, internal political decay, and further Japanese territorial expansion in China. Efforts at reform failed in the context of China's weakness and internal division.

Chinese intellectuals searched for new ideas from abroad to strengthen their nation during the vibrant May Fourth period and New Culture Movement (spanning the period from roughly 1917 through the mid-1920s). In the process, the Chinese invited influential foreigners such as the English mathematician, philosopher, and socialist Bertrand Russell, and the American philosopher and educator John Dewey, to lecture in China. Thousands of Chinese traveled, worked, and studied abroad. It was during this period that ideas such as liberal democracy, syndicalism, guild socialism, and communism were put forth as possible solutions to China's many problems.

The Founding of the Chinese Communist Party

It was during that period, in 1921, that a small Marxist study group in Shanghai founded the Chinese Communist Party (CCP). The Moscow-based Comintern (Communist International) advised this highly intellectual but politically and militarily powerless group to join with the more militarily powerful Kuomintang (KMT or Nationalist Party, led first by Sun Yat-sen and, after his death in 1925, by Chiang Kai-shek) until it could gain strength and break away to establish themselves as an independent party. Thus it was with the support of the Communists in a "united front" with the Nationalists that Chiang Kai-shek conquered the warlords and reunified China under one central government. Chiang felt threatened by the Communists' ambitions to gain political power, however, so in 1927 he executed all but the few Communists who managed to escape.

Members of the CCP continued to take their advice from Moscow; and they tried to organize an orthodox Marxist, urban-based movement of industrial workers. Because the cities were completely controlled by the KMT, the CCP found it difficult to organize the workers, and ultimately the KMT's police and military forces decimated the ranks of the Communists. It is a testimony to the appeal of communism in that era that the CCP managed to recover its strength each time. Indeed, the growing power of the CCP was such that Chiang Kai-shek considered the CCP, rather than the invading Japanese, to be the main threat to his complete control of China.

Eventually the Chinese Communist leaders agreed that an urban strategy could not succeed. Lacking adequate military power to confront the Nationalists, however, they retreated. In what became known as the Long March (1934–1935), they traveled more than 6,000 miles from the southeast, through the rugged interior and onto the windswept, desolate plains of Yan'an in northern China.

It was during this retreat, in which as many as 100,000 followers perished, that Mao Zedong staged his contest for power within the CCP. With his victory, the CCP reoriented itself toward a rural strategy and attempted to capture the loyalty of China's peasants, then comprising some 85 percent of China's total population. Mao saw the downtrodden peasantry as the major source of support for the revolutionary overthrow of Chiang Kai-shek's government. Suffering from an oppressive and exploitative system of landlord control, disillusioned with the government's unwillingness to carry out land reform, and desirous of owning their own land, the peasantry looked to the CCP for leadership. Slowly the CCP started to gain control over China's vast countryside.

United against the Japanese

In 1931, Japan invaded China and occupied Manchuria, the three northeastern provinces. In 1937, Japan attacked again, advancing southward to occupy China's heartland. Although the CCP and KMT were determined to destroy each other, Japan's threat to spread its control over the rest of China caused them to agree to

MAO ZEDONG: CHINA'S REVOLUTIONARY LEADER

Mao Zedong (1893–1976) came from a moderately well-to-do peasant family and, as a result, received a very good education, as compared to the vast majority of the Chinese of his time. Mao was one of the founders of the Chinese Communist Party in 1921, but his views on the need to switch from an orthodox Marxist strategy, which called for the party to seek roots among the urban working class, to a rural strategy centered on the exploited peasantry were spurned by the leadership of the CCP and its sponsors in Moscow.

Later, it became evident that the CCP could not flourish in the Nationalist-controlled cities, as time and again the KMT quashed the idealistic but militarily weak CCP. Mao appeared to be right: "Political power grows out of the barrel of a gun."

The Communists' retreat to Yan'an in northern China at the end of the Long March was not only for the purpose of survival but also for regrouping and forming a stronger Red Army. There the followers of the Chinese Communist Party were taught Mao's ideas about guerrilla warfare, the importance of winning the support of the people, principles of party leadership, and socialist values. Mao consolidated his control over the leadership of the CCP during the Yan'an period and led it to victory over the Nationalists in 1949.

From that time onward, Mao became a symbol of the new Chinese government, of national unity, and of the strength of China against foreign humiliation. In later years, although his real power was eclipsed, the party maintained the public illusion that Mao was the undisputed leader of China.

In his declining years, Mao waged a struggle, in the form of the "Cultural Revolution," against those who followed policies antagonistic to his own—a struggle that brought the country to the brink of civil war and turned the Chinese against one another. The symbol of Mao as China's "great leader" and "great mentor" was used by those who hoped to seize power after him: first the minister of defense, Lin Biao, and then the "Gang of Four," which included Mao's wife.

(Public Domain)
Mao Zedong.

Mao's death in 1976 ended the control of policy by the Gang of Four. Within a few years, questions were being raised about the legacy that Mao had left China. By the 1980s, it was broadly accepted throughout China that Mao had been responsible for a full 20 years of misguided policies. Since the Tiananmen Square protests of 1989, however, there has been a resurgence of nostalgia for Mao. This nostalgia is reflected in such aspects of popular culture as a tape of songs about Mao entitled "The Red Sun"—a best-selling tape in China, at over 5 million copies—that captures the Mao cult and Mao mania of the Cultural Revolution; and in a small portrait of Mao that virtually all car owners and taxi drivers hang over their rear-view mirrors for "good luck." Many Chinese long for the "good old days" of Mao's rule, when crime and corruption were at far lower levels than today and when there was a sense of collective commitment to China's future. But they do not long for a return to the mass terror of the Cultural Revolution, for which Mao also bears responsibility. In the commercialized twenty-first century China, attaching the name of Mao to a product is a good way to sell it.

a second "united front," this time for the purpose of halting the Japanese advance. Both the KMT and the CCP had ulterior motives, but according to most accounts, the Communists contributed more to the national wartime efforts. The Communists organized guerrilla efforts to peck away at the fringes of Japanese-controlled areas while Chiang Kai-shek, head of the KMT, retreated to the wartime capital of Chongqing (Chungking). His elite corps of troops and officers kept the best of the newly arriving American supplies for themselves, leaving the rank-and-file Chinese to fight against the Japanese in cloth boots and with inferior equipment. It was not the Nationalist Army but, rather, largely the unstinting efforts and sacrifices of the Chinese people and the American victory over Japan that brought World War II to an end in 1945. With the demobilization of the Japanese, however, Chiang Kai-shek was free once again to focus on defeating the Communists.

The Communists Oust the KMT

It seemed as if the Communists' Red Army had actually been strengthened through its hard fighting during World War II, turning itself into a formidable force. Meanwhile, the relatively soft life of the KMT military elite during the war did not leave it well prepared for civil war against the Red Army. Chiang Kai-shek relied on his old strategy of capturing China's cities, but the Communists, who had gained control over the countryside by winning the support of the vast peasantry, surrounded the cities. Like besieged fortresses, the cities eventually fell to Communist control. By October 1949, the CCP could claim control over all of China, except for the island of Taiwan. It was there that the Nationalists' political, economic, and military elites, with American support, had fled.

Scholars still dispute why the Red Army ultimately defeated the Nationalist Army. They cite as probable reasons the CCP's promises to undertake land reform; the Communists' more respectful treatment of the peasantry as they marched through the countryside (in comparison to that of the KMT soldiers); the CCP's more successful appeal to the Chinese sense of nationalism; and Chiang Kai-shek's unwillingness to undertake reforms that would benefit the

RED GUARDS: ROOTING OUT THOSE "ON THE CAPITALIST ROAD"

During the Cultural Revolution, Mao Zedong called upon the country's young people to "make revolution." Called "Mao's Red Guards," their ages varied, but for the most part they were teenagers.

Within each class and school, various youths would band together in a Red Guard group that would take on a revolutionary-sounding name and would then carry out the objective of challenging people in authority. But the people in authority—especially school-teachers, school principals, bureaucrats, and local leaders of the Communist Party—initially ignored the demands of the Red Guards that they reform their "reactionary thoughts" or eliminate their "feudal" habits.

The Red Guards initially had no real weapons and could only threaten. Since they were considered just misdirected children by those under attack, their initial assaults had little effect. But soon the frustrated Red Guards took to physically beating and publicly humiliating those who stubbornly refused to obey them. Since Mao had not clearly defined precisely what should be their objectives or methods, the Red Guards were free to believe that the ends justified extreme and often violent means. Moreover, many Red Guards took the opportunity to take revenge against authorities, such as teachers who had given them bad grades. Others (at right) would harangue crowds on the benefits of Maoism and the evils of foreign influence.

The Red Guards went on rampages throughout the country, breaking into people's houses and stealing or destroying their property, harassing people in their homes in the middle of the night, stopping girls with long hair and cutting it off on the spot, destroying the files of ministries and industrial enterprises, and clogging up the transportation system by their travels throughout the country to "make revolution." Different Red Guard factions began to fight with one another, each claiming to be the most revolutionary.

Mao eventually called on the army to support the Red Guards in their effort to challenge "those in authority taking the capitalist road." This created even more confusion, as many of the Red Guard groups actually supported the people they were supposed to be attacking. But their revolutionary-sounding names and their pretenses at being "Red" (Communist) confused the army. Moreover, the army was divided within itself and did not particularly wish to overthrow the Chinese Communist Party authorities, the main supporters of the military in their respective areas of jurisdiction.

Since the schools had been closed, the youth of China were not receiving any formal education during this period. Finally, in 1969, Mao called a halt to the excesses of the Red Guards. They were disbanded and sent home. Some were sent to work in factories or out to the countryside to labor in the fields with the peasants. But the chaos set in motion during the Cultural Revolution did not come to a halt until the arrest of the Gang of Four, some 10 years after the Cultural Revolution had begun.

During the "10 bad years," when schools were either closed or operating with a minimal program, children received virtually no formal education beyond an elementary school level. As a result, China's development lagged behind its neighbors such as Japan and South Korea.

(New York World-Telegram and Sun Newspaper Photograph Collection/ Library of Congress LC-USZ62-134168)
"Anti-revoluntionary" leaders in dunce caps for public shame by Red Guards in Beijing, January 26, 1967.

peasantry, advance economic development, and control corruption. Still, even had the KMT made greater efforts to reform, any wartime government confronted with the demoralization of the population ravaged by war, inflation, economic destruction, and the humiliation of a foreign occupation would have found it difficult to maintain the loyal support of its people. Even the middle class eventually deserted the KMT. Many of those industrial and commercial capitalists who had supported the Nationalists now joined with the CCP to rebuild China. Others, however, stayed behind only because they were unable to flee to Hong Kong or Taiwan.

One thing is clear: The Chinese Communists did not gain victory because of support from the Soviet Union; for the Soviets, who were anxious to be on the winning side in China, chose to give aid to the KMT until it was evident that the Communists would win. Furthermore, the Communists' victory rested not on superior weapons but, rather, on a superior strategy, support from the Chinese people, and (as Mao Zedong believed) a superior political "consciousness." It was because of the Communist victory over a technologically superior army that Mao thereafter insisted on the superiority of "man over weapons" and the importance of the support of the people for an army's victory. The relationship of the soldiers to the people is, Mao said, like the relationship of fish to water—without the water, the fish will die.

THE PEOPLE'S REPUBLIC OF CHINA

The Red Army's final victory came rapidly—far faster than anticipated. Suddenly China's large cities fell to the Communists, who now found themselves in charge of a nation of more than 600 million people. They had to make critical decisions about how to unify and rebuild the country. They were obligated, of course, to fulfill their promise to redistribute land to the poor and landless peasantry in return for the peasants' support of the Communists during the Civil War. The CCP leaders were, however, largely recruited from among the peasantry; and like revolutionary fighters everywhere, knew how to make a revolution but had little experience

(International Institute of Social History/Stefan R. Landsberger Collection (www.iisg.nl/~landsberger))
A Cultural Revolution—style poster. Below the poster, the inscription reads, "We must definitely carry forward the great struggle in thoroughly exposing and criticizing the Gang of Four."

THE GANG OF FOUR

The current leadership of the Chinese Communist Party views the Cultural Revolution of 1966–1976 as having been a period of total chaos that brought the People's Republic of China to the brink of political and economic ruin. While Mao Zedong is criticized for having begun the Cultural Revolution with his ideas about the danger of China turning "capitalist," the major blame for the turmoil of those years is placed on a group of extreme radicals labeled the "Gang of Four."

The Gang of Four consisted of Jiang Qing, Mao's wife, who began playing a key role in China's cultural affairs during the early 1960s; Zhang Chunqiao, a veteran party leader in Shanghai; Yao Wenyuan, a literary critic and ideologue; and Wang Hongwen, a factory worker catapulted into national prominence by his leadership of rebel workers during the Cultural Revolution. By the late 1960s, these four individuals were among the most powerful leaders in China. Drawn together by common political interests and a shared belief that the Communist Party should be relentless in ridding China of suspected "capitalist roaders," they worked together to keep the Cultural Revolution on a radical course. One of their targets had been Deng Xiaoping, who emerged as China's paramount leader in 1978, after the members of the Gang of Four had been arrested.

Although they had close political and personal ties to Mao and derived many of their ideas from him, Mao became quite disenchanted with them in the last few years of his life. He was particularly displeased with the unscrupulous way in which they behaved as a faction within the top levels of the party. Indeed, it was Mao who coined the name Gang of Four, as part of a written warning to the radicals to cease their conspiracies and obey established party procedures.

The Gang of Four hoped to take over supreme power in China following Mao's death, on September 9, 1976. However, their plans were upset less than a month later, when other party and army leaders had them arrested—an event that is now said to mark the formal end of the "10 bad years" or "Cultural Revolution." Removing the party's most influential radicals from power set the stage for the dramatic reforms that have become the hallmark of the post-Mao era in China. In November 1980, the Gang of Four were put on trial in Beijing. They were charged with having committed serious crimes against the Chinese people and accused of having had a hand in "persecuting to death" tens of thousands of officials and intellectuals whom they perceived as their political enemies. All four were convicted and sentenced to long terms in prison.

with governance. So, rejected by the Western democratic/capitalist countries because of their embrace of communism, and desperate for aid and advice, the Communists turned to the Soviet Union for direction and support. They did this in spite of the Soviet leader Joseph Stalin's fickle support of the Chinese Communists throughout the 1930s and '40s.

The Soviet Model

In the early years of CCP rule, China's leaders "leaned to one side" and followed the Soviet model of development in education, the legal system, the economic system, and elsewhere. The Soviet economic model favored capital-intensive industrialization, but all the Soviet "aid" had to be repaid. Furthermore, following the Soviet model required a reliance on Soviet experts and well-educated Chinese, whom the Communists were not sure they could trust. Without Soviet support in the beginning, however, it is questionable whether the CCP would have been as successful as it was in developing China in the 1950s.

The Maoist Model

China soon grew exasperated with the limitations of Soviet aid and the inapplicability of the Soviet model to Chinese circumstances. China's preeminent leader, Mao Zedong, proposed a Chinese model of development more appropriate to Chinese circumstances. What came to be known as the "Maoist model" took account of China's low level of development, poverty, and large population. Mao hoped to substitute China's enormous manpower for expensive capital equipment by organizing people into ever larger working units.

In 1958, in what became known as the "Great Leap Forward," Mao Zedong launched his model of development. It was a bold scheme to rapidly accelerate the pace of industrialization so that China could catch up with the industrialized states of the West. In the countryside, land was merged into large communes, untested and controversial planting techniques were introduced, and peasant women were engaged fully in the fields in order to increase agricultural production. The communes became the basis for industrializing the countryside through a program of peasants building their own "backyard furnaces" to smelt steel. The Maoist model assumed that those people possessing a proper revolutionary, or "red" (communist), consciousness—that is, a commitment to achieving communism—would be able to produce more than those who were "expert" but lacked revolutionary consciousness. In the cities, efforts to increase industrial production through longer work days, and overtaxing industrial equipment, likewise led to a marked decline in production and industrial wastage.

The Maoist model of extreme "egalitarianism"—captured in the Chinese expression "all eat out of the same pot"—and "continuous revolution," was a rejection of the Soviet model of development, which Mao came to see as an effort to hold the Chinese back from more rapid industrialization. In particular, the Soviets' refusal to give the Chinese the most advanced industrial-plant equipment and machinery, or to share nuclear technology with them, made Mao suspicious of their intentions.

Sino–Soviet Relations Sour

For their part, the Soviets believed that the Maoist model was doomed to failure. The Soviet leader Nikita Khrushchev denounced the Great Leap Forward as "irrational"; but he was equally distressed at what seemed a risky scheme by Mao Zedong to bring the Soviets and Americans into direct conflict over the Nationalist-controlled Offshore Islands in the Taiwan Strait. The combination of what the Soviets viewed as Mao's irrational economic policy and his risk-taking confrontation with the United States prompted the Soviets to abruptly withdraw their experts from China in 1959. They packed up their bags, along with spare parts for Soviet-supplied machinery and blueprints for unfinished factories, and returned home.

The Soviets' withdrawal, combined with the disastrous decline in production resulting from the policies of the Great Leap Forward and several years of bad weather, set China's economic development back many years. Population figures now available indicate that millions died in the years from 1959 to 1962, mostly from starvation and diseases caused by malnutrition. The catastrophic consequences of the Great Leap Forward resulted in the leadership paying no more than lip service to Mao Zedong's ideas. The Chinese people were not told that Mao Zedong bore blame for their problems, but the Maoist model was abandoned for the time being. More pragmatic leaders took over the direction of the economy, but without further support from the Soviets. Not until 1962 did the Chinese start to recover their productivity gains of the 1950s.

By 1963, the Sino–Soviet split had become public, as the two Communist powers found themselves in profound disagreement over a wide range of issues: whether socialist countries could use capitalist methods, such as free markets, to advance economic development; appropriate policies toward the United States; whether China or the Soviet Union could claim to follow Marxism-Leninism more faithfully, entitling it to lead the Communist world. By the mid-1960s, the Sino–Soviet relationship had deteriorated to the point that the Chinese were worried that the Soviets might launch a military attack on them.

The Cultural Revolution

In 1966, Mao launched what he termed the "Great Proletarian Cultural Revolution." Whether Mao Zedong hoped to provoke an internal party struggle and regain control over policy, or (as he alleged) to re-educate China's exploitative, corrupt, and oppressive officials in order to restore a revolutionary spirit to the Chinese people and to prevent China from abandoning socialism, is unclear. He called on China's youth to "challenge authority," particularly "those revisionists in authority who are taking the capitalist road." If China continued along its "revisionist" course, he said, the achievements of the Chinese revolution would be undone. China's youth were therefore urged to "make revolution."

Such vague objectives invited abuse, including personal feuds and retribution for alleged past wrongs. Determining just who was "Red" and committed to the Communist revolution, and who was "reactionary" itself generated chaos, as people tried to protect themselves by attacking others—including friends and relatives. During that period, people's cruelty was immeasurable. People were psychologically, and sometimes physically, tortured until they "admitted" to their "rightist" or "reactionary" behavior. Murders, suicides, ruined careers, and broken families were the debris left behind by this effort to "re-educate" those who had strayed from the revolutionary path. It is estimated that approximately 10 percent of the population—that is, *80 million people*—became targets of the Cultural Revolution, and that tens of thousands lost their lives during these years of political violence.

The Cultural Revolution attacked Chinese traditions and cultural practices as being feudal and outmoded. It also destroyed the authority of the Chinese Communist Party, through prolonged public attacks on many of its most respected leaders. Policies changed frequently in those "10 bad years" from 1966 to 1976, as first one faction and then another gained the upper hand. Few leaders escaped unscathed. Ultimately, the Chinese Communist Party and Marxist-Leninist ideology were themselves the victims of the Cultural Revolution. By the time the smoke cleared, the legitimacy of the CCP had been destroyed, and the people could no longer accept the idea that the party leaders were infallible. Both traditional Chinese morality and Marxist-Leninist values had been thoroughly undermined.

Reforms and Liberalization

With the death of Mao Zedong and the subsequent arrest of the politically radical "Gang of Four" (which included Mao's wife) in 1976, the Cultural Revolution came to an end. Deng Xiaoping, a veteran leader of the CCP who had been purged twice during the "10 bad years," was "rehabilitated" in 1977.

By 1979, China once again set off down the road of construction and put an end to the radical Maoist policies of "continuous revolution" and the idea that it was more important to be "red" than "expert." Saying that he did not care whether the cat was black or white, as long as it caught mice, Deng Xiaoping pursued more pragmatic, flexible policies in order to modernize China. In other words, Deng did not care if he used capitalist methods, as long as they helped modernize China. He deserves credit for opening up China to the outside world and to reforms that led to the liberalization of both the economic and the political spheres. When he died in 1997, Deng left behind a country that, despite some setbacks and reversals, had already traveled a significant distance down the road to liberalization and modernization.

In spite of Deng Xiaoping's pragmatic policies, and Mao Zedong's clear responsibility for precipitating policies that were devastating to the Chinese people, Mao has never been defrocked in China; for to do so would raise serious questions about the legitimacy of the CCP. China's leaders have admitted that, beginning with the Anti-Rightest Campaign of 1957 and the Great Leap Forward of 1958, Mao made "serious mistakes"; but the CCP insists that these errors must be seen within the context of his many accomplishments and his commitment, even if sometimes misdirected, to Marxism-Leninism. In contrast to the Gang of Four and others who were condemned as "counter-revolutionaries," Mao has been called a "revolutionary" who made "mistakes." As recently as the 17th National Congress of the CCP in 2007, Mao Thought (the Chinese adaptation of Marxism-Leninism to Chinese conditions) remained enshrined in the party's constitution as providing the foundation for continued CCP rule.

The Challenge of Reform

The erosion of traditional Chinese values, then of Marxist-Leninist values and faith in the Chinese Communist Party's leadership, and finally of Mao Thought, left China without any strong belief system. Such Western values as materialism, capitalism, individualism, and freedom swarmed into this vacuum to undermine both Communist ideology and the traditional Chinese values that had provided the glue of society. Deng Xiaoping's prognosis had proven correct: The "screen door" through which Western science and technology (and foreign investments) could flow into China was unable to keep out the annoying "insects" of Western values. The screen door had holes that were too large to prevent this invasion.

China's leadership in the reform period has not been united. The less pragmatic, more ideologically oriented "conservative" or "hard-line" leadership (who in the new context of reforms could be viewed as ideologues of a Maoist vintage) challenged the introduction of liberalizing economic reforms precisely because they threatened to undo China's earlier socialist achievements and erode Chinese culture. To combat the negative side effects of introducing free-market values and institutions, China's leadership launched a number of "mass campaigns": the campaign in the 1980s against "spiritual pollution"—the undermining of Chinese values;[2] a repressive campaign following the brutal crackdown against those challenging the leadership in Tiananmen Square in 1989; ongoing campaigns against corruption; and campaigns to "strike hard" against crime and to "get civilized."[3]

Since 1979, in spite of setbacks, China's leadership has been able to keep the country on the path of liberalization. As a result, the economy has had an average annual growth rate of 9.5 percent for the last 30 years. It is now ranked as the fourth largest economy in the world. China has dramatically reformed the legal and political system as well, even though much work remains to be done. The third generation of the PRC leadership was headed by Jiang Zemin, who led China throughout the 1990s and into the twenty-first century. In turn, Jiang stepped down from his position as party leader in 2001, as president in 2002, and as the head of the Military Affairs Commission in 2004. China's leaders now operate within what is a younger and increasingly well-institutionalized and better-educated system of collective leadership. The problems that the leadership of President Hu Jintao and Premier Wen Jiabao faces as a consequence of China's rapid modernization and liberalization are formidable: massive and growing unemployment; increasing crime, corruption, and social dislocation; a lack of social cohesion; and challenges to the CCP's monopoly on power put into play by its policies of liberalization, pluralization, and modernization. The forces of rapid growth and social and economic modernization have taken on a momentum of their own. China's increasing involvement in the international community has also put into motion seemingly uncontrollable forces, some of which are destablizing, and others that are contributing to demands for political reform. As will be noted below in the discussion of these issues, the real concern for China is not whether China will engage in further reform and democratization, but whether it can maintain stability in the context of this potentially destabilizing international and domestic environment.

The Student and Mass Movement of 1989

Symbolism is very important in Chinese culture; the death of a key leader is a particularly significant moment. In the case of Hu Yaobang, the former head of the CCP, his sudden death in April 1989 became symbolic of the death of liberalizing forces in China. The deceased leader's career and its meaning were touted as symbols of liberalization, even though his life was hardly a monument to liberal thought. More conservative leaders in the CCP had removed him from his position as the CCP's general-secretary in part because he had offended their cultural sensibilities. Apart from everything else, Hu's suggestion that the Chinese turn in their chopsticks for knives and forks, and not eat food out of a common dish because it spread disease, were culturally offensive to them.

Hu's death provided students with a catalyst to place his values and policies in juxta-position with those of the then increasingly conservative leadership.[4] The students' reassessment of Hu Yaobang's career, in a way that rejected the party's evaluation, was in itself a challenge to the authority of the CCP's right to rule China. The students' hunger strike in Tiananmen Square—essentially in front of party headquarters—during the visit of the Soviet Union President, Mikhail Gorbachev, to China was, even in the eyes of ordinary Chinese people, an insult to the Chinese leadership. Many Chinese later stated that the students went too far, as by humiliating the leadership, they humiliated *all* Chinese.

Part of the difficulty in reaching an agreement between the students and China's leaders was that the students' demands changed over time. At first they merely wanted a reassessment of Hu Yaobang's career. But quickly the students added new demands: dialogue between the government and the students (with the students to be treated as equals with top CCP leaders), retraction of an offensive *People's Daily* editorial, an end to official corruption, exposure of the financial and business dealings of the central leadership, a free press, the removal of the top CCP leadership, and still other actions that challenged continued CCP rule.

The students' hunger strike, which lasted for one week in May, was the final straw that brought down the wrath of the central leadership. Martial law was imposed in Beijing. When the citizens of Beijing resisted its enforcement and blocked the armies' efforts to reach Tiananmen Square to clear out the hunger strikers, both students and CCP leaders dug in; but both were deeply divided bodies. Indeed, divisions within the student-led movement caused it to lose its direction; and divisions within the central CCP leadership incapacitated it. For two weeks, the central leadership wrangled over who was right and the best course of action. On June 4, the "hard-liners" won out, and they chose to use military power rather than a negotiated solution with the students.

Did the students make significant or well-thought-out statements about "democracy" or realistic demands on China's leaders? The short and preliminary answer is no; but then, is this really the appropriate question to ask? One could argue that what the students *said* was less important than what they *did:* They mobilized the population of China's capital and other major cities to support a profound challenge to the legitimacy of the CCP's leadership. Even if workers believed that "You can't eat democracy," and even if they participated in the demonstrations for their *own* reasons (such as gripes about inflation and pensions), they did support the students' demand that the CCP carry out further political reforms. This was because the students successfully promoted the idea that if China had had a democratic system rather than authoritarian rule, the leadership would have been more responsive to the workers' bread-and-butter issues.

Repression within China Following the Crackdown

By August 1989, the CCP leadership had established quotas of "bad elements" for work units and identified 20 categories of people to be targeted for punishment. But people were more reluctant than in the past to follow orders to expose their friends, colleagues, and family members, not only because such verdicts had often been reversed at a later time, but also because many people questioned the CCP's version of what happened in Beijing on June 4. Although the citizenry worried about informers, there seemed to be complicity from top to bottom, whether inside or outside the ranks of the CCP, in refusing to go along with efforts to ferret out demonstrators and sympathizers with the prodemocracy, antiparty movement. Party leaders below the central level appeared to believe that the central government's leadership was doomed; for this reason, they dared not carry out its orders. Inevitably, there would be a reversal of verdicts, and they did not want to be caught in it.

As party leaders in work units droned on in mandatory political study sessions about Deng Xiaoping's important writings, workers wondered how long it would be before the June 4 military crackdown would be condemned as a "counterrevolutionary crime against the people." Individuals in work units had to fill out lengthy questionnaires. A standard one had 24 questions aimed at "identifying the enemy." Among them were such questions as, "What did you think when Hu Yaobang died?" "When Zhao Ziyang went to Tiananmen Square, what did you think? Where were you?" At one university, each faculty member's questionnaire had to be verified by two people (other than one's own family) or the individual involved would not be allowed to teach.[5]

As part of the repression that followed the military crackdown in June 1989, the government carried out arrests of hundreds of those who participated in the demonstrations. During the world's absorption with the Persian Gulf War in 1991, the government suddenly announced the trials and verdicts on some of China's best known leaders of the 1989 demonstrations. Of those who were summarily executed, available information indicates that almost all were workers trying to form labor unions. All the other known 1989 student and dissident leaders were eventually released, although some were deported to the West as a condition of their release. The government has also occasionally re-arrested 1989 protesters for other activities. In 1998, for example, some former protesters made bold attempts to establish a new party to challenge Chinese Communist Party rule. Although their efforts to register this new party were at first tolerated, several were later arrested, tried, and sentenced to prison. Finally, as discussed below, the government has attempted to ferret out and arrest activist leaders of the Falun Gong.

In spite of these important exceptions, many repressive controls were relaxed, and China's mass media have steadily expanded the parameters of allowable topics and opinions. Today, although there are occasional arrests of individuals who are blatantly challenging CCP rule, and although the establishment of a competing party is not tolerated, the leadership is more focused on harnessing the talents of China's best and brightest for the country's modernization than it is on controlling dissent. No longer a revolutionary party, the CCP is intent on effectively governing and developing China.

THE PEOPLE OF CHINA

Population Control

In 2010, China's population was estimated to be over 1.3 billion. In the 1950s, Mao had encouraged population growth, as he considered a large population to be a major source of strength: Cheap human labor could take the place of expensive technology and equipment. No sustained attempts to limit Chinese population occurred until the mid-1970s. Even then, because there were no penalties for those Chinese who ignored them, population control programs were only marginally successful.

In 1979, the government launched a serious birth-control campaign, rewarding couples giving birth to only one child with work bonuses and priority in housing. The only child was later to receive preferential treatment in university admissions and job assignments (a policy eventually abandoned). Couples who had more than one child, on the other hand, were to be penalized by a decrease in their annual wages, and their children would not be eligible for free education and healthcare benefits.

The one-child policy in China's major cities was rigorously enforced, to the point where it was almost impossible for a woman to get away with a second pregnancy. Who was allowed to have a child, as well as when she could give birth, was rigidly controlled by the woman's work unit. Furthermore, with so many state-owned enterprises paying close to half of their entire annual wages as "bonuses," authorities came up with additional sanctions to ensure compliance. Workers were usually organized in groups of 10 to 30 individuals. If any woman in the group gave birth to more than one child, *the entire group* would lose its annual bonus. With such overwhelming penalties for the group as a whole, pressures for a couple not to give birth to a second child were enormous.

To ensure that any unauthorized pregnancy did not occur, women who had already given birth were required to stand in front of x-ray machines (fluoroscopes) to verify that their IUDs (intrauterine birthcontrol devices) were still in place. Abortions could and would be performed throughout the period of a woman's unsanctioned pregnancy. (The moral issues that surround abortions for some Christians are not concerns for the Chinese.)

The effectiveness of China's family planning policy in the cities has been due not merely to the surveillance by state-owned work units, neighborhood committees, and the "granny police" who watch over the families in their residential areas. Changed social attitudes also play a critical role, and urban Chinese now accept the absolute necessity of population control in their overcrowded cities.

The one-child policy in China's cities has led to a generation of remarkably spoiled children. Known as "little emperors," these only children are the center of attention of six anxious adults (the parents and two sets of grandparents), who carefully scrutinize their every movement. It has led to the overuse of medical services by these parents and grandparents, who rush their only child/grandchild to the doctor at the first signs of a sniffle or sore throat. It has also led to overfed, even obese, children. Being overweight used to be considered a hedge against bad times, and the Chinese were initially pleased that their children were becoming heavier. A common greeting showing admiration had long been, "You have become fat!" But as contemporary urban Chinese adopt many of the values associated in the developed world with becoming wealthier, they are changing their perspectives on weight. Jane Fonda–style exercise programs are now a regular part of Chinese television, and weight-loss salons and fat farms are coming into vogue for China's well-fed middle class. Still, most people view the major purpose of exercise as staying healthy and keeping China a strong nation, not looking attractive.

The strictly enforced and administered family planning program has been undermined by a number of trends. First, in the cities, there are large migrant populations who are really under no one's control: the villages from which they fled have no responsibility for them; and the cities to which they migrate rarely issue them a "household registration" certificate, so they really belong to no official's jurisdiction. Yet, rural migrant families usually prefer to have only one child, as in their tenuous economic circumstances, taking care of more than one would make survival in a new city far more difficult. Second, well-to-do entrepreneurs who live in private housing can avoid population-control measures because they are not part of any public housing or work unit. They are willing to pay all the relevant fines and bribes necessary to have as many children as they want. Nevertheless, even they are unlikely to have more than two children. Indeed, some members of China's growing middle class are deciding not to have children, for the same reasons as in other more developed societies: they want to pursue careers and spend time and money in ways that leave little room for children. There are more and more such "DINK" (double income no kids) families now. Third, in recent years, there has been a more relaxed enforcement of the one-child policy because of the demographic crisis on the horizon: too few young people to support the large number of elderly people in future years.

One step the government has taken to address this inverted population pyramid is to allow those married couples who both come from one-child families to have two children. The government has also tried to grapple with one of the unintended side-effects of the one-child policy: the aborting of female fetuses. Given the cultural preference for males, a certain percentage of female fetuses are aborted. (This has resulted in a lopsided male–female ratio of at least 105:100, although in some areas it is said to be as high as 125:100. These are, however, much debated figures. The problem of a lopsided sex ratio is true in many Asian countries, including India and South Korea.) The sex of fetuses is usually known because of the widespread use in China of ultrasound machines. To use these machines to reveal the sex of the child is illegal, but for a very small bribe, doctors will usually do so. In addition, although female infanticide is illegal, it sometimes happens, especially in rural areas. So the government has promulgated several new laws and has investigated several thousand cases of alleged abuse of sex-identification of fetuses.

In the meantime, China's orphanages have absorbed some of the unwanted girl babies, now much sought after in the West. In 2007, however, China instituted new regulations that make it harder for foreigners to adopt Chinese babies (almost all of whom are girls—boys are usually only put up for adoption if there is a physical or mental defect). Arguably, one of the reasons for this change in policy is because of the other demographic crisis resulting from the one-child policy—tens of millions of men coming of marriage age without women to marry. Apart from societal unhappiness, the lack of brides has led to a sharp increase in the kidnapping of young women, as well as the practice of selling girls as brides in rural marketplaces when they reach marriageable age (usually to men who live in remote villages that have little to offer a new bride).

In the vast rural areas of China, where some three-quarters of the population still live, efforts to enforce the one-child policy have met with less success than in the cities, because the benefits and punishments are not as relevant for peasants. After the communes were disbanded in the early 1980s and families were given their own land to till, peasants wanted sons, to do the heavy farm labor. As a result, the government's policy in the countryside became more flexible. In some villages, if a woman gives birth to a girl and decides to have another child in hopes of having a boy, she may pay the government a substantial fee (usually an amount more than the entire annual income of the family) in order to be allowed to do so. Yet, in an ironic reflection of this still very male-dominant society, today's farming, which is far more physically demanding and less lucrative than factory jobs, is increasingly left to the women, while the men go off to towns and cities to make their fortunes.

Some analysts suggest that at least several million peasants have taken steps to ensure that their female offspring are not counted toward their one-child (and now, in some places, two-child) limit: One strategy is for a pregnant woman simply to move to another village to have her child. Since the local village leaders are not responsible for women's reproduction when they are not from their own village, women are not harassed into getting an abortion in other villages. If the child is a boy, the mother can simply return to her native village, pay a fine, and register him; if a girl, the mother can return and not register the child. Thus a whole generation of young girls is growing up in the countryside without ever having been registered with the government. Since, except for schooling, peasants have few claims to state-supplied benefits anyway, they may consider this official nonexistence of their daughters a small price to pay for having as many children as necessary until giving birth to a boy. And if this practice is as common as some believe, it may mean that China will not face quite such a large demographic crisis in the ratio between males and females as has been projected.

Males continue to be more valued in Chinese culture because only sons are permitted to carry on traditional Chinese family rituals and ancestor worship. This is unbearably painful for families without sons, who feel that their entire ancestral history, often recorded over several hundred years on village-temple tablets, is coming to an end. As a result, a few villages have changed the very foundations of ancestral worship: They now permit daughters to continue the family lineage down the female line. The government itself is encouraging this practice, and it is also changing certain other family-related policies. For example, it used to be the son who was responsible under the law for taking care of their parents. This meant that parents whose only child was a girl could not expect to be supported in their old age. Now, both sons and daughters are legally responsible. Furthermore, it is hoped that a new system of social security and pensions for retired people will gradually lead the state and employers to absorb the responsibility for caring for the elderly.

China's strict population-control policies have been effective: Since 1977, the population has grown at an *average* annual rate of below one percent, one of the lowest in the developing world. According to official statistics, the policy has prevented about 400 million births. While the overall population is still growing, the annual population growth rate has declined over the years. In 2008 it was 0.629 percent; in 2010 it dropped to 0.494 percent. Yet, even this low rate works out to an average annual population *increase* for several years to come. (This is because previous generations had large numbers of offspring.) The dilemma is this: on the one hand, the growing population is a drain on China's

(Alastair Drysdale (DAL/mhhe010359))
An old couple in China. China's aging population is expanding rapidly due to improved healthcare and declining birth rate. The country is getting older before getting richer.

limited resources and poses a threat to its environment and economic development. As China continues to provide a substantial percentage of its citizens with a higher standard of living, the continuing pollution of China's air, water, and land, and the depletion of nonrenewable resources and energy are leading to ecological crises that are increasingly difficult to redress. On the other hand, there are concerns that the population replacement rate is too low to provide enough workers to support the growing elderly population. Yet, arguably the productivity of new generations of workers will be far higher; for as farmers abandon the land and move into cities, their productivity (that is, their return on labor) increases; and so does the productivity of workers who move into less labor-intensive jobs, or move from manufacturing into the service sector. As a result, it could well turn out that it will take far fewer younger workers to support China's larger retired community than it did in the past. And, as is the case in developed countries now facing problems of a burgeoning retired population, those beyond the retirement age who are still able may simply keep working.

The one-child policy has led to problems such as abortion of female fetuses, abduction and trafficking of women in areas with an excess number of men, illegal marriages, and forced prostitution. In certain regions, the male:female ratio is higher than 125:100. According to *China Daily*, some 24 million Chinese men of marrying age will find themselves lacking wives in 2020.

Officials from the Ministry of Civil Affairs reveal that China's aging population reached 12.79 percent (169 million) of the total population at the end of 2008. More than 8.3 percent of the Chinese population was above 65, and in most cities, more than 50 percent of the elderly people live without the company of their children.

The fact that the one-child policy has not been strictly enforced in minority areas and the countryside creates a long-term challenge for China.[6] The population base in the countryside and minority areas is becoming relatively bigger while that in the cities is shrinking. If this trend continues, the quality of China's population is going to deteriorate because education and healthcare are worse in rural areas. The one-child policy needs to be examined by the government to arrest the quick graying of the society and the sharp decline of newborns in big cities. Already, officials in Shanghai and Beijing have begun encouraging newlyweds who are the only children in their families to have two children. Some scholars predict that by 2020 or so, China will have a two-child policy.

Women

It is hardly surprising that overlaying (but never eradicating) China's traditional culture with a communist ideology in which men and women are supposed to be equal has generated a bundle of contradictions. Under Chinese Communist Party rule, women have long had more rights and opportunities than women in almost any other developing country, and in certain respects, more than women in some developed countries. For example, Chinese women were expected to work, not stay at home. And, in state-owned enterprises, they received from three to twelve months of paid maternity leave and child care in the workplace, decades before this became common practice in the Western countries. Although Chinese women rarely broke through the "glass ceiling" to the highest levels of the workplace or the ruling elite, and were often given "women's work," their pay scale was similar to that of men. Furthermore, an ideological morality that insisted on respect for women as equals (with both men and women being addressed as "comrades") combined with a de-emphasis on the importance of sexuality, resulted in at least a superficial respect for women that was rare before the Communist period.

The economic reforms that began in 1978, however, precipitated changes in the manner in which women are treated, and in how women act. While many women entrepreneurs and workers benefit as much as the men from economic reforms, there have also been certain throwbacks to earlier times that have undercut women's equality. Women are now treated much more as sex objects than they used to be; and while some women revel in their new freedom to beautify themselves, some companies will hire only women who are perceived as physically attractive, and many enterprises are now using women as "window dressing." For example, women dressed in *qipao*—the traditional, slim-fitting Chinese dress slit high on the thigh—stand outside restaurants and other establishments to entice customers. At business meetings, many women have become mere tea-pourers. In newspapers, employment ads for Chinese enterprises often state in so many words that only young, good-looking women need apply.

The emphasis on profits and efficiency since the reforms has also made state-run enterprises reluctant to hire women because of the costs in maternity benefits and because mothers are still more likely than fathers to be in charge of sick children and the household. Under the socialist system, where the purpose of an enterprise was not necessarily to make profits but to fulfill such socialist objectives as the equality of women and full employment, women fared better. Economic reforms, which emphasize profitability, have provided enterprise

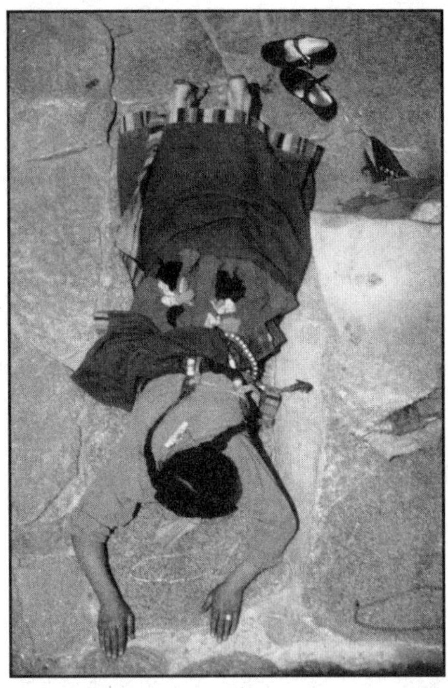

(© Glowimages, Inc./Punchstock RF)
Buddhist worshipper in Tibet.

(© Glowimages, Inc./Punchstock RF)
Jokhang Temple, a Buddhist temple in Lhasa, Tibet.

managers with the excuse they need not to hire women. Whatever the real reason, they can always claim that their refusal to hire more women or to promote them is justified: Women are more costly, or less competent, or less reliable.

National Minorities

Ninety-two percent of the population is Han Chinese. Although only 8 percent is classified as "national minorities," they occupy more than 60 percent of China's geographical expanse. These minorities inhabit almost all of the border areas, including Tibet, Inner Mongolia, and Xinjiang. The stability and allegiance of the border areas are important for China's national security. Furthermore, China's borders with the many neighboring countries are poorly defined, and members of the same minority usually live on both sides of the borders.

To address this issue, China's central government pursued policies designed to get the minorities on the Chinese side of the borders to identify with the Han Chinese majority. Rather than admitting to this objective of undermining distinctive national identities, the CCP leaders phrased the policies in terms of getting rid of the minorities' "feudal" customs, such as religious practices, which are contrary to the "scientific" values of socialism. Teaching children their native language was often prohibited. At times these policies have been brutal and have caused extreme bitterness among the minorities, particularly the Tibetans and the Uighurs (who practice Islam) in the northwest border province of Xinjiang. The extreme policies of the "10 bad years" that encouraged the elimination of the "four olds" led to the wanton destruction of minority cultural artifacts, temples, mosques, texts, and statuary.

In the 1980s, the Deng Xiaoping leadership conceded that Beijing's harsh assimilation policies had been ill-conceived, and it tried to gain the loyalty of the national minorities through more culturally sensitive policies. Minority children are now taught their own language in schools, alongside the "national" language (Mandarin). By the late 1980s, however, the loosening of controls had led to further challenges to Beijing's control, so the central government tightened up security in Xinjiang and reimposed marital law in Tibet in an effort to quell protests and riots against Beijing's discriminatory policies. Martial law was lifted in Tibet in 1990, but security has remained tight ever since. The terrorist attacks on the United States on September 11, 2001, led to even greater surveillance and controls on those minority groups that practice Islam and are believed to have ties with terrorist organizations in the Middle East.

Tibet

The Dalai Lama is the most important spiritual leader of the Tibetans, but he lives in exile in India, where he fled after a Chinese crackdown on Tibetans in 1959. He has stepped up his efforts to reach some form of accommodation with China. The Dalai Lama insists that he is not pressing for independence, only for greater autonomy; and that as long as he is in charge, Tibetans will use only nonviolent methods to this end. The Dalai Lama believes that more Tibetan control over their own affairs is necessary to protect their culture from extinction. Nevertheless, the people who surround the Dalai Lama are far more militant and see autonomy as only the first step toward independence from China.

In the past, the Chinese government made a concerted effort to assimilate Tibetans by eradicating Tibetan cultural practices and institutions that differentiated them from the majority Han culture. These policies were largely abandoned during the 1980s. At this point, the major threat to Tibetan culture comes from globalization, and from Chinese entrepreneurs who, thanks to economic liberalization policies, have taken over many of the commercial and entrepreneurial activities of Tibet. Ironically, the Tibetan feelings about the Chinese mirror the feelings of the Chinese toward the West: The Tibetans want Chinese technology and commercial goods, but not the values that come with the people providing those goods and technology. And among Tibetans, as among the Chinese, the young are more likely to want to become part of the modern world, to be modern and hip, and to leave behind traditional culture and values. Young Tibetans in Lhasa have been swept up in efforts to make money, a pleasure somewhat reduced by the fact that the increasingly large number of Chinese entrepreneurs in Lhasa usually make higher profits than they do. If Tibetan culture is to survive, Tibetans need to take on a modern identity, one that allows them to be both Tibetan and modern at the same time. Otherwise, the sheer dynamism of Han Chinese culture and globalization may well overwhelm Tibetan culture.[7]

Not all Tibetans accept the Dalai Lama's preferred path of nonviolence. In 1996, Beijing revealed that there were isolated bombing incidents and violent clashes between anti-Chinese Tibetans (reportedly armed) and Chinese authorities. The government, in response, sealed off most monasteries in Lhasa, the capital of Tibet. Beginning in the late 1990s, however, China decided to restore many of Tibet's monasteries—in part to placate Tibetans, in part to attract tourists. China has also made it far easier for foreigners to travel to Tibet. By 2006 it had completed a new highway across the length of the vast Qinghai Province plateau to connect Tibet with the rest of China; and another engineering feat—a railroad through the Tibetan mountains to the capital, Lhasa.

As a result of a policy to alleviate poverty in Tibet, the autonomous region now receives more financial aid from the central Chinese government than any other province or autonomous region in China. Tibetans are better fed, clothed, and housed than in the past; but Tibet still remains China's poorest administrative region. Generous state subsidies have not generated development. This is largely because of disastrous, centrally conceived policies, a bloated administrative structure, a large Chinese military presence to house and feed, disdain for Tibetan culture, and incompetent Han cadres who have little understanding of local issues and rarely speak Tibetan.[8] Tibet's landlocked, remote location and its lack of arable land certainly exacerbate problems in development.

Nevertheless, in recent years, Tibetan farmers have discovered that they can produce a valuable caterpillar fungus, *Cordyceps sinensis* (also called "winter worm, summer grass"). The fungus consumes caterpillars, then produces a columnar growth that shoots through the soil. It can be harvested and dried. It is much sought after by the Chinese as an herbal remedy for enhanced health, longevity, and male potency. Overnight, some of Tibet's poorest farmers have gained considerable wealth from cultivation of the caterpillar. Tibet's tourist industry is also suddenly booming, thanks to an expanded airport, new hotels, fewer restrictions by the Chinese government, and the determination of both Chinese and Western travel agencies to provide opportunities for tourists to visit "Shangrila."

None of this is meant to suggest that Tibet is going to leave poverty behind any time soon, even though the Chinese government is fully engaged in an anti-poverty program in Tibet. Regardless of what actions the Chinese government takes to develop Tibet, moreover, it is viewed with suspicion: New roads or railroads? The better for the Chinese to exploit Tibetan resources and even to invade. Encouraging tourism in Tibet or sending Tibetans to higher-quality Chinese schools outside of Tibet? The better to destroy Tibetan culture. Allowing Chinese entrepreneurs to do business in Tibet or desperately poor youth from neighboring provinces to go to Tibet for work? The better to take away jobs from Tibetans. Projects to develop Tibet's infrastructure for development? The better to destroy its environment. In short, Tibetans tend to regard all of Beijing's policies, and Han cadres in Tibet, with suspicion.

Tibet's anger toward China's central government was exacerbated by Beijing's decision in 1995 not to accept the Tibetan Buddhists' choice (chosen according to traditional Tibetan Buddhist ritual) of a young boy as the reincarnation of the second-most important spiritual leader of the Tibetans, the Panchen Lama, who died in 1989. Instead, Beijing substituted its own six-year old candidate. The Tibetans' choice, meanwhile, is living in seclusion somewhere in Beijing, under the watchful eye of the Chinese. China's concern is that any new spiritual leader could become a focus for a new push for Tibetan independence—an eventuality it wishes to avoid.

Inner Mongolia

Inner Mongolia (an autonomous region under Beijing's control) lies on the southern side of Mongolia, which is an independent state. Beijing's concern that the Mongolians in China would want to unite with Mongolia led to a policy that diluted the Mongol population with what has grown to be an overwhelming majority of Han Chinese. According to the 2000 national census, the national minority (largely Mongol) population was only 4.93 million—a mere 20.76 percent of the total population of 23.76 million. Inner Mongolia's capital, Huhhot, is essentially a Han city, and assimilation of Mongols into Han culture in the capital is almost complete. Mongolians are dispersed throughout the vast countryside as shepherds, herdsmen, and farmers and retain many of their ethnic traditions and practices.

Events in (Outer) Mongolia have led China's central leadership to keep a watchful eye on Inner Mongolia. In 1989, Mongolia's government—theoretically independent but in fact under Soviet tutelage—decided to permit multiparty rule at the expense of the Communist Party's complete control; and in democratic elections held in 1996, the Mongolian Communist Party was ousted from power.

Beijing has grown increasingly concerned that these democratic inklings might spread to their neighboring cousins in Inner Mongolia, with a resulting challenge to one-party CCP rule. China's leadership worries that the Mongols in Inner Mongolia may try to secede from China and join with the independent state of Mongolia because of their shared culture. So far, however, those Inner Mongolians who have traveled to Mongolia have been surprised by the relative lack of development there and have shown little interest in drumming up a secessionist movement. Nevertheless, privatization of the economy, combined with an insensitivity to Mongolian culture have led to periodic demonstrations against the Han Chinese-dominated government.

Muslim Minorities

In the far northwest, the predominantly Muslim population—particularly the Uyghur minority—of Xinjiang Autonomous Region continues to challenge the authority of China's central leadership. The loosening of policies aimed at assimilating the minority populations into the Han (Chinese) culture has given a rebirth to Islamic culture and practices, including prayer five times a day, architecture in the Islamic style, traditional Islamic medicine, and teaching Islam in the schools. With the dissolution of the Soviet Union in 1991 into 15 independent states, the ties between the Islamic states on China's borders (Kazakhstan, Kyrgyzstan, and Tajikistan, as well as Afghanistan and Pakistan) have accelerated rapidly.

Beijing is concerned that China's Islamic minorities may find that they have more in common with these neighboring Islamic nations than with the Chinese Han majority and may attempt to secede from China. Signs of a growing, worldwide Islamic movement have exacerbated Beijing's anxieties about controlling China's Islamic minorities. The 9/11/01 attacks by Islamic terrorists, followed by the U.S.–led "war on terrorism" throughout the world, have led China to intensify its efforts to root out Islamic radicals. Although some analysts argue that the war on terrorism has given Beijing an ideal pretext for cracking down on what is a legitimate desire for national independence by Uyghur Muslims in Xinjiang, others accept Beijing's view that those using violence (including bombs) are "terrorists," not "freedom fighters."

Uyghurs who have engaged in terrorism are not motivated by religious fanaticism, but rather, by a desire to achieve a concrete, pragmatic goal: Xinjiang's secession from China. Still, in the last decade, they have received funding from the Islamic world, including, it is believed, from terrorist groups located therein. The Uyghurs do not, however, accept the tenets of Islamic fundamentalists, nor that they view their struggle against Chinese rule as a struggle of good against evil. (Indeed, it has often been noted that Islam is much more moderate, tolerant, and progressive as it spreads eastward.) Evidence that the Uyghurs are engaged in an out-and-out political struggle for independence from Chinese rule would be that Uyghur violence in Xinjiang has not been in form of terrorist attacks on the local Han population but, rather, on the state structure of the governing Han.

China's nearly 9 million Hui—Han Chinese who practice Islam—are also classified as a "national minority," but over many centuries, they have become so integrated into mainstream Chinese culture that at this point in history their only remaining distinct characteristic is their practice of Islam. They speak standard Chinese and live together with other Han. Although a large number of Hui live in one autonomous region, Ningxia, they are also spread out throughout China. In general, in spite of shared Islamic beliefs, they do not identify with Uyghur nationalism, which is seen as particular to Uyghur ethnicity, and not to a broader Islamic identity.

Religion

Confucianism

Confucianism is the "religion" most closely associated with China. It is not, however, a religion in Western terms, as there is no place for gods, faith, or many other beliefs associated with formal religions. But like most religions, it does have a system of ethics for human relationships; and it adds to this what most religions do not have, namely, principles for good governance that include the hierarchical ordering of relationships, with obedience and subordination of those in lower ranks to those in higher ranks.

The Chinese Communists rejected Confucianism until the 1980s, but not because they saw it as an "opiate of the masses." (That was Karl Marx's description of religion, which he viewed as a way of trapping people in a web of superstitions, robbing them of their money, and causing them to passively endure their miserable lives on Earth.) Instead, they denounced Confucianism for providing the ethical rationale for a system of patriarchy that allowed officials to insist on obedience from subordinates. During the years in which "leftists" such as the Gang of Four set the agenda, moreover, the CCP rejected Confucianism for its emphasis on education as a criterion for joining the ruling elite. Instead, the CCP favored ideological commitment—"redness"—as the primary criterion for ruling. The reforms since 1979, however, have emphasized the need for an educated elite, and Confucian values of hard work and the importance of the family are frequently referred to. The revival of Confucian values has, in fact, provided an important foundation for China's renewed emphasis on national identity and Chinese culture as a substitute for the now nearly defunct values of Marxism-Leninism-Mao Thought.

Buddhism

Buddhism has remained important among some of the largest of the national minorities, notably the Tibetans and Mongols. The CCP's efforts to eradicate formal Buddhism have been interpreted by the minorities as national oppression by the Han Chinese. As a result, the revival of Buddhism since the 1980s has been associated with efforts by Tibetans and Mongolians to assert their national identities and to gain greater autonomy in formulating their own policies. Under the influence of the more moderate policies of the Deng Xiaoping reformist leadership, the CCP reconsidered its efforts to eliminate religion. The 1982 State Constitution permits religious freedom, whereas previously only atheism was allowed. The state has actually encouraged the restoration of Buddhist temples, in part because of Beijing's awareness of the continuing tensions caused by its efforts to deny minorities their respective religious practices, and in part because of a desire to attract both tourists and money to the minority areas.

But Buddhism is far more widespread than in just Tibet and Inner Mongolia. Indeed, popular Buddhism, which is full of stories and Buddhist mythology, is pervasive throughout the rural population—and even among some urban populations. Popular Buddhist beliefs are even worked into many of the sects, cults, and folk religions in China. Today, Buddhist temples are frequented by increasingly large numbers of Chinese, who go there to propitiate their ancestors and to pray for good health and more wealth.

Folk Religions

For most Chinese, folk religions are far more important than any organized religion.[9] The CCP's best efforts to eradicate folk religions and to impart in their place an educated "scientific" viewpoint have failed. Animism—the belief that nonliving things have spirits that should be respected through worship—continues to be practiced by China's vast peasantry. Ancestor worship—based on the belief that the living can communicate with the dead and that the dead spirits to whom sacrifices are ritually made have the ability to bring a better (or worse) life to the living—once again absorbs much of the income of China's peasants. The costs of offerings, burning paper money, and using shamans and priests to perform rituals that will heal the sick, appease the ancestors, and exorcise ghosts (who are often poorly treated ancestors returned to haunt their descendants) at times of birth, marriage, and death, can be financially burdensome. But the willingness of peasants to spend money on traditional religious folk practices is contributing to the reconstruction of practices prohibited in earlier decades of Communist rule.

Taoism, Qigong, and Falun Gong

Taoism, which requires its disciples to renounce the secular world, has had few adherents in China since the early twentieth century. But during the repression that followed the crackdown on Tiananmen Square's prodemocracy movement in 1989, many Chinese who felt unable to speak freely turned to mysticism and Taoism. *Qigong,* the ancient Taoist art of deep breathing, had by 1990 become a national pastime. Some 30 Taoist priests in China took on the role of national soothsayers, portending the future of everything from the weather to China's political leadership. What these priests said—or were believed to have said—quickly spread through a vast rumor network in the cities. Meanwhile, on Chinese Communist Party–controlled television, qigong experts swallowed needles and thread, only to have the needles subsequently come out of their noses perfectly threaded. It is widely believed that, with a sufficient concentration of *qi* (vital energy or breath), a practitioner may literally knock a person to the ground.[10] The revival of Taoist mysticism and meditation, folk religion, and formal religions suggests a need to find meaning from religion to fill the moral and ideological vacuum created by the near-collapse of Communist values.

Falun Gong ("Wheel of Law"), which the government has declared a "sect"—and hence not entitled to claim a constitutional right to practice religion freely—has been

DENG XIAOPING = TENG HSIAO-P'ING. WHAT IS PINYIN?

Chinese is the oldest of the world's active languages and is now spoken and written by more people than any other modern language. Chinese is written in the form of characters, which have evolved over several thousand years from picture symbols (like ancient Egyptian hieroglyphics) written on oracle bones to the more abstract forms now in use. Although spoken Chinese varies greatly from dialect to dialect (for example, Mandarin, Cantonese, Shanghai-ese), the characters used to represent the language remain the same throughout China. Dialects are really just different ways of pronouncing the same characters.

There are more than 50,000 different Chinese characters. A well-educated person may be able to recognize as many as 25,000 characters, but basic literacy requires familiarity with only a few thousand.

Since Chinese is written in the form of characters rather than by a phonetic alphabet, Chinese words must be transliterated so that foreigners can pronounce them. This means that the sound of the character must be put into an alphabetic approximation.

Since English uses the Roman alphabet, Chinese characters are Romanized. (We do the same thing with other languages that are based on non-Roman alphabets, such as Russian, Greek, Hebrew, and Arabic.) Over the years, a number of methods have been developed to Romanize the Chinese language. Each method presents what the linguists who developed it believe to be the best way of approximating the sound of Chinese characters. *Pinyin* (literally, "spell sounds"), the system developed in the People's Republic of China, has gradually become the most commonly accepted system of Romanizing Chinese.

日 **rì** sun
月 **yuè** moon
人 **rén** person
木 **mù** tree

Chinese characters are the symbols used to write Chinese. Modern Chinese characters fall into two categories: one with a phonetic component, the other without it. Most of those without a phonetic component developed from pictographs. From ancient writing on archaeological relics we can see their evolution, as in the examples shown (from left to right) above.

However, other systems are still used in areas such as Taiwan. This can cause some confusion, since the differences between Romanization systems can be quite significant. For example, in pinyin, the name of China's former leader is spelled Deng Xiaoping. But the Wade-Giles system, which was until recently the Romanization method most widely used by Westerners, transliterates his name as Teng Hsiao-p'ing. Same person, same characters, but a difference in how to spell his name in Roman letters.

charged with involvement in a range of illegal activities. Falun Gong is a complex mixture of Buddhism, Taoism, and qigong practices—the last relying on many ideas from traditional Chinese medicine. According to its adherents, the focus is on healing and good health, but it also has a millennial component, predicting the end of the world and a bad ending for those who are not practitioners. According to the government, the sect's practices can endanger people's health and have in fact caused the deaths of hundreds. It also accuses the sect of being a front for antigovernment political activities.

In 1999, thousands of Falun Gong adherents, some from distant provinces surrounded CCP headquarters, on the edge of Tiananmen Square in Beijing. Hundreds were arrested, but most were soon released and sent back to their home provinces. Others were sent to labor camps or jailed, and some died while incarcerated.[11] In many state-owned work units, officials continue to meet regularly to discuss the dangers of Falun Gong, to encourage followers to end their participation in Falun Gong, and to root out its leaders.

Religious practice often provides the foundation for illegal or "black" societies. Falun Gong and a number of other sects have been accused of using their organizations as fronts for drugs, smuggling, prostitution, and other illegal activities. Religious sects and black societies are widely believed to provide the basis of power for candidates for office. In the countryside, religion can become a tool of the family clans, who sometimes use it to pressure villagers to vote for their candidates.

Christianity

Christianity, which was introduced in the nineteenth and early twentieth centuries by European missionaries, has several million known adherents; and its churches, which were often used as warehouses or public offices after the Communist victory in 1949, have been reopened for religious practice. Bibles in several editions are available for purchase in many large-city bookstores. A steady stream of Christian proselytizers flow to China in search of new converts. Today's churches are attended as much by the curious as by the devout. As with eating Western food in places such as McDonald's and Kentucky Fried Chicken, attending Christian churches is a way that some Chinese feel they can participate in Western culture. Some Chinese want to become Christians because they see that in the West, Christians are rich and powerful. They believe that Christianity helps explain the wealth and power of Western capitalists, and hope that converting to Christianity will do the same for them.

The government generally permits mainstream Christian churches to practice in China, but it continues to exercise one major control over Roman Catholics: Their loyalty must be declared to the state, not to the pope. The Vatican is prohibited from involvement with China's priesthood, and Beijing does not recognize the validity of the Vatican's appointment of bishops and cardinals for China. Underground "house churches," primarily for smaller Christian sects, offshoots of mainstream Protestant religions, and papal Catholics, are forbidden. Nevertheless, they seem to flourish as officials busy themselves with addressing far more pressing social isues.

Since the mid-1990s, the government has tried to clamp down on non-mainstream Christian churches as well as religious sects, arresting and even jailing some of their leaders. They have justified their actions on the grounds that, as in the West, some of the churches are involved in practices that endanger their adherents; some are actually involved in seditious activities against the state; and some are set up as fronts for illegal activities, including gambling, prostitution, and drugs.

Marxism-Leninism-Mao Zedong Thought

In general, Marxists are atheists. They believe that religions hinder the development

COMMUNES: PEASANTS WORK OVERTIME DURING THE GREAT LEAP FORWARD

In the socialist scheme of things, communes are considered ideal forms of organization for agriculture. They are supposed to increase productivity and equality, reduce inefficiencies of small-scale individual farming, and bring modern benefits to the countryside more rapidly through rural industrialization.

These objectives are believed to be attained largely through the economies of scale of communes; that is, it is presumed that things done on a large scale are more efficient and cost-effective. Thus, using tractors, harvesters, trucks, and other agricultural machinery makes sense when large tracts of land can be planted with the same crops and plowed at one time. Similarly, a communal unit of 30,000 to 70,000 people can support small-scale industries, since, in such a large work unit, not everyone has to work in the fields.

Because of its size, a commune can support small-scale industries, as well as other types of organizations that smaller work units could not. A commune, for example, can support a hospital, a high school, an agricultural-research organization, and, if the commune is wealthy enough, even a "sports palace" and a cultural center for movies and entertainment.

During the Great Leap Forward, launched in 1958, peasants were—much against their will—forced into these larger agricultural and administrative units. They were particularly distressed that their small remaining private plots were, like the rest of their land, collectivized. Communal kitchens were to prepare food for everyone. Peasants were told that they had to eat in the communal mess halls rather than in the privacy of their own homes. And, they were ordered to build "backyard furnaces" to smelt steel and bring the benefits of industry to the countryside—part of Mao Zedong's idea of "closing the gap" between agriculture and industry, and between countryside and city.

When the combination of bad policies and bad weather led to a severe famine, widespread peasant resistance forced the government to retreat from the Great Leap Forward policy and abandon the communes. But a modified commune system remained intact in much of China until the late 1970s, when the government ordered communes to be dissolved. A commune's collective property was then distributed to the peasants within it, and a system of "contract responsibility" was launched. Individual households are again, as before 1953, engaged in small-scale agricultural production on private plots of land.

of "rational" behavior and values that are so important to modernization. Yet societies seem to need some sort of spiritual, moral, and ethical guidance. For Communist party-led states, Marxism was believed to be adequate to fulfill this role. In China, however, Marxism-Leninism was reshaped by Mao Zedong Thought to accommodate for Chinese culture and conditions. Paramount among these conditions was that China was a predominantly peasant society, not a society in which there was a capitalist class exploiting large numbers of urban workers. The repackaged ideology became known as Marxism-Leninism-Mao Zedong Thought. The Chinese leadership believed that it provided the ethical values necessary to guide China toward communism; and it was considered an integrated, rational thought system.

Nevertheless, this core of China's Communist political ideology exhibited many of the trappings of religions. It included scriptures (the works of Marx, Lenin, and Mao, as well as party doctrine); a spiritual head (Mao); and ritual observances (particularly during the Cultural Revolution, when Chinese were forced to participate in the political equivalent of Bible study each day). Largely thanks to the shaping of this ideology by Maoism, it included moral axioms that embodied traditional Chinese—and, some would say, Confucian—values that resemble teachings in other religions. Thus the moral of Mao's story of "The Foolish Old Man Who Wanted to Remove a Mountain" is essentially identical to the Christian principle "If you have faith you can walk on water," based on a story in the New Testament. Like this teaching, the essence of Mao Zedong Thought was concerned with the importance of a correct moral (political) consciousness.

In the 1980s, the more pragmatic leadership focused on liberalizing reforms and encouraged the people to "seek truth from facts" rather than from Marxism-Leninism-Mao Zedong Thought. As a result, the role of ideology declined, in spite of efforts by more conservative elements in the political leadership to keep it as a guiding moral and political force. Participants in the required weekly "political study" sessions in most urban work units abandoned any pretense of interest in politics. Instead, they focused on such issues as "how to do our work better" (that is, how to become more efficient and make a profit) that were in line with the more pragmatic approach to the workplace. Nevertheless, campaigns like the "get civilized" and anticorruption ones retain a strong moralistic tone.

Ideology has not been entirely abandoned. In the context of modernizing the economy and raising the standard of living, the current leadership is still committed to building "socialism with Chinese characteristics." Marxist-Leninist ideology is still being reformulated in China; but it is increasingly evident that few true believers in communism remain. Rarely does a Chinese leader even mention Marxism-Leninism in a speech. Leaders instead focus on modernization and becoming more efficient; they are more likely to discuss interest rates and trade balances than ideology. Fully aware that they need something to replace their own nearly defunct guiding ideological principles, however, and fearing that pure materialism and consumerism are inadequate substitutes, China's leaders seem to be relying on patriotism, nationalism, and national identity as the key components of a new ideology. Its primary purpose is very simple: economic modernization and support of the leadership of the Chinese Communist Party.

Undergirding China's nationalism is a fierce pride in China's history, civilization, and people. Insult, snub, slight, or challenge China, and the result is certain to be a country united behind its leadership, against the offender. To oppose the CCP or its objective of modernization is viewed as "unpatriotic." China's nationalism, on the other hand, is fired by antiforeign sentiments. These sentiments derive from the belief that foreign countries are—either militarily, economically, or through insidious cultural invasion—attempting to hurt China or to intervene in China's sovereign affairs by telling China's rulers how to govern properly. This is most notably the case whenever the Western countries condemn China for its human-rights record. U.S. support for Taiwan, and the U.S. bombing of the Chinese Embassy in Belgrade, Yugoslavia, during the Kosovo War in May 1999, fueled Chinese nationalism and injected even more tension into Sino–American relations. China's decision to join with the United States in its war on terrorism since 9/11 has, however, resulted in a toned-down nationalism and a less strident approach to international relationships. China's growing entanglement in a web of international economic and political relationships has also contributed to a softening of its nationalistic stance.

Language

By the time of the Shang Dynasty, which ruled in the second millennium B.C., the Chinese had a written language based on "characters." Over 4,000 years, these characters, or "ideographs," have evolved from being pictorial representations of objects or ideas into their present-day form. Each character usually contains a phonetic element and one (or more) of the 212 symbols called "radicals" that help categorize and organize them.[12] Before the New Culture Movement of the 1920s, only a tiny elite of highly educated men could read these ideographs, which were organized in the difficult grammar of the classical style of writing, a style that in no way reflected the spoken language. All this changed with language reform in the 1920s: The classical style was abandoned, and the written language became almost identical in its structure to the spoken language.

Increasing Literacy

When the Chinese Communists came to power in 1949, they decided to facilitate the process of becoming literate by allowing only a few thousand of the more than 50,000 Chinese characters in existence to be used in printing newspapers, official documents, and educational materials. However, since a word is usually composed of a combination of two characters, these few thousand characters form the basis of a fairly rich vocabulary: Any single character may be used in numerous combinations in order to form many different words. The Chinese Communists have gone even further in facilitating literacy by simplifying thousands of characters, often reducing a character from more than 20 strokes to 10 or even fewer.

In 1979, China adopted a new system, *pinyin,* for spelling Chinese words and names. This system, which uses the Latin alphabet of 26 letters, was created largely for foreign consumption and was not widely used within China. The fact that so many characters have the same Romanization (and pronunciation), plus cultural resistance, have thus far resulted in ideographs remaining the basis for Chinese writing. There are, as an example, at least 70 different Chinese ideographs that are pronounced *zhu,* but each means something different. Usually the context is adequate to indicate which word is being used. But when it may not be clear which of many homonyms is being used, Chinese often use their fingers to draw the character in the air.

Something of a national crisis has emerged in recent years over the deleterious effect of computer use on the ability of Chinese to write Chinese characters from memory. Computers are set up to write Chinese characters by choosing from multiple Chinese words whose sound is rendered into a Latin alphabet. As a result, computer users no longer need to remember how to write the many strokes in Chinese characters—they simply scroll down to the correct Chinese character under the sound of, say, "Zhen." Then they press the "enter" key. The problem is that, without regular practice writing out characters, it is easy to forget how to write them—even for Chinese people. It is a problem akin to the loss of mathematical skills due to the use of calculators and computers.

Spoken Chinese

The Chinese have shared the same written language over the last 2,000 years, regardless of which dialect of Chinese they spoke. (The same written characters were simply pronounced in different ways, depending on the dialect.) Building a sense of national unity was difficult, however, when people needed interpreters to speak with someone living even a few miles away. After the Communist victory in 1949, the government decided that all Chinese would speak the same dialect in order to facilitate national unity. A majority of the delegates to the National People's Congress voted to adopt the northern dialect, Mandarin, as the national language, and required all schools to teach in Mandarin (usually referred to as "standard Chinese").

In the countryside, however, it has been difficult to find teachers capable of speaking and teaching Mandarin; and at home, whether in the countryside or the cities, the people have continued to speak their local dialects. The liberalization policies that began in 1979 have had as their by-product a discernible trend back to speaking local dialects, even in the workplace and on the streets. Whereas a decade ago a traveler could count on the national language being spoken in China's major cities, this is no longer the case. As a unified language is an important factor in maintaining national cohesion, the re-emergence of local dialects at the expense of standard Chinese threatens China's fragile unity.

One force that is slowing this disintegration is television, for it is broadcast almost entirely in standard Chinese. As there is a growing variety of interesting programming available, it may be that most Chinese will make an effort to maintain or even acquire the ability to understand standard Chinese. Many television programs have Chinese characters (representing the words being spoken) running along the bottom of the screen. This makes it less necessary for viewers to understand spoken standard Chinese; but it makes it more necessary for those who do not speak standard Chinese to be literate in order to enjoy television programs.

Education

The People's Republic of China has been remarkably successful in educating its people. Before 1949, less than 20 percent of the population could read and write. Today, nine years of schooling are compulsory. In the larger cities, 12 years of schooling is becoming the norm, with children attending either a vocational middle school or a college-preparatory school. Computer use is increasingly common in urban schools.

It is difficult to enforce the requirement of nine years of school in the impoverished countryside. Still, close to 90 percent of those children living in rural areas attend at least primary school. Village schools, however, often lack rudimentary equipment such as chairs and desks. Rural education also suffers from a lack of qualified teachers, as any person educated enough to teach can probably get a better-paying job in the industrial or commercial sector. But the situation is in flux, because as rural families are having fewer children than before, more can now afford the cost of educating their children. In 2006, the government abolished school fees, but there remain significant costs for poor villagers to educate their children. Still, as the goals of rural familes have changed from preparing their children for farming to preparing them for factory work and office jobs in the towns and cities, education is seen as all the more important. So, at the same time that the collective basis for funding schools has deteriorated, in some villages many more farmers are able and willing to pay the necessary costs for schooling their children.

At the other end of the spectrum, many more students are now pursuing a college-oriented curriculum than will ever go on to college. From 1998 to 2004, China doubled the number of students it admitted to universities, from 2.1 to 4.2 million students.[13] Nevertheless, only about 5 percent of the senior middle school graduates will pass the university entrance examinations and be admitted. As a result, many who had prepared for a college curriculum are

((AP Photo) 060305058541)

Rapid industrialization has not only created an income gap but also shrunk China's already limited arable land.

inappropriately educated for the workplace. The government is attempting to augment vocational training for high school students, but it is also increasing the number of slots available in colleges and universities. Private high schools and colleges are becoming increasingly popular as parents try to optimize the chances for their only child to climb the academic ladder in order to gain social and economic success. Enrollments in on-line education courses, especially in business, have also soared in numbers as China's economy becomes increasingly specialized and demands greater skills and expertise for jobs.

Political Education

Until the reforms that began in 1979, the content of Chinese education was suffused with political values and objectives. A considerable amount of school time—as much as 100 percent during political campaigns—was devoted to political education. Often this amounted to nothing more than learning by rote the favorite axioms and policies of the leading faction in power. When that faction lost power, the students' political thought had to be reoriented in the direction of the new policies and values. History, philosophy, literature, and even foreign languages and science were laced with a political vocabulary.

The prevailing political line has affected the balance in the curriculum between political study and the learning of skills and scientific knowledge. Beginning in the 1960s, the political content of education increased dramatically until, during the Cultural Revolution, schools were shut down. When they reopened in the early 1970s, politics dominated the curriculum. When Deng Xiaoping and the "modernizers" consolidated their power in the late 1970s, this tendency was reversed. During the 1980s, in fact, schools jettisoned the study of political theory because both administrators and teachers wanted their students to do well on college-entrance examinations, which by then focused on academic subjects. As a result, students refused to clog their schedules with the study of political theory and the CCP's history. The study of Marxism and party history was revived in the wake of the events of Tiananmen Square in 1989, with the entering classes for many universities required to spend the first year in political study and indoctrination, sometimes under military supervision; but this practice was abandoned after two years. Today, political study has again been confined to a narrow part of the curriculum, in the interest of giving students an education that will help advance China's modernization.

Study Abroad

Since 1979, when China began to promote an "open door" policy, more than 1.6 million PRC students have been sent for a university education to the United States, and tens of thousands more have gone to Europe and Japan. In the 2009–2010 academic year alone, nearly 128,000 Chinese students (excluding those from Hong Kong and Taiwan) were studying in the United States. China has sent so many students abroad in part because the quality of education had seriously deteriorated during the Cultural Revolution, and in part because China's limited number of universities can accommodate only a tiny percentage of all high school graduates. Although an increasingly large number of Chinese universities are able to offer graduate training, talented Chinese students still travel abroad to receive advanced degrees.

Chinese students who have returned home have not always met a happy fate. Many of those educated abroad who were in China at the time of the Communists' victory in 1949, or who returned to China thereafter, were not permitted to hold leadership positions in their fields. Ultimately they were the targets of class-struggle campaigns and purges in the 1950s, '60s, and '70s, precisely because of their Western education. For the most part, those students who returned to China in the 1980s found that they could not be promoted because of the continuation of a system of seniority.

Since 1992, however, when Deng Xiaoping announced a major shift in economic and commercial policy to support just about anything that would help China become rich and powerful, the government has offered students significant incentives to return to China, including excellent jobs, promotions, good salaries, and even the chance to start new companies. Chinese students educated abroad are also recruited for their expertise and understanding of the outside world by the rapidly multiplying number of joint ventures in China, and by universities that are establishing their own graduate programs. Today, fully one third of students educated abroad return to live in China.

The Chinese government also now sees those Chinese who do stay abroad as forming critical links for China to the rest of the world. They have become the bridges over which contracts, loans, and trade flow to China, and are viewed as a positive asset. Finally, like immigrants elsewhere, Chinese who settle abroad tend to send remittances back to their families in China. These remittances amount to hundreds of millions of U.S. dollars in foreign currency each year and are valuable not just to the family recipients but also to the government's bank reserves.

Chinese studying abroad learn much about liberal democratic societies. Those who have returned to China bring with them the values at the heart of liberal-democratic societies. While this does not necessarily mean that they will demand the democratization of the Chinese political system, they do bring pluralistic liberal-democratic ideas to their own institutions. Some have been instrumental in setting up institutions such as "think tanks," have encouraged debate within their own fields, and have been insistent that China remain open to the outside world through the Internet, travel, conferences, and communications.

THE ECONOMIC SYSTEM

A Command Economy

Until 1979, the Chinese had a centrally controlled command economy. That is, the central leadership determined the economic policies to be followed and allocated all of the country's resources—labor, capital, land, and raw materials. It also determined how much each enterprise, and even each individual, would be allocated for production and consumption. Once the Chinese Communist Party leadership determined the country's political goals and the correct ideology, the State Planning Commission and the State Economic Commission would then decide how to implement these objectives through specific policies for agriculture and industry and the allocation of resources. This is in striking contrast to a capitalist laissez-faire economy, in which government control over both consumers and producers is minimal and market forces of supply and demand play the primary role in determining the production and distribution of goods.

The CCP leadership adopted the model of a centralized planned economy from the Soviet Union. Such a system was not only in accord with the Leninist model of centralized state governance; it also made sense for a government desperate to unify China after more than 100 years of internal division, instability, and economic collapse. Historically, China suffered from the ability of large regions to evade the grasp of central control over such matters as currency and taxes. The inability of the Nationalist government to gain control over the country's economy in the 1930s and early 1940s undercut its power and contributed to its failure to win control over China. Thus, the Chinese Communist Party's decision to centralize economic decision making after 1949 helped the state to function as an integrated whole.

Over time, however, China's highly centralized economy became inefficient and too inflexible to address the complexity of the country's needs. Although China possesses a large and diverse economy, with a broad range of resources, topography, and climate, its economic planners made policy as if it were a uniform, homogeneous whole. Merely increasing production was itself considered a contribution to development, regardless of whether a market for the products existed or whether the products actually helped advance modernization.

State planning agencies, without the benefit of market research or signals from the marketplace, determined whether or not a product should be manufactured, and in what quantity. For example, the central government might set a goal for a factory to manufacture 5 million springs per year—without knowing if there was even a market for them. The factory management did not care, as the state was responsible for marketing the products and paid the factory's bills. If the state had no buyer for the springs, they would pile up in warehouses; but rarely would production be cut back, much less a factory be closed, as this would create the problem of employing the workers cut from the factory's payroll. Economic inefficiencies of this sort were often justified because socialist political objectives such as full employment were being met. Even today the state worries about shutting down a state-owned factory that is losing money, because it creates unemployment. In turn, unemployment leads to popular anger and provides a volatile, unstable environment, ripe for public political protest. Quality control was similarly not as important an issue as it should have been for state-run industries in a centrally planned economy. Until market reforms began in 1979, the state itself allocated all finished products to other industries that needed them. If a state-controlled factory made defective parts, the industry using them had no recourse against the supplier, because each factory had a contract with the state, not with other factories. It was the state that would pay for additional parts to be made, so the enterprises did not bear the costs.

As a result, China's economic development under the centralized political leadership of the CCP occurred by fits and starts. Much waste resulted from planning that did not take into account market factors of supply and demand. Centrally set production quotas took the place of efficiency and profitability in the allocation of resources. Although China's command economy was able to meet the country's most important industrial needs, problems like these took their toll over time. Enterprises had little incentive to raise productivity, quality, or efficiency when doing so did not affect their budgets, wages, or funds for expansion.

Agricultural Programs

By the late 1950s, central planning was causing significant damage to the agricultural sector. Regardless of geography or climate, China's economic planners

(Courtesy of Ron Church)

Chinese street shops in the 1980s.

repeatedly ordered the peasants to restructure their economic production units according to one centralized plan. China's peasants, who had supported the CCP in its rise to power before 1949 in order to acquire their own land, had enthusiastically embraced the CCP's fulfillment of its pledge of "land to the tillers" after the Communists took over in 1949. But in 1953, the leadership, motivated by a belief that small-scale agricultural production could not meet the production goals of socialist development, ordered all but 15 percent of the arable land to be pooled into "lower-level agricultural producer cooperatives" of between 300 and 700 workers. The remaining 15 percent of land was to be set aside as private plots for the peasants, and they could market the produce from these plots in private markets throughout the countryside. Then, in 1956, the peasants throughout the country were ordered into "higher-level agricultural producer cooperatives" of 10 times that size, and the size of the private plots allotted to them was reduced to 5 percent of the cooperatives' total land.

Many peasants felt cheated by these wholesale collectivization policies. When in 1958 the central leadership ordered them to move into communes 10 times larger still than the cooperatives they had just joined, they were irate. Mao Zedong's Great Leap Forward policy of 1958 forced all peasants in China to become members of communes: enormous economic and administrative units consisting of between 30,000 and 70,000 people. Peasants were required to relinquish their private plots, and turn over their private utensils, as well as their household chickens, pigs, and ducks, to the commune. Resisting this mandate, many peasants killed and ate their livestock. Since private enterprise was no longer permitted, home industries ground to a halt.

CCP chairman Mao's vision for catching up with the West was to industrialize the vast countryside. Peasants were therefore ordered to build "backyard furnaces" to smelt steel. Lacking iron ore, much less any knowledge of how to make steel, and under the guidance of party cadres who themselves were ignorant of steelmaking, the peasants tore out metal radiators, pipes, and fences. Together with pots and pans, they were dumped into their furnaces. Almost none of the final smelted product was usable. Finally, the central economic leadership ordered all peasants to eat in large, communal mess halls. This was reportedly the last straw for a people who valued family above all else. Being deprived of time alone with their families for meals, the peasants refused to cooperate further in agricultural collectivization.

When the catastrophic results of the Great Leap Forward policy poured in, the CCP retreated—but it was too late. Three subsequent years of bad weather, combined with the devastation wreaked by these policies and the Soviet withdrawal of all assistance, brought economic catastrophe. Demographic data indicate that in the "three bad years" from 1959 to 1962, more than 20 million Chinese died from starvation and malnutrition-related diseases.

By 1962, central planners had condoned peasants returning to production units that were smaller than communes. Furthermore, peasants were again allowed to farm a small percentage of the total land as private plots, to raise domestic animals for their own use, and to engage in household industries. Free markets, at which the peasantry could trade goods from private production, were reopened. The commune structure was retained throughout the countryside, however; and until the CCP leadership introduced the contract responsibility system in 1979, it provided the infrastructure of rural secondary school education, hospitals, and agricultural research.

Other centrally determined policies, seemingly oblivious to reality, compounded the P.R.C.'s difficulties in agriculture. Maoist policies carried out during both the Great Leap Forward and renewed during the Cultural Revolution included attempts to plant three crops per year in areas that for climatic reasons can only support two (as the Chinese put it, "Three times three is not as good as two times five"); and "close planting," which often more than doubled the amount of seed planted, with the result that all of it grew to less than full size or simply wilted for lack of adequate sunshine and nutrients.

A final example of centrally determined agricultural policy bringing catastrophe was the decision during the Cultural Revolution that "the whole country should grow grain." The purpose was to establish China's self-sufficiency in grain. Considering China's immense size and diverse climates, soil types, and topography, a policy ordering everyone to grow the same thing was doomed to failure. Peasants were ordered to plow under fields of cotton and cut down rubber plantations and fruit orchards, planting grain in their place. China's central planning was largely done by the CCP, not by economic or agricultural experts, and they ignored overwhelming evidence that grain would not grow well, if at all, in some areas. These policies ignored the fact that China would have to import everything that it had replaced with grain, at far greater costs than it would have paid for importing just grain. Peasant protests were futile in the face of local-level Communist Party leaders who hoped to advance their careers by implementing central policy. The policy of self-sufficiency in grain was abandoned only with the arrest of the Gang of Four in 1976.

Economic Reforms: Decentralization and Liberalization

In 1979, Deng Xiaoping undertook reform and liberalization of the economy, a critical component of China's modernization program. The government tried to maintain centralized state control of the direction of policy and the distribution and pricing of strategic and energy resources, while decentralizing decision making down to the level of the township and village. Decentralization was meant to facilitate more rational decision making, based on local conditions, needs, and efficiency. Although the state has retained the right to set overall economic priorities, township and village enterprises (TVEs), as well as larger state-owned enterprises, have been encouraged to respond to local market forces of supply and demand. Centrally determined quotas and pricing for most products have been phased out, and enterprises now contract with one another rather than with the state. Thus the government's role as the go-between in commercial transactions—and as the central planner for everything in the economy—is gradually disappearing.

Since the introduction of market capitalism in the 1980s, the economy has become increasingly privatized. Some 70 percent of China's gross national product (GNP) is now produced by these nonstate enterprises. They compete with state-run enterprises to supply goods and services, and if they are not profitable, they go bankrupt.

On the other hand, some state-run enterprises that operate at a loss continue to be subsidized by the government rather than shut down. One reason is that the enterprises are in a sector over which the state wants to keep control, such as energy, raw materials, and steel production. Another reason is fear of the destabilizing impact of a high level of unemployment if the state closes down large state-run enterprises. Subsidies and state-owned bank loans (which will not be repaid) to unprofitable enterprises consume a significant portion of the state's budget. China's membership in the World Trade Organization (WTO) since 2001 has, however, forced China's large state-owned enterprises to become efficient, or else perish in the face of foreign competition. In the meantime, the government continues to close down the most unprofitable state-run enterprises, and it counts on the entrepreneurial sector and foreign investors to fuel economic growth and absorb the unemployed.

In addition, under a carefully managed scheme, the government has started to sell off some state-run industries to TVEs, the private sector, and foreign investors. Whoever buys them must in most cases guarantee some sort of livelihood, even if not full employment, to the former employees of the state-run enterprises. The new owners, however, have far more freedom to make a profit than did the enterprises when they were state-owned. The state is also slowly introducing state-run pension and unemployment funds to care for those workers who lose their jobs when state-owned enterprises are shut down.

Today, the agricultural sector is almost fully market-driven. The "10,000 yuan" household (about U.S. $1,200), once a measure of extraordinary wealth in China, has become a realizable goal for most peasants. Free markets are booming in China, and peasants may now produce whatever they can get the best price for in the market, once they have filled their grain quotas to the state. Wealthy rural towns are springing up throughout the agriculturally rich and densely populated east coast, as well as along China's major transportation routes.

The achievements of economic reforms are impressive. China's GDP has grown at an average of about 10 percent since 1979, a growth that has been higher and longer than that of any other economy. Some 400 million Chinese have been lifted out of abject poverty. Living standards for most people continue to rise despite the growing income gaps.

Even during the 2008–2010 global financial crisis, China's economy continued to charge ahead. The government announced a $586 billion economic stimulus package at the end of 2008 to maintain a high growth rate. It has continued to invest in infrastructure. Today, China is home of the most modern airports and railways, the fastest supercomputer, and the fastest passenger train, among others. In 2010, China surpassed Japan to become the second largest economy in the world.

Problems Created by Economic Reforms Disputes over Property

One downside of privatization and marketization is a distinct increase in disputes among villagers. There are disputes over who gets to use the formerly collectively owned goods, such as tractors, harvesters, processing equipment, and boats for transporting goods at harvest time. And with formerly collective fields turned into a patchwork of small private plots, villagers often protest when others cross their land with farm equipment. Theoretically, the land that is "leased" to the peasants is collectively owned by the village, but in practice the land is treated as the private property of the peasants. Those who choose to leave their land may contract it out to others, so some peasants have amassed large tracts of land suitable for large-scale farm machinery. To encourage development, the government has permitted land to be leased for as long as 30 years and for leased rights to be inherited. Furthermore, peasants have built houses on what they consider their own land, itself a problem because it is usually arable land. Nevertheless, the village councils also have some ability to reallocate land so that soldiers and others who return to settle in the villages can receive adequate land to farm.

With the growth of free enterprise in the rural towns since 1979, some 60 million to 100 million peasants have left the land to work for higher pay in small-scale rural industry or in cities. It is not just that the wages are higher, but also that the peasants are burdened by the arbitrary imposition of local taxes and fees. Combined with the low profits on agricultural products and the unpredictability of weather, farmers seek better wages in local enterprises or migrate to the cities. Others leave because their land has been confiscated by the local government for development. Rarely compensated adequately, if at all, for their land, they are forced to leave the villages for towns and cities. Parents often migrate together, but are unable to bear the costs of taking their children with them, so they are left with their relatives or friends—or to fend for themselves, alone. To address the problems being created by the confiscation of agricultural land, as well as to stop the boom in building "development zones" for no specific need or purpose, the government cancelled some 4,800 of the 6,866 development zones that had sprung up throughout China. The Ministry of Land and Resources has discovered that, of the total zones, 70 percent had illegally acquired land, or had confiscated the land and then left it unused. The measure shut down about 65 percent of the total land area planned for development zones, and returned some of the land to agricultural use.[14]

Instability and Crime

For some, especially those able to find employment in the booming construction industry in many small towns and cities, migration has brought a far better standard of living. But tens of millions of unemployed migrants clog city streets, parks, and railroad stations. They have been joined by equally large numbers of workers from bankrupt state-owned enterprises. Together, they have contributed to a vast increase in criminality and social instability. According to Chinese

CHINA'S SPECIAL ECONOMIC ZONES

In 1980, China opened four "Special Economic Zones" (SEZs) within the territory of the People's Republic of China as part of its program to reform the socialist economy and to encourage foreign investment. The SEZs were allowed a great deal of leeway in experimenting with new economic policies. For example, Western management methods, including the right to fire unsatisfactory workers (something unknown under China's centrally planned economy), were introduced into SEZ factories. Laws and regulations on foreign investment were greatly eased in the SEZs in order to attract foreign capital. Export-oriented industries were established with the goal of earning large amounts of foreign exchange in order to help pay for importing technology for modernization. To many people, the SEZs looked like pockets of capitalism inside the socialist economy of the P.R.C.; indeed, they are often referred to as "mini-Hong Kongs." In 1984, 14 coastal cities were allowed to open up development zones to help attract foreign investment.

The largest of the Special Economic Zones is Shenzhen, which is located just across the border from the Hong Kong New Territories. Over several decades, Shenzhen was transformed from a sleepy little rural town to a large, modern urban center and one of China's major industrial cities. The city boasts broad avenues and skyscrapers, and a standard of living that is among the highest in the country.

But with growth and prosperity have come numerous problems. The pace of construction has outstripped the ability of the city to provide adequate services to the growing population. Speculation and corruption have been rampant, and crime is a more serious problem in Shenzhen than in most other parts of China. Strict controls on immigration help stem the flood of people who are attracted to Shenzhen in the hopes of making their fortune.

Nevertheless, the success of Shenzhen and other Special Economic Zones led the leadership to expand the concept of SEZs throughout the country, most notably Shanghai's new Pudong district. The special privileges, such as lower taxes, that foreign businesses and joint ventures could originally enjoy only in these zones were expanded first along the coast and then to the interior, so that it too could benefit from foreign investment. By 2004, there were 6,866 "development zones," formed largely by confiscating agricultural land from the peasantry.

(© IMS Communications, Ltd/Capstone Design/FlatEarth Images RF)
A Shenzhen street at night.

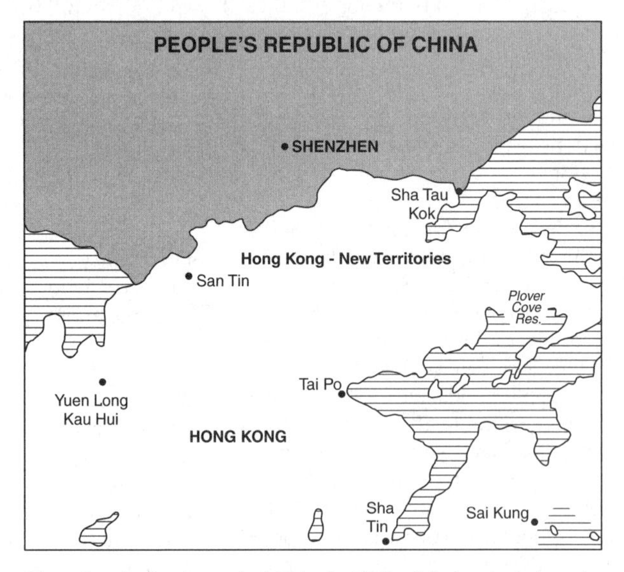

Shenzhen is the largest of China's SEZs. It is located close to the Hong Kong New Territories.

government reports, there were more than 74,000 "mass incidents" (protests and riots) in 2004, a 30 percent increase over the 58,000 reported in 2003, and more than six times the number it reported 10 years earlier.[15] In 2005 over 87,000 such demonstrations occurred. By the late 2000s, the number had hit 100,000.

Problems Arising from a Mixed Economy

Still other problems arose from the post–1979 reform policies that led to a mixed socialist and capitalist economy. For example, with decentralization, industrial enterprises tried to keep their profits but "socialize" their losses. That is, profitable enterprises (whether state-owned, collective TVEs, or private) try to hide their profits to avoid paying taxes to the state. State-owned enterprises that are losing money ask the state for subsidies to keep them in business. For this reason, the amount of profits turned over to the state was not commensurate with the dramatic increase in the value of industrial output since 1979. In the last decade, however, this situation has been ameliorated as China has become a primarily market-driven economy. Almost all prices are now market-determined, except for energy, water, and natural resources. Although it is common in almost all market economies for the state to regulate the prices of energy and utilities, the fact that certain other commodities and products in short supply are still regulated by the state continues to create problems in China.[16]

Products such as rolled steel, glass, cement, or timber are either hoarded as a safeguard against future shortages, or resold at unauthorized higher prices. By doing so, these enterprises make illegal profits for themselves and deprive other enterprises of the materials they need for production. In short, the remnants of a mixed economy, combined with the lack of a strong regulatory state to administer the economy, provide many opportunities for corruption and abuse of the system.

Further, while private and collective enterprises rely on market signals to determine whether they will expand production facilities, the state continues to centrally allocate resources based on a national plan. A clothing factory that expands its production, for instance, requires more energy (coal, oil) and water, and more cotton. The state, already faced with inadequate energy resources to keep most industries operating at full capacity, continues to allocate the same amount to the now-expanded factory. Profitable enterprises want a greater share of centrally allocated scarce resources, but find they cannot acquire them without the help of "middlemen" and a significant amount of under-the-table dealing. Corruption has, therefore, become rampant at the nexus where the capitalist and socialist economies meet.

Widespread corruption in the economic sector has led the Chinese government to wage a series of campaigns against economic crimes such as embezzlement and graft. An increasing number of economic criminals are going to prison, and serious offenders are sometimes executed. Until energy and transportation bottlenecks and the scarcity of key resources are dealt with, however, it will be extremely difficult to halt the bribery, smuggling, embezzlement, and extortion now pervasive in China. The combination of relaxed centralized control, the mandate for the Chinese people to "get rich," and a mixed economy have exacerbated what was already a problem under the socialist system. In a system suffering from serious scarcities but controlled by bureaucrats, it is political power, not the market, that determines who gets what. This includes not only goods, but also opportunities, licenses, permits, and approvals.

Although the Chinese may now purchase in the market many essential products previously distributed through bureaucratically controlled channels, there are still many goods available only through the "back door"—that is, through people they know and with whom they exchange favors or money. Scarcity, combined with bureaucratic control, has led to "collective corruption": Individuals engage in corrupt practices, even cheat the state, in order to benefit the enterprise for which they work. Since enterprises not owned by the state will not be bailed out if they suffer losses, the motivation for corrupt activities is stronger than under the previous system.

Liberalization of the economy is providing a massive number and variety of goods for the marketplace. The Chinese people may buy almost any basic consumer goods in stores or the open markets. But the nexus between continued state control and the free economy still fuels a rampant corruption that threatens the development of China's economy.

The mixed economy, then, provides the environment in which corruption has increased; but it is further exacerbated by the sheer amount of money in China today. Before 1979, corruption tended to involve exchanges of favors, many of which amounted to giving access, a privilege, or a necessary document to someone in exchange for another favor. Today, corruption is about getting large amounts of money through various deals. Further, unlike in most developing countries, corruption is not just at the top. It is widespread, all the way to the bottom of the economic ladder. Some argue that the fact that so many people at all rungs of the economic ladder can share in the cashing out of corrupt activities allows it to be tolerated.[17]

Still, there is a huge amount of wealth concentrated in the elite, if for no other reason than that they have the "connections" and access to gain wealth. Figures indicate that 90 percent of China's billionaires are, in fact, children of the senior party-state elite.[18] The ability of corrupt individuals to abscond to a foreign country has made capital flight an increasingly worrisome issue. China's Ministry of Commerce reported that thousands of corrupt officials (many of whom are said to be the children and relatives of political leaders) have illegally transferred China's wealth to companies registered in offshore tax shelters and financial havens.[19]

Unequal Benefits

Not all Chinese have benefited equally from the 30 years of economic reforms. Those who inhabit cities in the interior, and peasants living far from cities and transportation lines or tilling less arable land, have reaped far fewer rewards. And although the vast majority of Chinese have seen some improvement in their standard of living, short-term gains in income are threatened in the long term by the deterioration of the infrastructure for education and medical care in large parts of China's hinterland. This deterioration is due to the elimination of the commune as the underlying institution for funding these services. Wealthier peasants send their children into the larger towns and cities for schooling.

In healthcare, the Communist government has overseen dramatic improvements in the provision of social services for the masses. Largely because of its emphasis on preventive medicine and sanitation, life expectancy has increased from 45 years in 1949 to 73 years (overall) today. The government has successfully eradicated many childhood diseases and has made great strides against other diseases, such as malaria and tuberculosis. The privatization of medicine in recent years has, however, meant that many no longer have guaranteed access to medical treatment. The press has highlighted many cases in which, if they cannot pay, they will not be treated. Wealthier families, on the other hand, can travel to the more comprehensive health clinics and hospitals farther away, although in some areas, they have actually built local schools and private hospitals. In recent years, under the leadership of Hu Jintao and Wen Jiabao, the state has stepped back in to address these issues of unequal access, in part because they are giving rise to social instability. Among their many policies has been the elimination of school fees for children who attend rural schools.

Nevertheless, the visible polarization of wealth, which had been virtually eradicated in the first 30 years of Communist rule, continues to deepen in the context of capitalism and a poorly regulated free market. The creation of a crassly ostentatious wealthy class and a simply ostentatious middle class, in the context of high unemployment, poverty, and a mobile population, is breeding the very class conflict that the Chinese Communists fought a revolution to eliminate. When reforms began 30 years ago, street crime was almost unheard of in China's cities. Now it is a growing problem, one that the government fears may erode the legitimacy of CCP rule if left uncontrolled.

Mortgaging the Future

One of the most damaging aspects of the capitalist "get-rich" atmosphere prevailing in China is the willingness to sacrifice the future for short-term profits. The environment continues to be deeply degraded by uncontrolled pollution; the rampant growth of new towns, cities, and highways; the building of houses on arable land; the overuse of nonrenewable water resources; and the destruction of forests. Some state institutions, including schools, have turned the open spaces (such as playgrounds) within their walls into parking lots in China's crowded cities to provide spaces for the rapidly growing number of privately owned cars.

To sustain its rapid economic growth, China has had to add new power-generating capacity, 90 percent of which is from new coal-fired plants, at the *annual* rate (as of 2006) of "the entire capacity of the UK and Thailand combined, or about twice the generating assets of California."[20] China's primary source of energy is coal, followed by oil, both of which emit substantial greenhouse gases (such as carbon dioxide) to the environment. In 2008, China became the world's largest producer of greenhouse gases. But regardless of how much it contributes to global warming, China is already suffering from its effects. The government is concerned about the impact of global warming on China's future economic growth and food security, as well as its role in generating severe weather patterns, desertification, and flooding. Environmental deterioration is, in turn, causing social instability. Indeed, each year there is an increasing number of major protests—often violent—from communities affected by environmental degradation. Since 1997, pollution disputes submitted to legal or administrative resolution have increased at an annual rate of about 25 percent. By 2002, the actual number of disputes reported as submitted for solution had reached 500,000.[21]

Health problems arising from polluted air and contaminated water are affecting both rural and urban populations. Cities that used to have clear blue skies before economic reforms began 30 years ago now infrequently see the sun. Coal-burning plants, which produce significant amounts of air particulates, supply two-thirds of China's energy needs. With close to 70 percent of China's rivers and lakes severely polluted, even boiled water is not fit to drink. In Beijing, 90 percent of the underground water supply on which it relies is contaminated. To add to its woes, a 60-billion-dollar pipeline is having to be built to transfer water from the south to the north, especially for Beijing.

So, even though China's government takes the same stance as most other developing countries—namely, that the *developed* countries should pay for their efforts to reduce greenhouse gases (and sulfur dioxide, which causes acid rain), China realizes it has to take its own actions before it is too late. Moreover, when Beijing hosted the 2008 Olympic Games, the government wanted to be sure that air quality was improved quickly. The problem is that the State Environment Protection Agency (SEPA) is one of China's weakest ministries. It must compete against other powerful ministries that want to promote economic development, regardless of the environmental costs. The government also promotes car ownership and consumption of durable goods such as air conditioners and washing machines, all of which consume considerable energy and thereby exacerbate the pollution. In the meantime, officials receive promotion based on how much they have advanced economic growth in their localities. This puts them at cross purposes with SEPA's goals of regulating pollution, and causes them to blatantly ignore state environmental regulations.

Added to the demand for energy created by economic growth is the hugely inefficient use of energy, with some products requiring as much as 10 to 20 times more energy to produce than in more efficient economies, such as Japan, the U.S., and Germany. So, in 2006, environmental policy, instead of capping greenhouse gas emissions, chose to reduce the amount of energy used in the production of a product. The central government is now demanding that major state enterprises as well as provincial governors sign contracts in which they pledge "to reduce the amount of energy consumed relative to economic output by 20 percent over five years." If China succeeds in doing so, even though economic output will continue to grow, the amount of energy used will grow at a much slower rate.[22]

THE CHINESE LEGAL SYSTEM

Ethical Basis of Law

In imperial China, the Confucian system provided the basis for the traditional social and political order. Confucianism posited that good governance should be based on maintaining correct personal relationships among people, and between people and their rulers. Ethics were based on these relationships. A legal system did exist, but the Chinese resorted to it in civil cases only in desperation, for the inability to resolve one's problems oneself or through a mediator usually resulted in considerable "loss of face"—a loss of dignity or standing in the eyes of others. (In criminal cases, the state normally became involved in determining guilt and punishment.)

This perspective on law carried over into the period of Communist rule. Until legal reforms began in 1979, most Chinese preferred to call in CCP officials, local neighborhood or factory mediation committees, family members, or friends to settle disputes. Lawyers were rarely used, and only when mediation failed did the Chinese resort to the courts. By contrast, the West lacks both this strong support for the institution of mediation and the concept of "face." So Westerners have difficulty understanding why China has had so few lawyers, and why the Chinese have relied less on the law than on personal relationships when problems arise.

Like Confucianism, Marxism-Leninism is an ideology that embodies a set of ethical standards for behavior. After 1949, it easily built on China's cultural predisposition toward ruling by ethics instead of law. Although Marxism-Leninism did not completely replace the Confucian ethical system, it did establish new standards of behavior based on socialist morality. These ethical standards were embodied in the works of Marx, Lenin, and Mao, and in the Chinese Communist Party's policies; but in practice, they were frequently undercut by preferential treatment for officials.

Legal Reforms

Before legal reforms began in 1979, the Chinese cultural context and the socialist system were the primary factors shaping the Chinese legal system. Since 1979, reforms have brought about a remarkable transformation in Chinese attitudes toward the law, with the result that China's laws and legal procedures—if not practices and implementation—look increasingly like those in the West. This is particularly true for laws that relate to the economy, including contract, investment, property, and commercial laws. The legal system has evolved to accommodate its more market-oriented economy and privatization. Chinese citizens have discovered that the legal system can protect their rights, especially in economic transactions. In 2004, 4.4 million civil cases were filed, double the number 10 years earlier. This is a strong indication that the Chinese now are more aware of their legal rights, and believe they can use the law to protect their rights and to hold others (including officials) accountable for their actions.[23] Still, in many civil cases (such as disputes with neighbors and family members), the Chinese remain inclined to rely on mediation and traditional cultural values in the settling of disputes.

Law and Politics

From 1949 until the legal reforms that began in 1979, Chinese universities trained few lawyers. Legal training consisted of learning law and politics as an integrated whole; for according to Marxism, law is meant to reflect the values of the "ruling class" and to serve as an instrument of "class struggle." The Chinese Communist regime viewed law as a branch of the social sciences, not as a professional field of study. For this reason, China's citizens tended to view law as a mere propaganda tool, not as a means for protecting their rights. They had never really experienced a law-based society. Not only were China's laws and legal education highly politicized, but politics also pervaded the judicial system.

With few lawyers available, few legally trained judges or prosecutors in the courts, and even fewer laws to refer to for standards of behavior, inevitably China's legal system has been subject to abuse. China has been ruled by people, not by law; by politics, not by legal standards; and by party policy, not by a constitution. Interference in the judicial process by party and local state officials has been all too common.

(United Nations photo by John Isaacs (UN161371))

Neighborhood mediation in China.

After 1979, the government moved quickly to write new laws. Fewer than 300 lawyers, most of them trained before 1949 in Western legal institutions, undertook the immense task of writing a civil code, a criminal code, contract law, economic law, law governing foreign investment in China, tax law, and environmental and forestry laws. One strong motivation for the Chinese Communist leadership to formalize the legal system was its growing realization, after years of a disappointingly low level of foreign investment, that the international business community was reluctant to invest further in China without substantial legal guarantees.

Even China's own potential entrepreneurs wanted legal protection against the *state* before they would assume the risks of developing new businesses. Enterprises, for example, wanted a legal guarantee that if the state should fail to supply resources contracted for, it could be sued for losses issuing from its nonfulfillment of contractual obligations. Since the leadership wanted to encourage investment, it needed to supplement economic reforms with legal reforms. Codification of the legal system fostered a stronger basis for modernization and helped limit the party-state's abuse of the people's rights.

In addition, new qualifications have been established for all judicial personnel. Demobilized military officers who became judges and prosecutors during the Cultural Revolution have been removed from the judiciary. Judges, prosecutors, and lawyers must now have formal legal training and pass a national judicial examination. It is hoped that judicial personnel endowed with higher qualifications and larger salaries, as well as judicial systems that are financially autonomous from local governments will diminish judicial corruption and enhance the autonomy of judicial decisions.[24]

Criminal Law

Procedures followed in Chinese criminal courts have differed significantly from those in the United States. Although the concept of "innocent until proven guilty" was introduced in China in 1996, it is still presumed that people brought to trial in criminal cases are guilty. This presumption is confirmed by the judicial process itself. That is, after a suspect is arrested by the police, the procuratorate (*procuracy,* the investigative branch of the judiciary system) will spend considerable time and effort examining the evidence gathered by the police and establishing whether the suspect is indeed guilty. This is important to understanding why 99 percent of all the accused brought to trial in China are judged guilty. Indeed, had the facts not substantiated their guilt, the procuracy would have dismissed their cases before going to trial.

In short, then, those adjudged to be innocent would never be brought to trial in the first place. For this reason, court trials function mainly to present the evidence upon which the guilty verdict is based—not to weigh the evidence to see if it indicates guilt—and to remind the public that criminals are punished. A trial is a "morality play" of sorts: the villain is punished, justice is done, the people's interests are protected. In addition, the trial process continues to emphasize the importance of confessing one's crimes, for those who confess and appear repentant in court will usually be dealt more lenient sentences. Criminals are encouraged to turn themselves in, on the promise that their punishment will be less severe than if they are caught. Those accused of crimes are encouraged to confess rather than deny their guilt or appeal to the next level, all in hopes of gaining a more lenient sentence from the judge.

This type of system, which tends to focus on confession, not on fact finding, is weighted against the innocent. The result is that police are more inclined to use brutal tactics in order to exact a confession; but of course, police in Western liberal-democratic countries also have been known to use brutality, and even torture, to get a confession.

Another serious problem with the Chinese system was that the procuracy, which investigated the case, also prosecuted the case. Once the procuracy established "the facts," they were not open to question by the lawyer or the representative of the accused. (In China, a person may be represented by a family member, friend, or colleague, largely because there are not enough lawyers to fulfill the guarantee of a person's "right to a defense.") The lawyer for the accused was not allowed to introduce new evidence, make arguments to dismiss the case based on technicalities or improper procedures (such as wire tapping), call witnesses for the defendant, or make insanity pleas for the client. Instead, the lawyer's role in a criminal case was simply to represent the person in court and to bargain with the court for a reduced sentence for the repentant client.

The 1996 legal reforms were aimed at improving the rights of the accused: They may now call their own witnesses and introduce their own evidence, they cannot be held for more than 30 days without being formally charged with a crime, and they are supposed to have access to a lawyer within several days of being formally arrested. But many suspects are still not accorded these rights.

The accused have the right to a defense, but it has always been presumed that a lawyer will not defend someone who is guilty. Most of China's lawyers are still employed and paid by the state. As such, a lawyer's obligation is first and foremost to protect the state's interests, not the individual's interests at the expense of the state. Lawyers who acted otherwise risked being condemned as "counter-revolutionaries" or treasonous. Small wonder that after 1949, the study of law did not attract China's most talented students.

Today, however, the law profession is seen as potentially lucrative and increasingly divorced from politics. Lawyers can now enter private practice or work for foreigners in a joint venture. The All-China Lawyers Association, established in 1995 by the Ministry of Justice to regulate the legal profession, also functions as an interest group to protect the rights of lawyers against the state. Thus, when in recent years lawyers have found themselves in trouble with the law because they have defended the political rights of their clients against the state, the association tries to protect them.

Lawyers in Civil and Commercial Law

In the areas of civil and commercial law, the role of the lawyer has become increasingly important since the opening of China's closed door to the outside world. Because foreign trade and investment have become crucial to China's development, the government has made an all-out effort to train many more lawyers. In today's China, upholding the law is no longer simply a matter of correctly understanding the party "line" and then following it in legal disputes. China's limited experience in dealing with economic, liability, corporate, and contractual disputes in the courts, as well as the insistence by foreign investors that Chinese courts be prepared to address such issues, have forced the leadership to train lawyers in the intricacies of Western law and to draft countless new laws and regulations. To protect themselves against what is difficult to understand in the abstract, the Chinese used to refuse to publish their newly written laws. Claiming a shortage of paper or the need to protect "state secrets," they withheld publication of many laws until their actual impact on China's state interests could be determined. This practice frustrated potential investors, who dared not risk capital investment in China until they knew exactly what the relevant laws were. Today, however, the complexity of both foreign-investment issues and the entrepreneurial activities of China's own citizens have led the Chinese government to publish most of its laws as quickly as possible.

THE POLITICAL SYSTEM

The Party and the State

In China, the Chinese Communist Party is the fountainhead of power and policy. But not all Chinese people are party members. Although the CCP has some 70 million members, this number represents only 5 percent of the population. Joining the CCP is a competitive, selective, rigorous process. Some have wanted to join out of a commitment to Communist ideals, others in hopes of climbing the ladder of success, still others to gain access to limited goods and opportunities. Ordinary Chinese are generally suspicious of the motives of those who do join the party. Many well-educated Chinese have grown cynical about the CCP and refused to join. Still, those who travel to China today are likely to find that many of the most talented people they meet are party members. Party hacks who are ideologically adept but incompetent at their work are gradually being squeezed out of a party desperate to maintain its leading position in a rapidly changing China.

Today's Party wants the best and brightest of the land as members, and it wants it to represent "the people" more broadly. In 2001, Jiang Zemin, then general-secretary of the CCP, laid the groundwork for this with his "theory of the three represents": The Party was from that point on to represent not just the workers and peasants, but *all* the people's interests, including both intellectuals and capitalists. This addition to party theory indicates that the Chinese Communist Party recognizes the important role that intellectuals and business people have played—and will play—in modernizing China; but it also acknowledges the reality that many individuals already within the Party have become capitalists. By enshrining this theory in the party constitution at the 16th Party Congress in 2002, the party was in effect attempting to shore up its legitimacy as China's ruling party. At the same time, it was in effect announcing that it had relinquished its role as a revolutionary party in favor of its new position as the *governing* party of China. Today's CCP adopts collective leadership. The days of dictatorial leadership under a strongman like Mao or Deng are gone. The party also promotes "inner-party democracy."

The CCP is still China's ultimate institutional authority. Although in theory the state is distinct from the party, in practice the two overlapped almost completely from the late 1950s to the early 1990s. Efforts to keep the party from meddling in the day-to-day work of the government and management of economic enterprises have had some effect; but in recent years, more conservative leaders within the CCP have exerted considerable pressure to keep the Party in charge.

The state apparatus consists of the State Council, headed by the premier. Under the State Council are the ministries and agencies and "people's congresses" responsible for the formulation of policy. The CCP has, however, exercised firm control over these state bodies through interlocking organizations. CCP branches exist within all government organizations; and at every administrative level from the

CENTRAL GOVERNMENT ORGANIZATION OF THE PEOPLE'S REPUBLIC OF CHINA

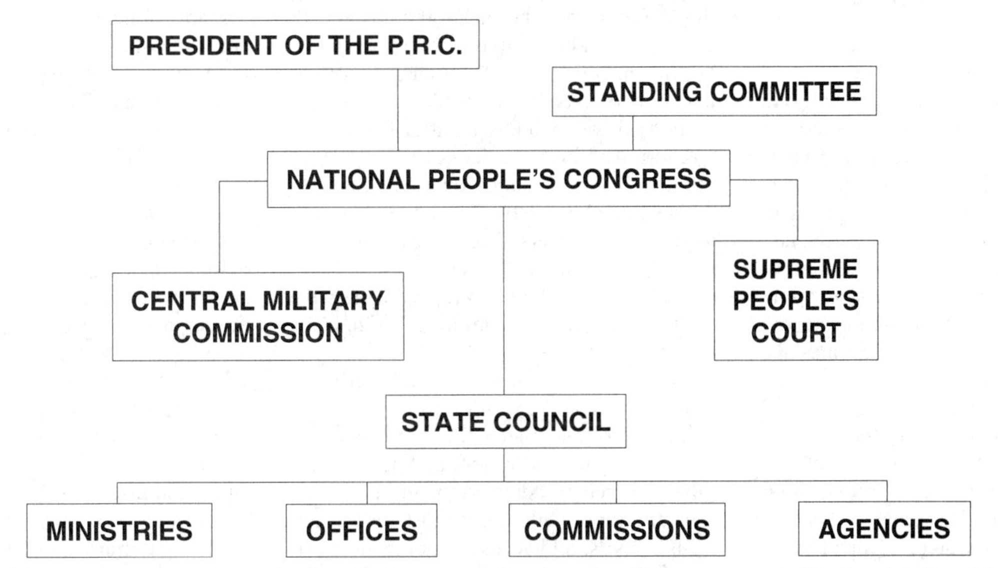

This central government organization chart represents the structure of the government of the People's Republic of China as it appears on paper. However since all of the actions and overall doctrine of the central government must be reviewed and approved by the Chinese Communist Party, political power ultimately lies with the party. To ensure this control, virtually all top state positions are held by party members.

THE CHINESE COMMUNIST PARTY (CCP)

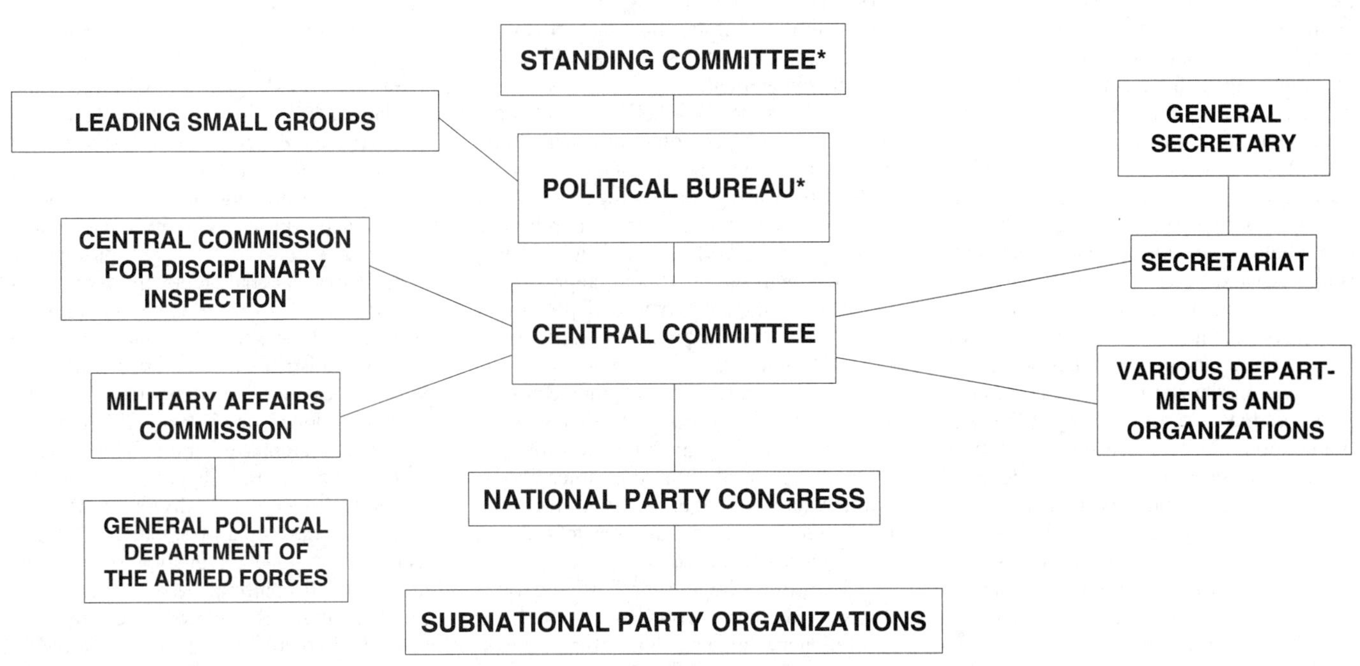

*The Political Bureau and its Standing Committee and the "Leading Small Groups" are the most powerful organizations within the Chinese Communist Party and are therefore the real centers of power in the P.R.C.

central government in Beijing down to the villages, almost everyone in a leadership position is also a party member.

Corruption in the Political System

China's political system is subject to enormous abuses of power. The lines of authority within both the CCP and the state system are poorly defined, as are the rules for selection to top leadership positions. In the past, this allowed individuals like Mao Zedong and the Gang of Four to usurp power and rule arbitrarily. By the late 1980s, China's bureaucracy appeared to have become more corrupt than at any time since the People's Republic of China came into being in 1949. Anger at the massive scale of official corruption was, in fact, the major factor unifying ordinary citizens and workers with students during the antigovernment protests in the spring of 1989.

Although campaigns to control official corruption continue, the problem appears to be growing even worse. Campaigns do little more than scratch the surface. This is in part because with so many opportunities to make money in China, especially for party and state officials whose positions give them the inside track for making profitable deals, the potential payoff for corruption can be huge—and the risks of getting caught appear small. This is particularly true in the case of selling off of state-owned assets. In most cases, state-owned enterprises that have been closed down, or privatized, have been turned over to the relatives of China's leaders for managing. The same is true in real estate: land, which used to be completely state-owned, is now controlled by partly privatized land companies, run by the relatives of leaders, who take huge profits for themselves when they make deals with investors and real estate developers. One problem is that Chinese institutions lack the transparency, acquired through financial checks within the system and open access to accounting books, that could help rein in corruption. The situation is exacerbated by a society that by its complicity encourages official corruption.

Individuals may write letters to the editors of the country's daily newspapers or to television stations to expose corruption. Many Chinese, especially those living in the countryside, feel that the only way in which local corruption will be addressed is if the media send reporters to investigate and publicly expose criminality. The China Central Television station has aired a popular daily program in prime time that records the successes of China's public security system in cracking down on official corruption and crime. The press also devotes substantial space to sensational cases of official corruption, in part because it helps sell newspapers. The media only has the resources, however, to address comparatively few of the numerous cases begging for investigation. Furthermore, those reporters who have threatened to expose scandals, abuses of authority, and inappropriate local policies are often harassed, or even had violence used against them by those who might suffer from such a report. And in 2007, the government was reported to have shut down the investigative branches of the papers with more aggressive reporters. There is evidence, on the other hand, that some reporters have actually used the threat of reporting a scandal to extort money from local officials in order to cover it up.

So far, most efforts to control official corruption have had little effect. Officials continue to use their power to achieve personal gain, trading official favors for others' services (such as better housing, jobs for their children, admission to the right schools, and access to goods in short supply), or for wining and dining. Getting things done in a system that requires layers of bureaucratic approval still depends heavily upon a complex set of personal connections and relationships, all reinforced through under-the-table gift giving. This stems in part from the still heavily centralized aspect of Chinese governance, and in part from the overstaffing of a bureaucracy that is plagued by red tape. Countless offices must sign off on requests for anything from installing a telephone to processing a request for a license or additional electrical outlets. This gives enormous power to individual officials who handle those requests, allowing them to ask for favors in return, or to stone-wall if the payoff is inadequate.

In today's more market-oriented China, officials have lost some of their leverage over the distribution of goods. Now, instead of waiting for a work unit official to decide whose turn it is to purchase a bicycle or who will have the right to live in a two-bedroom apartment, anyone with adequate funds may buy virtually anything they want. In a society where prostitution is banned brothels can be run in the open, virtually without interference from the police, who are bribed to look the other way. In short, officials may have lost control over the distribution of many consumer goods, but they have kept their ability to facilitate or obstruct access to many services, documents, licenses, and so on.

Controlling the abuse of official privilege is difficult in part because of the large discretionary budgets that officials have, and in part because the Chinese have made an art form out of going around regulations. For example, the government issued a regulation stipulating that governmental officials doing business could order only four dishes and one soup at a meal. But as most Chinese like to eat well, especially at the government's expense, the restaurants accommodated them by simply giving them much larger plates on which they put many different dishes, and then wrote it up as if it were just one dish.

The definition of corrupt behavior has also become more complex as the country moves from a socialist economy to a market economy. In the initial stages of introducing a market economy, selling imported goods in China for high profits was considered corrupt, as was paying middlemen to arrange business transactions. In the 1980s, some businesspeople were arrested, and even executed, for such activities. Now, it is assumed that those importing goods from abroad will make as large a profit as possible; and instead of looking at middlemen as the embodiment of corruption, government regulations allow them to be paid a transaction "fee." Yet, middlemen continue many activities considered corrupt, such as demanding a fee for introducing potential investors to appropriate officials—in part because the middlemen must in turn pay the officials for agreeing to meet with the potential investors—even though it is their job to do so.

Reform of party-state institutions and procedures has been an important avenue through which the government has attempted to curb corruption; but its broader goal in reforming the party state has been to improve the quality of China's leadership. Otherwise, China's leaders worry, the Chinese Communist Party may lose its legitimacy. Reforms have encouraged, even demanded, that the Chinese state bureaucracy reward merit more than mere seniority, and expertise more than political activism. And in 1996, the government's practice of allowing officials to stay in one ministry during their entire career was replaced by new regulations requiring officials from divisional chiefs up to ministers and provincial governors to be rotated every five years. Restrictions on tenure in office have brought a much younger generation of leaders into power; and they have placed a time limit on any one individual's access to power. The emphasis on a collective leadership since reforms began in 1979 has, moreover, made it virtually impossible for a leader to develop a personality cult, such as that which reached fanatical proportions around CCP chairman Mao Zedong during the Cultural Revolution. These reforms have dramatically reshaped the leadership structure and process. But other reforms, such as an anti-nepotism regulation that

(Embassy of the People's Republic of China)
Chinese Communist Party 17th national congress October 2007.

prohibits any high official from working in the same office as a spouse or direct blood relative, have seemingly had little effect on the overall pattern of officials using their power and access to put family members in positions where they can acquire significant wealth.

CONTEXT FOR DEMOCRACY

Cultural and Historical Authoritarianism

The Chinese political system reflects a history, political culture, and values entirely different from those in the West. For millennia, Chinese thought has run along different lines, with far less emphasis on such ideals as individual rights, privacy, and limits on state power. The Chinese political tradition is weighed down with a preference for authoritarian values, such as obedience and subordination of individuals to their superiors in a rigidly hierarchical system, and a belief in the importance of moral indoctrination. China's rulers have throughout history shown greater concerns for establishing their authority and maintaining unity in the vast territory and population they control than in protecting individual rights. Apart from some of China's intellectuals, the overwhelming majority of the Chinese people have appeared to be more afraid of chaos than an authoritarian ruler. Even today, the Chinese people seem more concerned that their leaders have *enough* power to control China than that the rights of citizens vis-à-vis their leaders be protected.

This is not to suggest that Confucianism and China's other traditions did not contain some mention of such rights. They did; but the *dominant* strand of Chinese political culture was authoritarian. It was critical in shaping the development of today's political system. As a result, when the Communists came to power in 1949, they were trying to operate within the context of an inherited patriarchal culture, in which the hierarchical values of superior–inferior and subordination, loyalty, and obedience prevailed over those of equality; and in which there was a historical predisposition toward official secrecy; a fear of officials and official power; and a traditional repugnance for courts, lawyers, and formal laws that protected individual rights. Thus, when Western democratic values and institutions were introduced, China's political culture and institutions were ill prepared to embrace them.

China's limited experience with democracy in the twentieth century was bitter. Virtually the entire period from the fall of China's imperial monarchy in 1911 to the Communist victory in 1949 was marred by warlordism, chaos, war, and a government masking brutality, greed, and incompetence under the label of "democracy." Although it is hardly fair to blame this period of societal collapse and externally imposed war on China's efforts to practice "democracy" under the "tutelage of the Kuomintang," the Chinese people's experience of democracy was nevertheless negative.

Moreover, the Chinese Communists condemned Western liberal democracy for being too weak a political system to prevent the Great Depression and two world wars in the twentieth century. In any event, foreign values were always suspect in China, as foreigners had repeatedly declared war on China in order to advance their own national interests. China was inclined to view the propagation of liberal democratic values as just one more effort by Western countries to enhance their own national power.

Experience with the Western powers and China's own efforts to implement democracy before 1949, together with China's traditional political culture, help explain the people's reluctance to embrace Western liberal democratic ideals. During the period of the Republic of China (1912–1949), China's "democratic" political and legal institutions proved inadequate to guarantee the nation's welfare, or to protect individual rights. Even under Communist rule, the one period described as "democratic mass rule" (the Cultural Revolution, from 1966 to 1976) was in fact a period of mass tyranny. For the Chinese, the experience of relinquishing power to "the masses" turned into the most horrific period of unleashed terrorism and cruelty they had experienced since the Communist takeover in 1949. Most analysts would argue that this was in no respect a "democracy," but rather the result of the Chinese people being manipulated by an ever-shifting nouveau elite, who were in a desperate competition with other pretenders to power. To those Chinese who experienced the Cultural Revolution firsthand, this was what could happen if China became democratic.

Socialist Democracy

When the CCP came to power in 1949, it inherited a country torn by civil war, internal rebellion, and foreign invasions for more than 100 years. The population was overwhelmingly illiterate and desperately poor, the economy in shambles. The most urgent need was for order and unity. China made great strides in securing its borders and ending internal fighting and chaos. Despite some serious setbacks and mistakes under the leadership of Mao Zedong, moreover, China also succeeded in establishing effective institutions of government and enhancing the material well-being of its people. But in the name of order and stability, China's leaders also severely limited the development of "democracy" in its Western liberal-democratic sense.

The Chinese people are accustomed to "eating bitterness," not to standing up to authority. The traditional Confucian emphases on the group rather than the individual and on respect for authority continue to this day, although modernization, internationalization, and disenchantment with the CCP leadership have diminished their powerful cultural hold. Rapid modernization has likewise undermined Marxist Leninist-Maoist values, the glue that, along with traditional values, had helped hold China together. None of this, however, necessarily bodes well for the propagation of democratic values; for the destabilizing social effects of a loss of values, and the concomitant rise of aggressive nationalism and materialism, hardly provide a receptive environment for liberal democratic values.

Nevertheless, since 1949, and especially since the reform period began in 1979, there has been a gradual accretion of individual rights for Chinese people. These include greater freedom of speech; access to far more information and a diversity of perspectives; the right to vote in local elections; and development of the rights to privacy, to choose one's own work (as opposed to being assigned by the government), to move and work in different locations, to own private property, and many more. Moreover, the impersonal market forces of supply and demand, combined with an abundant variety of consumer goods, have undercut the power of officials to control the distribution of resources and opportunities in the society. The result is that the Chinese are no longer beholden to officials to supply them in exchange for favors, gift giving, banqueting, and outright bribery. This equality of access in the marketplace contributes to a greater sense of control by the people over their daily lives.

Unfortunately, even as such rights grow, other very important rights previously enjoyed by the Chinese people—such as a job, healthcare, housing, and education—are being eroded. Such "welfare rights" have provided the context in which other rights have gained meaning. At the same time that it has led to greater political and societal rights, then, economic liberalization has contributed to the polarization of wealth and destabilization of Chinese society.

Patterns of political participation are also changing. Participation in the political process at the local level has already led to greater responsiveness by local officials to the common people's needs. Village officials are more inclined than in the past to seek out advice for improving the economic conditions in their localities, and incompetent village officials are usually unable to gain re-election. In spite of local elections and other efforts to advance village democracy, however, villagers in many parts of China do not believe that elections have made much of a difference. Even in those areas where there has been extraordinary economic development, villagers do not necessarily see the connection of elections and democratization with prosperity.[25]

China has experienced only limited open popular demand for democracy. When the student-led demonstrations in Beijing began in 1989, the demands for democratic reforms were confined largely to the small realm of the intellectual elite—that is, students and well-educated individuals, as well as some members of the political and economic ruling elite. The workers and farmers of China remained more concerned about bread-and-butter issues—inflation, economic growth, unemployment, and their own personal enrichment—not democratic ideals.

By the mid-1990s, many Chinese had discovered that they could get what they wanted through channels other than mass demonstrations, because of the development of numerous alternative groups, institutions, and processes. Many of these groups are not political in origin, but the process by which they are pressing for policy changes in the government is highly political. When reforms began in 1979, there was only a handful of such groups. By the end of 2006, there were 346,000 interest groups and associations, commonly called NGOs (non-governmental organizations) in the West, but labeled as "civilian organizations" in China.[26]

While all such organizations are headed by a member of the CCP (usually a retired person), they reflect and promote the interests of their membership. Indeed, it is in the self-interest of the appointed party head of the organization to advance its power and respond to the requests of the membership to represent their interests at higher levels. As an example, the party-state instituted an association to which small entrepreneurs in each city and town must belong. The association's purpose is to assert some level of control over a totally unsupervised arena of small business and commerce activity, but it also promotes the interests of its members; for entrepreneurs engaged in legitimate businesses do not want to have to compete against unregistered businesses that flog shoddy, illegal, and even dangerous products. Further, by helping crack down on those whose products violate copyright and trademark regulations, the association helps promote legitimate businesses.

In many areas, such as environmental protection, education, healthcare, and poverty reduction, interest groups, associations, foundations, and charities work together with the government to advance their causes. The fact that each group must be registered with the government and led by a party-state official need not, in most cases, be viewed as a negative factor. Indeed, the existence of a party-state official at the top ensures access to power that an organization would not otherwise have. Sometimes, the government has demanded that all "civilian associations," foundations, and charities re-register, in part because countless businesses have tried to avoid taxes by claiming to be "civilian organizations," charities, or foundations. The government likewise prohibits democracy-promoting organizations from registering. Organizations that are funded by foreigners are always watched carefully, and when they appear to cross the line and become involved in sponsoring political movements, the government shuts them down.

It is perhaps ironic that the Chinese Communist Party's penchant for organizing people has resulted in teaching them organizational skills that they now use to pressure the government to change policy.[27] They work through these organizations to protect and advance their members' interests within the framework of existing law and regulations.

Even those Chinese not working through officially organized associations ban together to petition local officials using the organizational techniques they learned from the CCP. For example, urban neighborhoods join forces to stop local noise pollution emanating from stereos blasting on the street where thousands of Chinese engage in ballroom dancing, or from the cymbals, tambourines, gongs, and drums of old ladies doing "fan dancing" on the city streets.[28] In turn, the dancers petition the local officials to maintain their "right" to express themselves through dancing in the streets, some of the few public spaces available to them in a crowded urban environment.

The tendency to organize around issues and interests in China today is more than a reflection of the decline of the role of Communist ideology in shaping policy. It also reflects the government's focus on problem solving and pragmatic concerns in policymaking. Many constituencies take advantage of this approach to issues and policy. Although China has never been homogeneous and uncomplicated, it is certainly a far more complex economy, society, and polity today than it was before 1980. Today's China has far more diverse needs and interests than was the case previously, and specialized associations and interest groups serve the need of articulating these interests.

Compared to Western liberal democratic systems, Chinese citizens are politically passive. But is this a sign of satisfaction

(Courtesy of Suzanne Ogden)

Villagers standing in line to vote, September 2001, near Shanghai.

with the CCP regime? Is Chinese submissiveness a sign of "collusion" with their oppressors? One could argue that, like Eastern Europeans, the Chinese have participated in their own political oppression simply by complying with the demands of the system. As the Czech Republic's then-president Vaclav Havel stated, "All of us have become accustomed to the totalitarian system, accepted it as an unalterable fact and therefore kept it running. . . . None of us is merely a victim of it, because all of us helped to create it together.[29]

Can we say that the Chinese, any more than the Czechs, collaborated with their authoritarian rulers if they did not flee into exile under Communist rule, or did not refuse to work? Is anyone who does not actively revolt against an oppressive system necessarily in collusion with it? In the case of China, one cannot assume that the major reason why people are not challenging the Communist system is out of fear of punitive consequences—although those who want to directly challenge CCP rule through the formation of new political parties or trade unions, or through the explicit public criticism of China's top political leaders do indeed risk punishment. The more logical explanation is that the Chinese have developed a completely different style from that of citizens in Western liberal democratic countries for getting what they want. This style is largely based on cultivating personal relationships rather than more formal and open institutionalized forms of political participation. More "democratic" behavioral skills can, of course, be acquired through practice; and as the Chinese political system gradually adopts more liberal democratic practices, the political culture is likely to evolve—indeed, it *is* evolving—in a more democratic direction.

Students, Intellectuals, and Business People as Potential Forces for Democratization

Democratization in China has been hampered by the people's inability to envision an alternative to CCP rule. What form would it take? How would it get organized? And if the CCP were overthrown, who would lead a new system? These questions are still far from being answered. So far, no dissident leadership capable of offering an alternative to CCP leadership, and laying claim to popular support, has formed. Even the mass demonstrations in 1989 were not led by either a worker, a peasant, or even an intellectual with whom the common people could identify:

> [C]ompared with the intellectuals of Poland and Czechoslovakia, for example, Chinese intellectuals have little contact with workers and peasants and are not sensitive to their country's worsening social crisis; they were caught unawares by the democratic upsurge of 1989, and proved unable to provide the people with either the theoretical or practical guidance they needed.[30]

In fact, during the Tiananmen protests in 1989, students were actually annoyed by the workers' participation in the demonstrations. They wanted to press their own political demands—not the workers' more concrete, work-related issues. Some Chinese believe that the students' real interest in setting forth their own demands was because they wanted to enhance their own prestige and power vis-àvis the regime: The students' major demands were for a "dialogue" with the government as "equals," and for free speech. These issues were of primary interest to them, but of little interest to the workers of China. Many Chinese believe that had the leaders of the 1989 demonstrations suddenly been catapulted to power, their behavior would have differed little from the ruling CCP elite. The student movement itself admitted to being authoritarian, of kowtowing to its own leaders, and of expecting others to obey rather than to discuss decisions. As one Beijing University student wrote during the 1989 Tiananmen Square protests:

> The autonomous student unions have gradually cut themselves off from many students and become a machine kept constantly on the run in issuing orders. No set of organizational rules widely accepted by the students has emerged, and the democratic mechanism is even more vague.[31]

In any event, few of those who participated in the demonstrations in 1989 are interested in politics or political leadership today. Most have thrown themselves into business and making money.

Apart from students and intellectuals, some of the major proponents of democratic reform today hail from China's newly emerging business circles; but these groups have not united to achieve reform, as they neither like nor trust each other. Intellectuals view venture capitalists "as uncultured, and business people as driven only by crass material interests." The latter in turn regard intellectuals and students as "well-meaning but out of touch with reality and always all too willing and eager to serve the state" when it suits their needs.[32] Moreover, although the business community is interested in pushing such rights as the protection of private property and the strengthening of the legal system in the commercial and economic spheres, it tends to be more supportive of the regime's "law and order" values than of democratic values; for an unstable social and political environment would not be conducive to economic growth.

Those who wanted reforms (or even the overthrow of Communist Party rule) but left China and remain abroad have lost their political influence. Apart from everything else, it is difficult for those living abroad to make themselves heard in China, even if their articles are published or appear on the Web. Although they may keep in touch with the dissident movement in China, their influence is largely limited to their ability to supply it with funds. Doing so, however, often gets the recipients in China in trouble. Today's college students in China are among the strongest supporters of the Chinese government.

The Impact of Global Interdependency on Democratization

Since the late 1970s, the cultural context for democracy in China has shifted. The expansion of the international capitalist economy and increasing cultural and political globalization have led to a social and economic transformation of China. For the first time in Chinese history, a significant challenge to the "we–they" dichotomy—of China against the rest of the world—is occurring. This in turn has led many Chinese to question the heretofore assumed superiority of Chinese civilization.

Such an idea does not come easily for a people long-accustomed to hearing about China's greatness. Hence the fuss caused by *River Elegy,* a TV documentary series first shown on Chinese national television in 1988. In this series, the film producers argued that the Chinese people must embrace the idea of global interdependency—technological, economic, and cultural. To insist at that time in history on the superiority of Chinese civilization, with the isolation of China from the world of ideas that this implied, would only contribute to China's continued stagnation. The series suggested that the Chinese must see themselves as equal, not superior to, others; and as interdependent with, not as victims of, others. Such concepts of equality, and the opening up of China to ideas from the rest of the world, have led to a remarkable transformation of the Chinese political, cultural, and economic landscape.

In 2007, a 12-part TV documentary series, *Rise of the Great Powers,* aired by CCTV (China Central Television) examined how great powers have risen in history, and whether and why rising powers became involved in war. The fact that a television documentary, nearly ten years after the *River Elegy* series, was focused on a different set of issues and questions for China's role in the world in itself reflects how much China's self-perceptions and role in the world have changed.

The Press and Mass Media

At the time that the student-led demonstrations for democracy began in Beijing's Tiananmen Square in the spring of 1989, China's press had grown substantially and become increasingly liberalized. By the end of 2010, China had about 800 million cell phones. China's Internet users reached 350 million in early 2010. Both numbers are still growing.With some 2,000 newspapers, 9,000 magazines, and 600 publishing houses, the Chinese are able to express a wider variety of viewpoints and ideas than at any time since the CCP came to power in 1949. The production of millions of television sets, radios, shortwave radios, cassette recorders, and VCRs also facilitated the growth of the mass media in China. They were accompanied by a wide array of foreign audio and video materials. The programs of the British Broadcasting Corporation (BBC) and the Voice of America (VOA), the diversification of domestic television and radio programs (a choice made by the Chinese government and facilitated by satellite communication), and the importation and translation of foreign books and magazines—all contributed to a more pluralistic press in China. In fact, by 1989, the stream of publications had so overwhelmed the CCP Propaganda Department that it was simply no longer able to monitor their contents.

During the demonstrations in Tiananmen Square, the international press in Beijing, unlike the Chinese press, was freely filming the events and filing reports on the demonstrations. The Chinese press then took a leap into complete press freedom. With camera operators and microphones in hand, Chinese reporters covered the student hunger strike that began on May 13 in its entirety; but with the imposition of martial law in Beijing on May 20, press freedom came to a crashing halt.

In the immediate aftermath of the brutal crackdown on Tiananmen Square demonstrators in June 1989, the CCP imposed a ban on a variety of books, journals, and magazines. The government ordered the "cleansing" of media organizations, with "bad elements" to be removed and not permitted to leave Beijing for reporting. All press and magazine articles written during the prodemocracy movement, as well as all television and radio programs shown during this period, were analyzed to see if they conformed to the party line. If they did not, those individuals responsible for editing were dismissed. And, as had been the practice in the past, press and magazine articles once again had to be on topics specified by the editors, who were under the control of the Propaganda Department of the CCP. In short, press freedom in China suffered a significant setback because of the pro-democracy demonstrations in the spring of 1989.

In the new climate of experimentation launched by Deng Xiaoping in 1992, however, the diversity of television and radio programs soared. China's major cities now have multiple channels carrying a broad range of programs from Hong Kong, Taiwan, South Korea, Japan, and the West. These programs, whether soap operas about the daily life of Chinese people living in Hong Kong and Taiwan, or art programs introducing the world of Western religious art through a visual art history tour of the Vatican, or in news about protests and problems faced by other nations in the world expose the Chinese to values, events, ideas, and standards of living previously unknown to them. They are even learning about the American legal process through China's broadcasting of American television dramas that focus on police and the judicial system. Chinese are fascinated that American police, as soon as they arrest suspects, inform them that they have "the right to remain silent" and "the right to a lawyer." Such programs may do more to promote reform of the Criminal Procedure Code than anything human rights groups do.

(Digital Vision/Getty Images RF)
Woman with personal organizer.

Today, television ownership is widespread. In addition to dozens of regular channels, China has numerous cable stations. And virtually all families have radios. Round-the-clock, all-news radio stations broadcast the latest political, economic, and cultural news and conduct live interviews; and radio talk shows take phone calls from anonymous listeners about everything from sex to political and economic corruption. There are even blatant critiques of police brutality,[33] and analyses of failed government policies on everything from trade policy to healthcare and unemployment. There is also far better coverage of social and economic news in

STOCK MARKETS, GAMBLING, AND LOTTERIES

China has had two stock markets since the late 1980s. One was created in the Special Economic Zone (SEZ) of Shenzhen; the other in Shanghai. With only seven industries originally registered on them, strict rules about how much daily profit or loss (1.2 percent for the Shanghai exchange until July 1992) a stock could undergo, and deep public suspicion that these original issues of stocks were worthless, these markets got off to a slow start. But when these same stocks were worth five times their original value just a year later, the public took notice. Rumors—as important in China as actual news—took over and exaggerated the likelihood of success in picking a winning stock. The idea of investors actually losing money, much less a stock-market crash, did not seem to be an idea whose time had come.

Because China is still largely a cash economy, a stock-market transaction can be complicated. Instead of just telephoning a stockbroker and giving an order, with a simple bank transfer or check to follow shortly, many Chinese must still appear in person, stand in line, and pay cash on the spot. Taiwan has added its own angle to the stock-mania by selling to the Mainlanders small radios that are tuned in to only one frequency—stock market news. Local branches of the stock market and brokerages are popular places for retired people to spend the day watching their stocks on the boards. In Shanghai, they can also tune into "Stock Market Today," a popular local television program. Issuing, buying, and selling stocks has at times been a near-national obsession. Not only do ordinary companies selling commercial goods, such as computers and clothing, issue stocks. So do taxicab companies and even universities. Thus far, few such stocks are actually listed on the national stock exchanges; but employees of these work units are eager to purchase the stocks. In most cases, the original issues are sold at far higher prices than their face value, as employees (and even nonemployees) eagerly buy up fellow employees' rights to purchase stocks, at grossly inflated prices. Presumably, the right of employees to own stock in their own work unit will make them eager to have it do well, and thus increase efficiency and profits.

Even as China's economy continued to roar along with double-digit growth in the twenty-first century, the ardor for wheeling and dealing in the stock market cooled considerably. Companies were making money, but stock prices were in the doldrums, with a 2003 poll indicating that 90 percent of investors had lost money.[34] Then, in 2006, investors fueled a 130 percent surge. The bull market was largely due to a massive inflow of speculative funds from both foreign investment, and domestic investment from Chinese investors hoping to profit more from the stock market than from leaving their money in banks, where they rarely earn more than 2 percent per annum.

The underlying problems in China's stock markets, however, remain a concern. First, they are not regulated by outside auditors, which means that investors cannot trust the information in company reports. Scandals involving stocks all too often erupt in the Chinese media. Second, the state itself usually owns the majority of voting shares—often as much as two-thirds of the shares of China's leading companies. (These shares were until recently also "untradable" shares, meaning the companies were not really private at all, and should not have been listed on a public stock exchange.) This results in the state's preferred policies, which are often based on bureaucratic and political concerns, rather than private investors and the market determining company strategies and profitability for, say, the technology, telecommunications, or heavy industrial sector. Because a state enterprise or state agency is usually the majority stockholder, moreover, it can by its own purchases and sales determine the price of a stock. The small investor is simply a pawn, and really has little chance to make a profit.

(© Photodisc/Getty Images RF)

Shanghai Pudong financial district.

Companies that are owned and run by state agencies and ministries, such as the Ministry of Defense, have used state funds to invest in both domestic and international stocks and real estate. They have sometimes lost millions of yuan in the process. As a result, the government now prohibits state-run units from investing state money in stocks and real estate. Nevertheless, some still do so indirectly, thereby evading the laws.

Learning from Western practices and catering to a national penchant for gambling (illegal, but indulged in nevertheless, in mahjong and cards), the Chinese have also begun a number of lotteries. Thus far, most of these have been for the purpose of raising money for specific charities or causes, such as for victims of floods and for the disabled. Recently, because the items offered, such as brand-new cars, have been so desirable, the lotteries have generated billions of yuan in revenue. The government has found these government-controlled lotteries to be an excellent way of addressing the Chinese penchant for gambling while simultaneously generating revenues to compensate for those it seems unable to collect through taxes.

Finally, companies follow such Western marketing gimmicks for increasing sales as putting Chinese characters on the inside of packages or bottle caps to indicate whether the purchasers have won a prize. With a little Chinese ingenuity, the world could witness never-before-imaged realms of betting and competitive business practices that appeal to people's desire to get something for nothing.

all the media than previously, and serious investigative reporting on corruption and crime. A popular half-hour program during evening prime time looks at cases of corruption and crime that are being investigated, or have been solved. Figures indicate the exponential growth of the media. The number of books published annually in China rose from about 5,000 in 1970 to about 104,000 in 1995. The number of newspapers grew from 42 in 1970 to over 2,000 by the year 2005, with a circulation of over 100 million. (In China, almost all work units post a number of newspapers in glass cases outside, so the readership is far higher than the circulation numbers.) Most of the newspapers also have their own websites. Similarly, the number of magazines grew from a handful in 1970 to more than 9,000 by 2005—an official number that excludes the large number of nonregistered magazines and illegal publications. Xinhua (New China) News Agency has more than 100 branches throughout the world that report on news for the Chinese people. There are hundreds of radio stations, and the number of television stations grew to 3,000 by 2005. Close to 90 percent of the population has access to television. CCTV-9, the "CNN" of China, is a slick, English-language program that is watched in many parts of the world, and actually has broader viewership in Asia than does CNN. By 2005, there were 750 cable-television stations and an estimated 125 million households with cable television access, with millions of new subscribers being added each year. The government's National Cable Company is forming a countrywide network to expand services geographically and to include new Internet and telecommunications connections.[35]

This effort to expand Internet availability indicates the government's dilemma: It wants China to modernize rapidly, and to have the scientific, economic, commercial, and educational resources on the Internet available to as many as possible. It has demanded that all government offices be updated to use the Internet in order to improve communication, efficiency, and transparency. At the same time, it wants to control which websites can be accessed, and what type of material may be made available to China's citizens on those sites. In spite of the government's best efforts, the Internet is virtually uncontrollable. Angry and inflammatory commentary on events in China, and even criticism of China's leadership, appear with increasing regularity. Some sites are closed down because they violate the unstated boundaries of acceptable commentary; but it is difficult in a computer-saavy world to block access to all sites that the government might find irritating, if not downright seditious. China's Internet users know what is going on in the world; and they can spend their time chatting to dissidents abroad and at home on e-mail if they so choose. But most prefer to use the Internet for business, news, sports, games, music, socializing, and, of course, pornography. The Chinese, in fact, seem to spend most of their time on the Internet for entertainment and playing games, in contrast to Americans' preference for using it primarily for information. China's largest Internet company, Ten-cent, dominates 80 percent of China's market, far more than Google dominates in the U.S. market. This is due largely to the complete package of entertainment it offers (with a preference for playing games, forming communities, and adopting virtual personas—avatars), as well as its mobile instant-messaging service. In China, 70 percent of the Internet's users are under the age of 30; whereas in the U.S., some 70 percent are *over* the age of 30.[36] With print and electronic media so prolific and diverse, much of it escapes any monitoring whatsoever, especially newspapers, magazines, and books. Even the weekend editions of the remaining state-run newspapers print just about any story that will sell. Often about a seamier side of Chinese life, they undercut the puritanical aspect of communist rule and expanding the range of topics available for discussion in the public domain. Were the state able to control the media, it would, at the very least, crack down on pornographic literature on the streets.

The sheer quantity of output on television, radio, books, and the press allows the Chinese people to make choices among the types of news, programs, and perspectives they find most appealing. Their choices are not necessarily for the most informative or the highest quality, but for the most entertaining. Because the media (with few exceptions) are market-driven, consumers' preferences, not government regulations or ideological values, shape programming and publishing decisions.

This market orientation results from economic reforms. By the 1990s, the government had cut subsidies to the media, thereby requiring that even the state-controlled media had to make money or be shut down. This in turn meant that the news stories it presented had to be more newsworthy in order to sell advertising and subscriptions. Even in the countryside, the end of government subsidies to the media has spurred publication. Thinking there is money to be made, township and village enterprises, as well as private entrepreneurs, have set up thousands of printing facilities during the last 10 to 15 years.[37] In short, "Even though China's media can hardly be called free, the emergence of divergent voices means the center's ability to control people's minds has vanished."[38]

The party tends, however, to concentrate its limited resources on the largest and influential journals, magazines, newspapers, and publishing houses. It seems to have written off the rest as the inevitable downside of a commercialized media market—and the part that it no longer supports, nor controls, financially. Funded by advertising and consumer demand, the media must now "march to the market." Nevertheless, Beijing is trying to tighten its control over the media system by instituting a penalty point system. In 2007, the Chinese Communist Party's propaganda department announced the new system of points-based penalties, whereby each media outlet will be allocated 12 points. If it uses up its 12 points, it may be closed. Point deductions are based on the seriousness of the media outlet's action. This will supplement the existing system in which the CCP's propaganda department and the government's media regulator have jointly decided when to issue warnings, remove offending executives or reports, or otherwise punish a media organization. The new approach is portrayed as an effort to increase "social harmony," and seemed to be increasing in 2007 intensity before the 17th National Party Congress and the Beijing 2008 Olympics.[39] As for China's film industry, because it produces relatively few films each year, it is more heavily censored than print media. Furthermore, all films are shot in a small number of studios, making control easier. Finally, a film is likely to have a much larger audience than most books, and so the censors are concerned that it be carefully reviewed before being screened.[40] Zhang Yimou, China's renowned film director and producer, and recipient of numerous international film awards, has had the screening in China of one after another of his films banned. A publication of the Party School of the Chinese Communist Party condemned his 2006 blockbuster, *Curse of the Golden Flower,* as "ugly" and "blood-thirsty," transgressing the moral limits of Chinese art. And yet, the film has not been banned in China, and has wracked up tens of millions of dollars in profits.[41]

Human Rights and Democracy: The Chinese View

Surveys indicate that most urban Chinese citizens believe the government has adopted policies that have greatly improved their daily lives. Many have seen the government's law-and-order campaigns—which sometimes involve crackdowns on perceived dissidents (such as Falun Gong activists)—as necessary to China's continued economic prosperity and political stability. They have tended to be far more interested in the prospect of a higher standard of living than in the rights of dissidents.

Even China's intellectuals no longer seem interested in protest politics. They do not "love the party," but they accept the status quo. Some just want a promotion and to make money. Others have become advisers to the government's think tanks and advisory committees. Many have gone into business. As one university professor put it, it is easy to be idealistic in one's heart; but to be idealistic in action is a sign of a true idealist, and there haven't been many of those in China since 1989. Today in China, it is difficult to find any student or member of the intellectual elite who demonstrated in Tiananmen Square in 1989 doing anything remotely political.

Most Chinese consider raising living standards and enjoying a better material life as their fundamental human rights. They give credit to the CCP for promoting economic growth and maintaining social stability. Even the middle class, those entrepreneurs and intellectuals who are most likely to lead any democratization movement, seem to be happy with the status quo and have become some of the strongest defenders of the current system from which they have benefited so much.

Still, those who are discontent with party rule have far more outlets today for their grievances: the mass media, and journals, as well as think tanks, policy advisory committees, and professional associations that actually influence policy. Street protests are no longer considered the best way for the educated and professional classes to change policy, although ordinary workers and peasants resort to them with increasing frequency.

Many members of China's elites are committed to reform, but the number of idealists committed to democracy—or communism—is limited. Few Chinese, including government officials, want to discuss communist ideology, and even fewer agree on what democracy means. They prefer to talk about business and development, and do so in terms familiar to capitalists throughout the world, but also in terms of the overall objective of strengthening China as a nation. In this respect, they are appealing to the strong nationalism that has virtually replaced communism as the normative glue holding the country together.

Apart from their changed perspectives on what really matters, many Chinese feel that they do not know enough to challenge government policy on human rights issues. Why should they risk their careers to fight for the rights of jailed dissidents about whom they know almost nothing, they ask. They know of the abuse of human rights in Western liberal democracies, such as the killing of student protesters at Kent State by the National Guard during the Vietnam War, the many deaths attributed to the British forces in Northern Ireland, the torture of American-held prisoners in Iraq and in Guantanamo Bay. When the U.S. State Department issues its annual human rights report, which inevitably makes harsh criticism of China, Beijing replies with an equally damning condemnation of human rights abuses by the United States. In 2007, it noted the "increased willingness by Washington to spy on its own citizens by monitoring telephone calls, computer connections, and travels."[42]

The Chinese people have heard of the unseemly behavior of several student leaders of the Tiananmen Square demonstrations, both during the movement in 1989 and after it. They wonder aloud if in the treatment of criminal suspects and those suspected of treason or efforts to harm the country—especially suspected terrorists—Western democratic states exhibit more virtuous behavior than does China. Some Chinese intellectuals argue that the recent difficulties in the United States and other Western democracies indicate that their citizens frequently elect the wrong leaders, people who not only make bad policies but who are also increasingly involved in corrupt money politics—the very issue of most concern to the Chinese. This, they suggest, indicates that democracy is no more able than socialism to produce good leaders. Furthermore, many support the view that the Chinese people are inadequately prepared for democracy because of a low level of education. The blatant political maneuverings and other problems surrounding the 2000 U.S. presidential election hardly offered reassurance as to the virtues of American democracy; nor did the American 2003 invasion and subsequent occupation of Iraq. The minimal response by the U.S. government to those challenging the administration, and the unwillingness of Congress to debate the war years after the invasion give the Chinese the impression that "the people" were not necessarily listened to, and that a democracy does not necessarily have an effective system of "checks and balances."

Within China, it is frequently the people, rather than the government, who demand the harshest penalties for common criminals, if not political dissidents. And it is China's own privileged urbanites who often demand that the government ignore the civil rights of other citizens. For example, urban residents in Beijing have repeatedly demanded that the government remove migrant squatters and their shantytowns, asserting that they are breeding grounds for criminality in the city. And they resist the government allowing migrants to attend the city's public schools or receive healthcare. Many ordinary people now seem to accept the government's overall assessment of the events of the spring of 1989, which is that the demonstrations in Tiananmen Square posed a threat to the stability and order of China and justified the military's crackdown. To many Chinese people, no less than to their government, stability and order are critical to the continued economic development of China. Advancement toward democracy and protection of human rights take a back seat.

INTERNATIONAL RELATIONS

From the 1830s onward, foreign imperialists nibbled away at China's territorial sovereignty, subjecting China to one national humiliation after another. As early as the 1920s, both the Nationalists and the Communists were committed to unifying and strengthening China in order to rid it of foreigners and resist further foreign incursions. When the Communists achieved victory over the Nationalists in 1949, they vowed that they would never again allow foreigners to tell China what to do. The "century of humiliation" beginning from the "Opium War" in 1839 is essential to understanding China's foreign policy in the period of Chinese Communist rule.

From Isolation to Openness

By the early 1950s, the Communists had forced all but a handful of foreigners to leave China. The target of the United States' Cold War policy of "isolation and containment," China under Chairman Mao Zedong charted an independent, and eventually an isolationist, foreign policy. Even China's relations with socialist bloc countries were suspended after the Cultural Revolution began in 1966. China took tentative steps toward re-establishing relations with the outside world with U.S. president Richard Nixon's visit to China in 1972, but it did not really pursue an "open door" policy until more pragmatic "reformers" gained control in 1978. By the 1980s, China was hosting several million tourists annually, inviting foreign investors and foreign experts to help with China's modernization, and allowing Chinese to study and travel abroad. Nevertheless, inside the country, contacts between Chinese and foreigners were still affected by the suspicion on the part of ordinary Chinese that ideological and cultural contamination comes from abroad and that association with foreigners might bring trouble.

These attitudes have moderated considerably, to the point where some Chinese are willing to socialize with foreigners, invite them to their homes, and even marry them. However, this greater openness to things foreign sits uncomfortably together with a new nationalism that has emerged since the West so heavily criticized, and punished, China for the military crackdown on Tiananmen Square demonstrators in June 1989. A strong xenophobia (dislike and fear of foreigners) and an awareness of the history of China's victimization by Western and Japanese imperialism mean that the Chinese are likely to rail at any effort by other countries to tell them what to do. The Chinese continue to exhibit this sensitivity on a wide variety of issues, from human rights to China's policies toward Tibet and Taiwan; from intellectual property rights to prison labor and environmental degradation.

(AP Photo/William Foreman)

People protest against the construction of a garbage incinerator in Guangzhou.

Ordinary Chinese people tended to concur with the government's anger over the U.S. threat of economic sanctions to challenge China's human rights policies (something that is no longer possible since China joined the World Trade Organization [WTO] in 2001). They were enraged by the American bombing of the Chinese Embassy in Belgrade during the war in Yugoslavia, which the U.S. government insisted was by error, not intent; the crash of a U.S. spy plane with a Chinese jet in 2001, which the Chinese believed to be in their own airspace; efforts to prevent China from entering the World Trade Organization or becoming a site for the Olympics; American accusations of Chinese spying and stealing of American nuclear-weapons secrets, charges that appeared to the Chinese to be motivated by hostility toward China, especially when they were ultimately dismissed for lack of evidence; American accusations of illegal campaign funding by the Chinese; the on-going human-rights barrage; the strengthening of U.S. military ties with Japan and Taiwan; and continued American interference in China's efforts to regain control of Taiwan.

Only a narrow line separates a benign nationalism essential to China's unity, however, and a popular nationalism with militant overtones threatens to careen out of control.[43] These aspects of nationalism worry China's government, whose primary concerns remain economic growth and stability. The government does not want to be forced into war by a militant nationalism. Nevertheless, China's xenophobia continues to show up in its efforts to keep foreigners isolated in certain living compounds; to limit social contacts between foreigners and Chinese; to control the import of foreign literature, films, and periodicals; and in general to limit the invasion of foreign political and cultural values in China. Since 1996, in an effort to protect China's culture, the government has ordered television stations to broadcast only Chinese-made programs during prime time. In 2007, the government limited television stations to showing "ethically inspiring TV series" during prime time. Programs broadcast during prime time must likewise "keep script and video records for future censorship against vulgarity."[44] The government has also attempted to enhance national pride through economic success; and by participation in international events, including the Olympic and Asian Games, music competitions, and film festivals.

Yet, in some respects, the resistance to the spread of foreign values in China is proving to be a losing battle, with growing numbers of foreigners in China; television swamped with foreign programs; Kentucky Fried Chicken, McDonald's, Starbucks, and Pizza Hut ubiquitous; "Avon calling" at several million homes;[45] bodybuilding, disco, cell phones, and Internet becoming part of the culture. Further, with millions of Chinese tourists and officials tromping through the world, vast numbers of foreign investors in China, China's hosting of trade fairs, international meetings, and the 2008 Olympics, as well as a heavy reliance on

(© blue jean images/Getty Images RF)

Students on grounds of Summer Palace in Beijing.

foreign experts to help China reform its economic, legal, financial, and banking systems, and set up a commodities and futures market, China is awash with values that contend with traditional Chinese ones. China is now a full participant in the international economic and financial system: it is the third-largest global trading country, and gives foreign aid to developing countries, especially in Africa. It is an important participant in the war on terrorism, and a key player in Interpol and other efforts to control international crime syndicates, smuggling of weapons and drugs, and international trafficking in children and women. China today is seen by most countries as a partner rather than an adversary, a part of the solution, not the problem (such as in negotiations with the North Koreans over their nuclear weapons program). Its powerful economy and investments abroad have earned it respect (leavened with fear) worldwide. Today, China is a key international actor. It cannot be dismissed as a poor country without the wherewithal to enter the modern world.

China is a much more open country than at any time since 1949. This is in spite of the efforts by the party's more conservative wing to limit the impact of foreign ideas about democracy and individual rights on the political system, as well as the impact from "polluting" values embedded in foreign culture. Over time, however, conservatives have lost the battle to limit foreign influence, due more to the inexorable forces of globalization, communication, and the Internet than to any struggle behind closed doors with political adversaries. At the same time, although China's flourishing business community and growing middle class have little interest in disrupting the emphasis on stability by calling for greater political democracy, they nevertheless encourage, and benefit from, China's greater openness. Foreign investment in China (annual total foreign direct investment in the last few years has averaged around US$60 billion) far outstrips foreign investment in any other developing country. This is due not just to China's potential market size but also to the favorable investment climate created by the party-state. Fully one-third of China's manufacturing output is produced by foreign companies in China (both in the form of joint ventures with the Chinese, and wholly owned foreign companies), indicating just how important foreign investment is to China's economy. (This is an important fact to remember when critics denounce China's trade surplus; that is, a considerable percentage of those exports are actually produced by companies wholly or partly owned by parent companies located in countries to which China exports its products.) China's openness, especially since joining the WTO in 2001, has forced its own enterprises to compete not just against imports, but also against foreign firms that have invested in China. China's service sector has been opened up to foreign competition. By 2007, foreign banks, insurance companies, securities firms, and telecommunications could compete on an increasingly equal basis.[46]

When it comes to the economy, then, the government has seemed less worried about the invasion of foreign values than anxious to attract foreign investment. China sees a strong economy as the key to both domestic stability and international respect. The government's view now seems to be: If it takes nightclubs, discos, exciting stories in the media, stock markets, rock concerts, the Internet, and consumerism to make the Chinese people content and the economy flourish under CCP rule, then so be it.

THE SINO–SOVIET RELATIONSHIP

While forcing most other foreigners to leave China in the 1950s, the Chinese Communist regime invited experts from the Soviet Union to China to give much-needed advice, technical assistance, and aid. This convinced the United States (already certain that Moscow controlled communism wherever it appeared) that the Chinese were puppets of the Soviets. Indeed, until the Great Leap Forward in 1958, the Chinese regime accepted Soviet tenets of domestic and foreign policy along with Soviet aid. But China's leaders soon grew concerned about the limits of Soviet aid and the relevance of Soviet policies to China's conditions—especially the costly industrialization favored by the Soviet Union. Ultimately, the Chinese questioned their Soviet "big brother" and turned to the Maoist model of development, which aimed to replace expensive Soviet technology with human labor. Soviet leader Nikita Khrushchev warned the Chinese of the dangers to China's economy in undertaking the Great Leap Forward; but Mao Zedong interpreted this as evidence that the Soviet "big brother" wanted to hold back China's development.

The Soviets' refusal to use their military power in support of China's foreign-policy objectives further strained the Sino–Soviet relationship. First in the case of China's confrontation with the United States and the forces of the "Republic of China" over the Offshore Islands in the Taiwan Strait in 1958, and then in the Sino–Indian border war of 1962, the Soviet Union backed down from its promise to support China. The Soviets also refused to share nuclear technology with the Chinese. The final blow to the by-then fragile relationship came with the Soviet Union's signing of the 1963 Nuclear Test Ban Treaty. The Chinese denounced this as a Soviet plot to exclude China from the "nuclear club" that included only Britain, France, the United States, and the Soviet Union. Subsequently, Beijing publicly broke party relations with Moscow. In 1964 China denotated its first nuclear bomb.

The Sino–Soviet relationship, already in shambles, took on an added dimension of fear during the Vietnam War, when the Chinese grew concerned that the Soviets (and Americans) might use the war as an excuse to attack China. China's distrust of Soviet intentions was heightened in 1968, when the Soviets invaded Czechoslovakia in the name of the "greater interests of the socialist community"—which, they contended, "overrode the interests of any single country within that community." Soviet skirmishes with Chinese soldiers on China's northern borders soon followed. Ultimately, it was the Chinese leadership's concern about the Soviet threat to China's national security that, in 1971, caused it to reassess its relationship with the United States and led to the establishment of diplomatic relations with China in 1979. Indeed, the real interest of China and the United States in each other was as a "balancer" against the Soviet Union. Thus, in the midst of the Cold War, which began in 1947 and did not end until the late 1980s, China had moved out of the Soviet-led camp; yet China did not begin benefiting from friendship with Western countries in the power balance with the Soviet Union until it gained the seat in the United Nations in 1971.

The Sino–Soviet relationship moved toward reconciliation only near the end of the Cold War. In 1987, the Soviets began making peaceful overtures: They reduced troops on China's borders and withdrew support for Vietnam's puppet government in neighboring Cambodia. Beijing responded positively to the new *glasnost* ("open door") policy of the Soviet Communist Party's General Secretary, Mikhail Gorbachev. Border disputes were settled and ideological conflict between the two Communist giants abated; for with the Chinese themselves shelving Marxist dogma in their economic policies, they could hardly continue to denounce the Soviet Union's "revisionist" policies and make self-righteous claims to ideological orthodoxy. With both the Soviet Union and China abandoning their earlier battle over who should lead the Communist camp, they shifted away from conflict over ideological and security issues to cooperation on trade and economic issues. Today, China and Russia have significant trade, and their relationship is based on national interests, not ideology.

From "People's War" to Cyber War

With the collapse of Communist party rule, first in the Central/Eastern European states in 1989, and subsequently in the Soviet Union, the dynamics of China's foreign policy changed dramatically. Apart from fear that their own reforms might lead to the collapse of CCP rule in China, the breakup of the Soviet Union into 15 independent states removed China's ability to play off the two superpowers against each other: The formidable Soviet Union simply no longer existed. Yet its fragmented remains had to be treated seriously, for the state of Russia still has nuclear weapons and shares a common border of several thousand miles with China, and the former Soviet republic of Kazakhstan shares a border of nearly 1,000 miles.

The question of what type of war the Chinese military might have to fight has affected its military modernization. For many years, China's military leaders were in conflict over whether China would have to fight a high-tech war or a "people's war," in which China's huge army would draw in the enemy on the ground and destroy it. In 1979, the military modernizers won out, jettisoning the idea that a large army, motivated by ideological fervor but armed with hopelessly outdated equipment, could win a war against a highly modernized military such as that of the Soviet Union. The People's Liberation Army (PLA) began by shedding a few million soldiers and putting its funds into better armaments. A significant catalyst to further modernizing the military came with the Persian Gulf War of 1991, during which the CNN news network broadcasts vividly conveyed the power of high-technology weaponry to China's leaders.

In the nationwide rush to become prosperous, the PLA plunged into the sea of business. China's military believed that it was allocated an inadequate budget for modernization, so it struck out on its own along the capitalist road to raise money. By the late 1990s, the PLA had become one of the most powerful actors in the Chinese economy. It had purchased considerable property in the Special Economic Zones near Hong Kong; acquired ownership of major tourist hotels and industrial enterprises; and invested in everything from golf courses, brothels, and publishing houses to CD factories and the computer industry, as a means for funding military modernization. In 1998, however, President Jiang Zemin demanded that the military relinquish its economic enterprises and return to its primary task of building a modern military and protecting China. The promised payoff was that China's government would allocate more funding to the PLA, making it unnecessary for it to rely on its own economic activities.

In recent years, China's military has purchased weaponry and military technology from Russia as Moscow scales back its own military in what sometimes resembles a going-out-of-business sale; but in doing so, China's military may have simply bought into a higher level of obsolescence, since Russia's weaponry lags years behind the technology of the West. China possesses nuclear weapons and long-distance bombing capability, but its ability to fight a war beyond its own borders is quite limited. Asian countries, torn between wondering whether China or Japan will be a future threat to their territory, do not seem concerned by China's military modernization, except when China periodically makes threatening statements about Taiwan or the Spratly Islands in the South China Sea. Even here, however, Beijing usually relies on economic and diplomatic instruments. In the case of Taiwan, it is essentially tying Taiwan's economy to the mainland by welcoming economic investment and trade; the hope is ultimately to bring Taiwan under the control of Beijing without a war. In the case of the Spratlys, under whose territorial waters there is believed to be significant oil deposits, Beijing has reached tentative agreement with the five governments involved in competing claims to the Spratlys to avoid the possibility of armed clashes.

Nevertheless, in spite of China's remarkable economic and diplomatic gains since the reform period began in 1979, the leadership continues to modernize China's military capabilities. Beijing is ever alert to threats to its national security, but there are no indications that it is preparing for aggression against any country. China's military modernization is primarily aimed at defensive capabilities and maintaining its deterrent capability against an American nuclear attack. It has also increased the number of missiles it aims at Taiwan in response to repeated suggestions by Taiwan's former President, Chen Shui-bian, that Taiwan would declare independence. With the arrival of the George W. Bush administration in Washington in early 2001, the American leadership began to seriously consider the possibility of deploying a limited "national missile defense" (NMD) in the United States, and even to deploy a "theater missile defense" (TMD) around Japan—and possibly Taiwan. Were a missile-defense system successfully deployed, it would limit the ability of China to prevent Japan or the United States from attacking it—or to prevent Taiwan from declaring independence. By providing a protective shield, TMD would allow Taiwan to declare independence with impunity. TMD is, then, perceived by Beijing as an aggressive move by the United States, and helps explain China's efforts to substantially increase the number of missiles it aims at Taiwan in order to overcome any defensive system (including "theater missile defense") that might be installed.

The PLA's doctrine has transformed from "the People's War" to "limited

(© AP Photo/Ed Wray)
Chinese acrobats perform to warm up the crowd at the Olympic weightlifting competition at the Beijing 2008 Olympics in Beijing, China Monday, Aug.11, 2008.

local war under information conditions," according to China's 2004 Defense White Paper.

China's leadership is, in any event, primarily concerned with economic development. China is working to become an integral part of the international economic, commercial, and monetary systems. It has rapidly expanded trade with the international community, even more so since joining the World Trade Organization in 2001. Today, China focuses its efforts primarily on infrastructure development and investment, not just in China but throughout the world. With the exponential growth in per capita income for more than 200 million Chinese, China is pressed to acquire natural resources to satisfy rocketing increases in consumer demand. It is investing heavily in resources for the future in Latin America, Southeast Asia, Africa, and the Middle East. It is ironic that China, considered "the sick man of Asia" in the early 20th century, should now in the 21st century be buying up companies not just in the developing world, but also in Europe and the United States; and that it has, along with Japan, become the "banker" for the United States, buying American debt and keeping the U.S. dollar from declining still further in value, especially when the United States is suffering from a severe financial crisis now. For these reasons, China holds substantial bargaining power vis-à-vis the United States and many other countries in the world.

THE SINO–AMERICAN RELATIONSHIP

China's relationship with the United States has historically been an emotionally turbulent one.[47] During World War II, the United States gave significant help to the Chinese, who at that time were fighting under the leadership of the Nationalist Party, headed by General Chiang Kai-shek. When the Americans entered the war in Asia, the Chinese Communists were fighting together with the Nationalists in a "united front" against the Japanese, so American aid was not seen as directed against communism.

After the defeat of Japan at the end of World War II, the Japanese military, which had occupied much of the north and east of China, was demobilized and sent back to Japan. Subsequently, civil war broke out between the Communists and Nationalists. The United States attempted to reconcile the two sides, but to no avail. As the Communists moved toward victory in 1949, the KMT leadership fled to Taiwan. Thereafter, the two rival governments each claimed to be the true rulers of China. The United States, already in the throes of the Cold War because of the "iron curtain" falling over Eastern Europe, viewed communism in China as a major threat to its neighbors.

Korea, Taiwan, and Vietnam

The outbreak of the Korean War in 1950 helped the United States to rationalize its decision to support the Nationalists, who had already lost power on the mainland and fled to Taiwan. The Korean War began when the Communists in northern Korea attacked the non-Communist south. When United Nations troops (mostly Americans) led by American general Douglas MacArthur successfully pushed Communist troops back almost to the Chinese border and showed no signs of stopping their advance, the Chinese—who had frantically been sending the Americans messages about their concern for China's own security, to no avail—entered the war. The Chinese forced the UN troops to retreat to what is today still the demarcation line between North and South Korea. Thereafter, China became a target of America's Cold War isolation and containment policies.

With the People's Republic of China condemned as an international "aggressor" for its action in Korea, the United States felt free to recognize the Nationalist government in Taiwan as the legitimate government to represent all of China. The United States supported the Nationalists' claim that the people on the Chinese mainland actually wanted the KMT to return to the mainland and defeat the Chinese Communists. As the years passed, however, it became clear that the Chinese Communists controlled the mainland and that the people were not about to rebel against Communist rule.

Sino–American relations steadily worsened as the United States continued to build up a formidable military bastion with an estimated 100,000 KMT soldiers in the tiny islands of Quemoy and Matsu in the Taiwan Strait, just off China's coast. Tensions were exacerbated by the steady escalation of U.S. military involvement in Vietnam from 1965 to the early 1970s. China, fearful that the United States was really using the war in Vietnam as the first step toward attacking China, concentrated on civil-defense measures: Chinese citizens used shovels and even spoons to dig air-raid shelters in major cities such as Shanghai and Beijing, with tunnels connecting them to the suburbs. Some industrial enterprises were moved out of China's major cities in order to make them less vulnerable in the event of a massive attack on concentrated urban areas. The Chinese received a steady barrage of what we would call propaganda about the United States "imperialist" as China's number-one enemy; but it is important to realize that the Chinese leadership actually *believed* what it told the people, especially in the context of the United States' steady escalation of the war in Vietnam toward the Chinese border, and the repeated "mistaken" overflights of southern China by American planes bombing North Vietnam. Apart from everything else, it is unlikely that China's leaders would have made such an immense expenditure of manpower and resources on civil-defense measures had they not truly believed that the United States was preparing to attack China.

Diplomatic Relations

By the late 1960s, China was completely isolated from the world community, including the Communist bloc. It saw itself as surrounded on all sides by enemies—the Soviets to the north and west, the United States to the south in Vietnam as well as in South Korea and Japan, and the Nationalists to the east in Taiwan. Internally, China was in such turmoil from the Cultural Revolution that it appeared to be on the verge of complete collapse.

In this context, Soviet military incursions on China's northern borders, combined with an assessment of which country could offer China the most profitable economic relationship, led China to consider the United States as the lesser of two evil giants and to respond positively to American overtures. In 1972, President Richard Nixon visited China. When the U.S. and China signed the Shanghai Communique at the end of his visit, the groundwork was laid for reversing more than two decades of hostile relations.

Thus began a new era of Sino–American friendship, but it fell short of full diplomatic relations, This long delay in bringing the two states into full diplomatic relations reflected not only each country's domestic political problems but also mutual disillusionment with the nature of the relationship. Although both sides had entered the relationship with the understanding of its strategic importance as a bulwark against the Soviet threat, the Americans had assumed that the 1972 opening of partial diplomatic relations would lead to a huge new economic market for American products; the Chinese assumed that the new ties would quickly lead the United States to end its diplomatic relations with Taiwan. Both were disappointed. Nevertheless, pressure from both sides eventually led to full diplomatic relations between the United States and the People's Republic of China on January 1, 1979.

The Taiwan Issue in U.S.–China Relations

Because the People's Republic of China and the Republic of China both claimed to be the legitimate government of the Chinese people, the establishment of diplomatic relations with the former necessarily entailed breaking them with the latter. Nevertheless, the United States continued to maintain extensive, informal economic and cultural ties with Taiwan. It also continued the sale of military equipment to Taiwan. Although these military sales are still a serious issue, American ties with Taiwan have diminished, while China's own ties with Taiwan have grown steadily closer since 1988. Taiwan's entrepreneurs have become one of the largest groups of investors in China's economy. More than one million people from Taiwan, about 5 percent of Taiwan's population, live on the mainland, with 500,000 living in Shanghai alone. Taiwan used to have one of the cheapest labor forces in the world; but because its workers now demand wages too high to remain competitive, Taiwan's entrepreneurs have dismantled many of its older industries and reassembled them on the mainland. With China's cheap labor, these same industries are now profitable, and both China's economy and Taiwan's entrepreneurs benefit. Taiwan's businesspeople are also investing in new industries and new sectors, and they are competing with other outside investors for the best and brightest Chinese minds, so the relationship has already moved beyond simply exploiting the mainland for cheap labor and raw materials.

Ties with the mainland have also been enhanced since the late 1980s by the millions of tourists from Taiwan, most of them with relatives on the mainland. They bring both presents and good will. Family members who had not seen one another since 1949 have reestablished contact, and "the enemy" now seems less threatening.

China hopes that its economic reforms and growth, which have substantially raised the standard of living, will make reunification more attractive to Taiwan. This very positive context has, however, been disturbed over the years by events such as the military crackdown on the demonstrators in Tiananmen Square in 1989, and efforts by Taiwan's leaders to move towards independence. In 1996, China responded to such efforts by "testing" its missiles in the waters around Taiwan. High-level talks to discuss eventual reunification were broken off and now occur on a sporadic basis. The 2000 election of the Democratic Progressive Party candidate, Chen Shui-bian as president (and his re-election in 2004), led to still more crises with Beijing. President Chen, who campaigned on the platform of an independent Taiwan, refused to acknowledge Beijing's "one China" principle, and he insisted that Taiwan negotiate with the P.R.C. as "an equal." This further strained the relationship and led to raising the bellicosity decibel level in Beijing. Nevertheless, both sides recognize it is in their interests for the foreseeable future to maintain the status quo—a peaceful and profitable relationship in which Taiwan continues to act as an independent state, but does not declare its independence. In 2008, the pro-independence Chen was replaced by KMT's Ma Ying-jeou, who favors a more conciliatory policy toward mainland China.

So far, the battle between Taipei and Beijing remains at the verbal level. At the same time, massive investments by Taiwan in the mainland and the 2000 opening of Taiwan's Offshore Islands of Quemoy and Matsu for trade with the P.R.C. are bringing the two sides still closer together. Their two economies are becoming steadily more intertwined, and both sides benefit from their commercial ties. This does not mean that they will soon be fully reunified in law. Furthermore, there remains the black cloud of Beijing possibly using military force against Taiwan if it declares itself an independent state. Beijing refuses to make any pledge never to use military force to reunify Taiwan with the mainland, on the grounds that what it does with Taiwan is China's internal affair. In Beijing's view, no other country has a right to tell China what to do about Taiwan.

Human Rights in U.S.–China Relations

Since U.S. president Jimmy Carter established diplomatic relations with China in 1979, each successive American president has campaigned on a platform that decried the abuse of human rights in China and vowed, if elected, to take strong action, including economic measures, to punish China. The Chinese people have been confused and distraught at this prospect. They do not see the point in punishing hundreds of millions of Chinese for human rights abuses committed not by the people, but by their leadership. Nor do they necessarily believe that their own government has been more abusive of human rights than other states that seem to escape scrutiny. In any event, within a few months (if not sooner) of being sworn in, each successive president has abandoned his campaign platform and taken a more moderate approach to China.

Why was this? Once inauguration day was over, it was quickly explained to the new president that the United States dare not risk jeopardizing its relations with an increasingly powerful state containing one-quarter of the world's population through punitive measures. Boycotts would probably give Japan and other countries a better trading position while undercutting the opportunity for Americans to do business with China. By 2000, President Bill Clinton had managed to get Congress to vote for "permanent normal trading relations" (PNTR). No longer would normal trade relations with China be subjected each spring to a congressional review of its human-rights record. This in turn cleared the way for China to join the World Trade Organization with an American endorsement in 2001. Under WTO rules, one country may

((Pool photo by Kyodo News)(Kyodo via AP Images) (100413134))
President Hu Jintao and President Barack Obama shaking hands. U.S.–China relations have experienced some ups and downs in recent years, but both sides emphasize cooperation now.

not use trade as a weapon to punish another for political reasons.

Clinton's China policy was also shaped by a new strategy of "agreeing to disagree" on certain issues such as human rights, while efforts continued to be made to bring the two sides closer together. This strategy came out of a belief that China and the United States had so many common interests that neither side could afford to endanger the relationship on the basis of a single issue. The American policy of "engagement" with China, which began with the Clinton administration, was based on the belief that isolating China had proven counterproductive. The administration argued that human rights issues could be more fruitfully addressed in a relationship that was more positive in its broader aspects. "Engagement" allowed the two countries to work together toward shared objectives, including the security of Asia.

Although President George W. Bush initially appeared intent on ending engagement, and treating China as a "strategic adversary," the Bush administration soon abandoned this policy—an act made complete by the September 11, 2001 terrorist attacks on the United States. After those events, President Bush told the world, "You are either with us or against us." China, not wishing to needlessly bring trouble on itself, immediately sided with the United States in the war on terrorism. This had important implications for the role that human rights could play in the U.S.–China relationship; for with the focus on terrorism, human rights took a back seat, even in the United States itself. With critics across the political spectrum raising countless

Timeline: PAST

1842
The Treaty of Nanking cedes Hong Kong to Great Britain

1860
China cedes the Kowloon Peninsula to Great Britain

1894–1895
The Sino–Japanese War

1895–1945
Taiwan is under Japanese colonial control

1898
China leases Northern Kowloon and the New Territories (Hong Kong) to Great Britain for 99 years

1900–1901
The Boxer Rebellion

1911
Revolution and the overthrow of the Qing Dynasty

1912–1949
The Republic of China

1921
The Chinese Communist Party (CCP) is established

1931
Japanese occupation of Manchuria (the northeast province of China)

1934–1935
The Long March

1937–1945
The Japanese invasion and occupation of much of China

1942–1945
The Japanese occupation of Hong Kong

1945–1949
Civil war between the KMT and CCP

The KMT establishes the Nationalist government on Taiwan. Keeps name of Republic of China

The People's Republic of China is established

1950
The United States recognizes the Nationalist government in Taiwan as the legitimate government of all China

1958
The "Great Leap Forward"; the Taiwan Strait crisis (Offshore Islands)

1963
The Sino–Soviet split becomes public

1966–1976
The "Cultural Revolution"

1971
The United Nations votes to seat the P.R.C. in place of the R.O.C.

1972
U.S. president Richard Nixon visits the P.R.C.; the Shanghai Communique

1976
Mao Zedong dies; removal of the Gang of Four

1977
Deng Xiaoping returns to power

1979
The United States recognizes the P.R.C. and withdraws recognition of the R.O.C.

1980s–1990s
Resumption of arms sales to Taiwan

The Shanghai Communique II: the United States agrees to phase out arms sales to Taiwan

China and Great Britain sign an agreement on Hong Kong's future Sino–Soviet relations begin to thaw

China sells Silkworm missiles to Iran and Saudi Arabia

Student demonstrations in Tiananmen Square; military crackdown; political repression follows

Deng encourages "experimentation" and the economy booms

The United States bombs the Chinese Embassy in Belgrade; says "an accident" Deng Xiaoping dies; Jiang Zemin assumes power

PRESENT

2000
China begins long-distance population resettlement for the Three Gorges Dam project

2001
China bans the Falun Gong movement

9/11 terrorist attacks lead to stronger ties between China and the United States, but China opposes the Iraq war

China becomes member of the WTO

2002
Hu Jintao becomes the leader of the CCP

SARS outbreak

2003
China sends first manned spacecraft (with astronaut Yang Liwei) into space

2004
The United Nations estimates 1 million Chinese are infected with HIV

China signs agreement with Russia settling their long-lasting border disputes

2005
Joint military maneuvers of China and Russia

2006
China surpasses the United States to be the world's largest emitter of carbon dioxide

China becomes the fourth largest economy and the third largest trading nation

2007
China shoots down old weather satellite, demonstrating capability of destroying hostile spy satellites

China launches the first domestically developed passenger jet Xiang Feng, or "Flying Phoenix"

2008
Violent ethnic protests erupt in Lhasa, Tibet

Earthquake strikes Sichuan province, with death toll surpassing 70,000

Summer Olympics in Beijing

China's first space walk (Zhai Zhigang)

2009
About 200 people die and over 1,700 are injured in ethnic violence in Xinjiang

China becomes the largest automobile market in the world

Mass celebrations to mark the 60th anniversary of the PRC

2010
China becomes the second largest economy in the world

Shanghai World Expo

China produces the fastest supercomputer in the world

Jailed dissident Liu Xiaobo receives Nobel Peace Prize

questions about the American government's treatment of suspected terrorists, the United States was hardly in a position to be pressing for improved human rights in China. Just as important, the United States did not want to raise gratuitous questions concerning China's alleged derogation of human rights when it needed China on its side, not just in the war on terrorism, but on almost every issue of international significance.

In spite of the White House's tendency to be more pragmatic about China and avoid the issue of human rights, the U.S. Congress has been a different matter. It was Congress that pressed the human rights agenda, especially under President Clinton. Indeed, during his second term in office, Congress used Clinton's favorable treatment of China as one more reason for trying to force him out of office. Coupled with 9/11, the curtailing of American liberties, and numerous accusations of the American abuse of the rights of those under detention in Guantanamo and Iraq, the Chinese human rights issue virtually disappeared from the congressional agenda. In the 2008 U.S. presidential election, China's human rights did not become a campaign issue. President Barack Obama has also emphasized the importance of cooperating with China in dealing with regional and global issues. U.S.–China relations today are more likely to suffer from trade and monetary issues than from human rights concerns.

(Getty Images 98767826 NSEA)
The night view of the imposing China Pavilion at the 2010 World Expo in Shanghai.

THE FUTURE

Since 1979, China has moved from being a relatively closed and isolated country to one that is fully engaged in the world. China's agenda for the future is daunting. It must avoid war; maintain internal political stability in the context of international pressures to democratize; continue to carry out major economic, legal, and political reforms without destabilizing society and endangering CCP control; and sustain economic growth while limiting environmental destruction.

Since the death of Deng Xiaoping in 1997, China has carried out smooth leadership transitions. The central party-state leadership has never deviated from the road of reform.

Strong economic growth has been crucial to the continuing legitimacy of the CCP leadership in the eyes of the Chinese people. The party may, however, some day change its name to one more reflective of its actual policies—not communism but capitalism, combined with socialist social policies. To some degree, it has adopted policies similar to those of European leftist parties, which tend to label themselves as "democratic socialist" or "socialist democratic" parties. Whatever name it adopts, the CCP is unlikely to allow the creation of a multiparty system any time soon that could challenge its leadership.

Governing the world's largest population is a formidable task, one made even more challenging by globalization. The integration of China into the international community has heightened the receptivity of China's leaders to pressures from the international system on a host of specific issues: human rights, environmental protection, intellectual property rights, prison labor, arms control, and legal codes. China's leadership insists on moving at its own pace and in a way that takes into account China's culture, history, and institutions; but China is now subject to globalizing forces, as well as internal social and economic forces, that have a momentum of their own.

In the meantime, China, like so many other developing countries, must worry about the polarization of wealth, high levels of unemployment, uncontrolled economic growth, environmental degradation, and the strident resistance by whole regions within China against following economic and monetary policies formulated at the center. It is also facing a major HIV/AIDS epidemic, a potential collapse of the banking and financial systems, the need to finance a social safety net and retirement pensions, a demographic crisis, and a looming threat to its state-owned enterprises as a result of China's entry into the WTO. To wit, the international community is pressing China to revalue the Chinese currency, the *yuan,* so that Chinese goods will be priced higher, thus making them less competitive internationally. Common criminality, corruption, and social instability provide additional fuel that could one day explode politically and bring down Chinese Communist Party rule. Overall, the CCP and the Chinese public are very optimistic about the future of China despite all the problems and challenges.

At this time there is no alternative leadership waiting in the wings to take up the burden of leading China and ensuring its stability. An unstable China would not be in anyone's interest, neither that of the Chinese people, nor of any other country. An insecure and unstable China would be a more dangerous China, and it would be one in which the Chinese people would suffer immeasurably.

NOTES

1. Pam Woodall, "The Real Great Leap Forward, *The Economist,* October 2, 2004, p. 6.
2. A concern about "spiritual pollution" is not unique to China. It refers to the contamination or destruction of one's own spiritual and cultural values by other values. Europeans are as concerned about it as the Chinese and have, in an effort to combat spiritual pollution, limited the number of television programs made abroad (i.e., in the United States) that can be broadcast in European countries.
3. The essence of the "get civilized" campaign was an effort to revive a value that had seemingly been lost: respect for others and common human decency. Thus, drivers were told to drive in a "civilized" way—that is, courteously. Ordinary citizens were told to act in a "civilized" way by not spitting or throwing garbage on the

ground. Students were told to be "civilized" by not stealing books or cheating, keeping their rooms and bathrooms clean, and not talking loudly.

4. The turmoil that ensued after his death had also ensued after the death of the former beloved premier, Zhou Enlai. Similar turmoil followed the death of another recently arrived hero of the students. Indeed, the central leadership was almost paralyzed when, in January 2005, Zhao Ziyang, the Secretary General of the CCP at the time of the Tiananmen demonstrations in 1989, died. Zhao, who was dismissed from his position because in the end he supported the students' demands for political reform, had been accused of trying to "split" the party. He spent the next 15 years, until his death, under house arrest in Beijing.
5. "Campaign to Crush Dissent Intensifies," *South China Morning Post* (August 9, 1989).
6. Zhiqun Zhu, *The People's Republic of China Today: Internal and External Challenges* (Singapore: World Scientific Publishing, 2010): pp. 431–433.
7. Susan K. McCarthy, "The State, Minorities, and Dilemmas of Development in Contemporary China," *The Fletcher Forum of World Affairs,* Vol. 26:2 (Summer/Fall 2002).
8. June Teufel Dreyer, "Economic Development in Tibet under the People's Republic of China," *Journal of Contemporary China,* Vol. 12, no. 36 (August 2003), pp. 411–430.
9. For excellent detail on Chinese religious practices, see Robert Weller, *Taiping Rebels, Taiwanese Ghosts, and Tiananmen* (Seattle: University of Washington Press, 1994); and Alan Hunter and Kimk-wong Chan, *Protestantism in Contemporary China* (Cambridge: Cambridge University Press, 1993). The latter notes that Chinese judge gods "on performance rather than theological criteria" (p. 144). That is, if the contributors to the temple in which certain gods were honored were doing well financially and their families were healthy, then those gods were judged favorably. Furthermore, Chinese pray as individuals rather than as congregations. Thus, before the Chinese government closed most temples, they were full of individuals praying randomly, children playing inside, and general noise and confusion. Western missionaries have found this style too casual for their own more structured religions (p. 145).
10. Professor Rudolf G. Wagner (Heidelberg University). Information based on his stay in China in 1990.
11. Richard Madesen, "Understanding Falun Gong," *Current History* (September 2000), Vol. 99, No. 638, pp. 243–247.
12. For a better understanding of how Chinese characters are put togther, see John DeFrancis, *Visible Speech: The Diverse Oneness of Writing Systems* (Honolulu: University of Hawaii Press, 1989); and Bob Hodge and Kam Louie, *The Politics of Chinese Language and Culture: The Art of Reading Dragons* (New York: Routledge, 1998).
13. "Number of University Students Recruited Doubles in Six Years," *People's Daily Online* (December 7, 2004).
14. Cao Desheng, "China Cancels 4,800 Development Zones," *China Daily,* August 24, 2004.
15. Howard W. French "Alarm and Disarray on Rise in China," *The New York Times,* August 24, 2005.
16. Nicholas R. Lardy, "China's Economy: Problems and Prospects," *Foreign Policy Research Institute,* Vol. 12, no. 4 (Feb. 2007), online at www.fpri.org
17. Yan Sun, "Corruption, Growth, and Reform," *Current History,* September 2005, pp. 257–263.
18. According to research done by the Research Office of the State Council, the Chinese Academy of Social Sciences, and the Party School's Research Office, the main source of the billionaires' wealth was, among other things, "Legal or illegal commissions from introducing foreign investments . . .; Importing facilities and equipment with . . . prices . . . usually 60 percent to 300 percent higher than market prices; . . . Developing and selling land with bank loans and zero costs; . . . Smuggling, tax evasion. . . . Obtaining and pocketing loans from banks without collateral." For this and other sources of wealth for Chinese billionaires, see Mo Ming, "90 Percent of China's Billionaires Are Children of Senior Officials," http://financenews.com/ausdaily/
19. Jonathan Watts, "Corrupt Officials Have Cost China 330 Million Pounds in 20 Years," *The Guardian* (August 20, 2004).
20. Richard McGregor, "China's Power Capacity Soars," *Financial Times,* February 6, 2007.
21. Yuanyuan Shen (Tsinghua University Law School), seminar, Harvard University Center for the Environment, China Project, October 20, 2005.
22. *Shai Oster,* "China Tilts Green: Climate Concerns Sway Beijing," *The Wall Street Journal,* February 13, 2007.
23. Philip P. Pan, "In China, Turning the Law into the People's Protector," *The Washington Post Foreign Service* (Dec. 28, 2004), p. A1.
24. Suzanne Ogden, *Inklings of Democracy in China* (Cambridge: Harvard University Asia Center and Harvard University Press, 2002), pp. 234–236.
25. Based on the author's trip to interview village leaders in 2000 and the author's visit with President Carter to monitor elections in a Chinese village in 2001. See Ogden, pp. 183–220.
26. "Chinese NGOs increase to 346,000 last year," posted February 4, 2007, at www.chinaelections.org
27. For an excellent analysis of how the "patterns of protest" in China have replicated the "patterns of daily life," see Jeffrey N. Wasserstrom and Liu Xinyong, "Student Associations and Mass Movements," in Deborah S. Davis, Richard Kraus, Barry Naughton, Elizabeth J. Perry, eds., *Urban Spaces in Contemporary China: The Potential for Autonomy and Community in Post-Mao China* (Cambridge: Cambridge University Press and Woodrow Wilson Center Press, 1995), pp. 362–366, 383–386. The authors make the point that students learned how to organize, lead, and follow in school. This prepared them for organizing so masterfully in Tiananmen Square. The same was true for the workers who participated in the 1989 protests "not as individuals or members of 'autonomous' unions but as members of *danwei* delegations, which were usually organized with either the direct support or the passive approval of work-group leaders, and which were generally led onto the streets by people carrying flags emblazoned with the name of the unit." p. 383.
28. In 1996–1997, the citizens of Beijing who were unable to sleep through the racket finally forced the government to pass a noise ordinance that lowered the decibel level allowed on streets by public performers, such as the fan and ballroom dancers.
29. Vaclav Havel, as quoted by Timothy Garton Ash, "Eastern Europe: The Year of Truth," *New York Review of Books* (February 15, 1990), p. 18, referenced in Giuseppe De Palma, "After Leninism: Why Democracy Can Work in Eastern Europe," *Journal of Democracy,* Vol. 2, No. 1 (Winter 1991), p. 25, note 3.
30. Liu Binyan, "China and the Lessons of Eastern Europe," *Journal of Democracy,* Vol. 2, No. 2 (Spring 1991), p. 8.
31. Beijing University student, "My Innermost Thoughts—To the Students of Beijing Universities" (May 1989), Document 68, in Suzanne Ogden, et al., eds., *China's Search for Democracy,* pp. 172–173.
32. Vivienne Shue, in a speech to a USIA conference of diplomats and scholars, as quoted and summarized in "Democracy Rating Low in Mainland," *The Free China Journal* (January 24, 1992), p. 7.
33. Joyce Barnathan, et al., "China: Is Prosperity Creating a Freer Society?" *Business Week* (June 6, 1994), p. 98.
34. Jim Yardley, "Chinese United by Common Goal: A Hot Stock Tip," *The New York Times* (January 30, 2007), pp. A1, A10.
35. These figures are a composite, taken from the Chinese government website on "Mass Media," www.china.org.cn/english/features/Brief/193358.htm (Feb. 10, 2007); and "Lexis-Nexis Country Report, 1999: China," www.lexis-nexis.com.
36. David Barboza, "Internet Boom in China is Built on Virtual Fun," *The New York Times* (Feb. 5, 2007), pp. A1, A4.
37. Wang, in Davis, pp. 170–171.
38. "Party Magazine Attacks Morality of Chinese Films," (article published on Feb. 9 2007, posted Feb. 10, 2007), www.chinaelections.org). Zhang Yimou's film *House of Flying Tigers,* and Chen

Kaige's film *The Promise,* were also condemned by the party periodical, *The Study Times.*

39. Cary Huang, "Beijing Tightens Media Grip with Penalty Points System," *South China Morning Post,* Feb. 9, 2007.
40. Wang Meng (former minister of culture and a leading novelist in China), speech at Cambridge University (May 23, 1996). An example of a movie banned in China is the famous producer Chen Kaige's *Temptress Moon.* This movie, which won the Golden Palm award at the Cannes Film Festival in 1993, is, however, allowed to be distributed abroad. The government has adopted a similar policy of censorship at home but distribution abroad for a number of films, including *Farewell My Concubine,* by China's most famous film directors.
41. Barnathan, et al., pp. 98–99.
42. For the Chinese response in 2007, see Edward Cody, "China: Bush Has No Right to Criticize on Human Rights," *Washington Post,* March 8, 2007.
43. For more on Chinese nationalism, see Suzanne Ogden, "Chinese Nationalism: The Precedence of Community and Identity Over Individual Rights," *Asian Perspective,* vol. 25, no. 4 (2001), pp. 157–185.
44. China to Show Only "Ethically Inspiring TV Series in Prime Time from Next Month" Published and posted Jan. 22, 2007 on www.chinaelections.org
45. In 1998, Avon was, at least temporarily, banned from China, as were other companies that used similar sales and marketing techniques. Too many Chinese found themselves bankrupted when they could not sell the products that they had purchased for resale.
46. Lardy, "China's Economy . . . ," (February 2007), *ibid.*
47. For excellent analyses of the Sino–American relationship from the nineteenth century, see Warren Cohen, *America's Response to China: A History of Sino–American Relations,* 3rd ed. (New York: Columbia University Press, 1990); Richard Madesen, *China and the American Dream: A Moral Inquiry* (Berkeley, CA: University of California Press, 1995); Michael Schaller, *The United States and China in the Twentieth Century,* 2nd ed. (New York: Oxford University Press, 1990); David Shambaugh, ed., *American Studies of Contemporary China* (Armonk, NY: M.E. Sharpe, 1993); and David Shambaugh, *Beautiful Imperialist: China Perceives America, 1972–1990* (Princeton, NJ: Princeton University Press,

Hong Kong Map

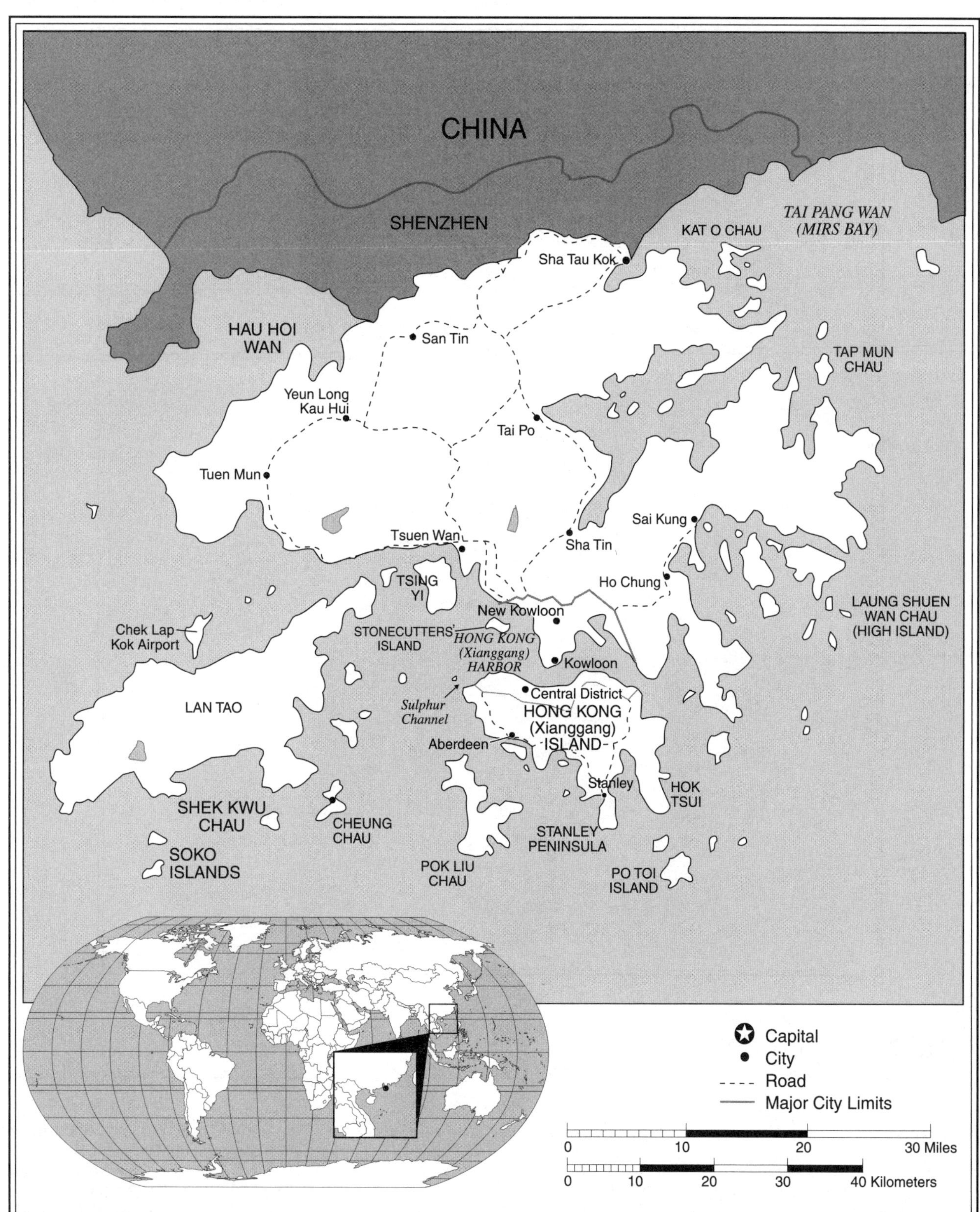

Hong Kong is comprised of the island of Hong Kong (1842), the Kowloon Peninsula and Stonecutters' Island (1860), the New Territories (1898) that extend from Kowloon to the Chinese land border; and 230 adjacent islets. Land is constantly being reclaimed from the sea, so the total land area of Hong Kong is continually increasing by small amounts. All of Hong Kong reverted to Chinese sovereignty on July 1, 1997. It was renamed the Hong Kong Special Administrative Region.

Hong Kong (Hong Kong Special Administrative Region)

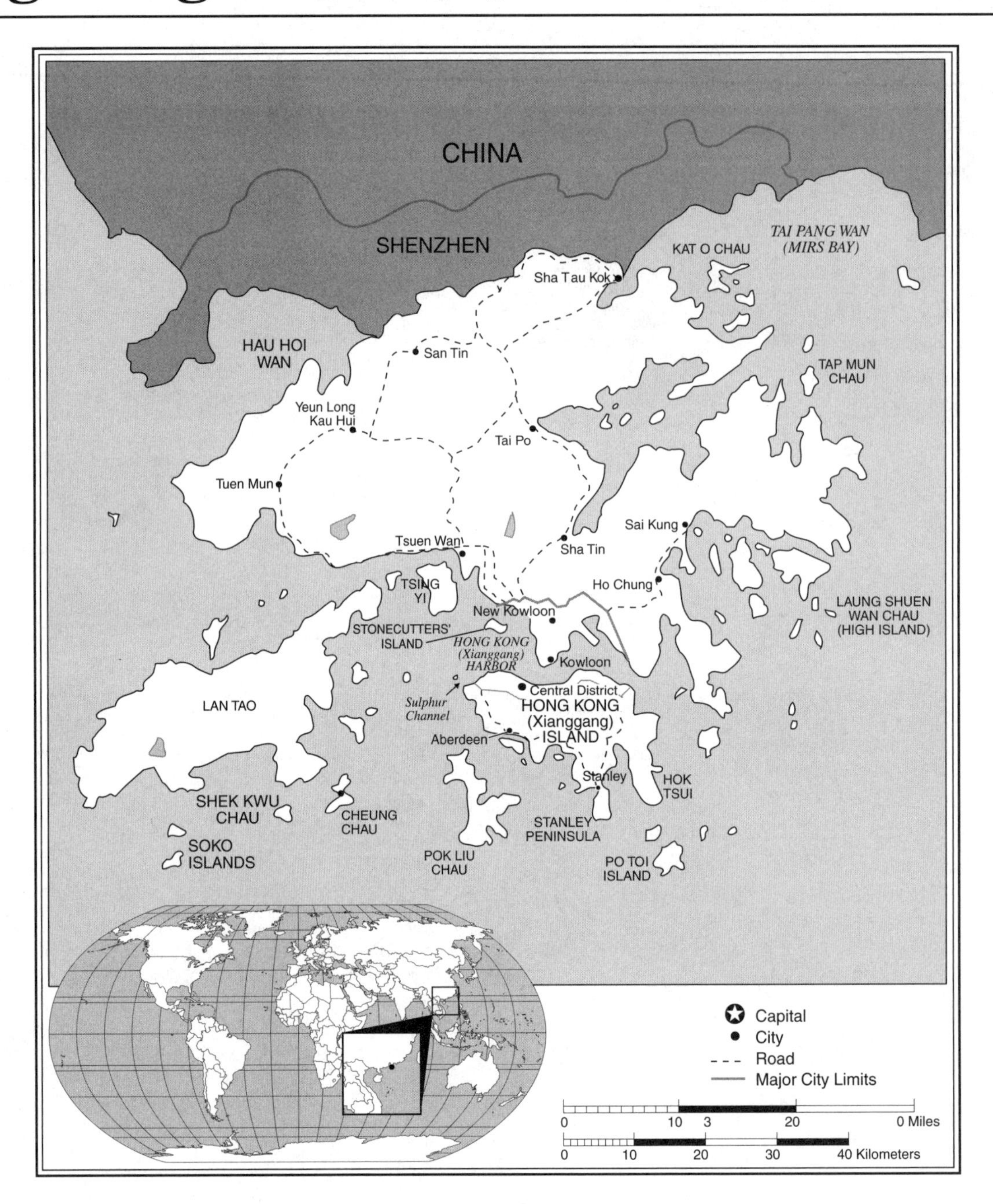

Hong Kong Statistics

GEOGRAPHY

Area in Square Miles (Kilometers): 421 (1,092) (about six times the size of Washington, D.C.)
Geographical Features: hilly to mountainous, with steep slopes; lowlands in the north; more than 200 islands
Climate: tropical
Environmental Concerns: air and water pollution

PEOPLE

Population

Total: 7,089,705 (July 2010 est.)
Annual Growth Rate: 0.476% (2010 est.)
Ethnic Makeup: 95% Chinese (mostly Cantonese); 5% others
Religions: local religions (Buddhism and Taoism) 90%; Christian 10%
Major Languages: Chinese (Cantonese) 90.89% (official), Mandarin, 0.9%, other Chinese dialects 4.4%; English 2.8% (official).

Health

Life Expectancy at Birth: 81.96 years (*male:* 79.24 years; *female:* 84.88 years) (2010)

Infant Mortality Rate: 2.91 deaths/1,000 live births
Total Fertility Rate: 1.04 child born/woman (2008 est.)
Physicians Available: 1.3/1,000 people
HIV/AIDS—Adult Prevalence Rate: 0.1% (2003 est.)

EDUCATION

Adult Literacy Rate: 93.5%
Education expenditures: 3.3% of GDP (2008)

COMMUNICATION

Telephones—main lines in use: 4.188 million (2009)
Telephones—mobile cellular: 10.55 million (2007)
Internet Users: 4.124 million (2008)
Television Broadcast Stations: 55 (2 TV networks, each broadcasting on 2 channels) (2007)

TRANSPORTATION

Airports: 2 (2010)
Heliports: 9 (2010)
Paved Roadways: 2,050 km (2009)

GOVERNMENT

Type: Special Administrative Region (SAR) of China
Head of State: President of China Hu Jintao (since March 2003)
Head of Government: Chief Executive Donald Tsang (since June 2005)
National Holiday: National Day (Anniversary of the Founding of the People's Republic of China), October 1. July 1 is celebrated as Hong Kong SAR Establishment Day
Constitution: The Basic Law, approved in March 1990 by China's National People's Congress, is Hong Kong's "mini-constitution"
Legal system: based on English common law
Political Parties: Association for Democracy and People's Livelihood or ADPL; Citizens Party; Civic Party; Democratic Alliance for the Betterment and Progress of Hong Kong or DAB; Democratic Party; Frontier Party; League of Social Democrats; Liberal Party
Suffrage: direct election—18 years of age for a number of non-executive positions; universal for permanent residents living in the territory of Hong Kong for the past seven years; indirect election—limited to about 220,000 members of functional constituencies and an 800-member election committee drawn from broad regional groupings, central government bodies, and municipal organizations

MILITARY

Military Expenditures (% of GDP): None. Defense is the responsibility of China.
Current Disputes: none

ECONOMY

Currency: Hong Kong dollar (HKD)
Exchange Rate: U.S.$1 = 7.77 HKD (December 2010)
Per Capita Income (Purchasing Power Parity): $42,700 (2009 est.)
GDP (official exchange rate): $210.6 billion (2009 est.)
GDP - Real Growth Rate: 6.4% (2007 est.)
Unemployment Rate: 5.2% (2009 est.)
Agriculture: fresh vegetables; poultry; pork; fish
Industries: textiles; clothing; tourism; banking; shipping; electronics; plastics; toys; watches; clocks
Exports: $321.8 billion f.o.b., including re-exports (2009 est.)
Exports Partners: China 51.2%; U.S. 13.7%; Japan 4.5% (2009)
Exports Commodities: electrical machinery and appliances; textiles; apparel; footwear; watches and clocks; toys; plastics; precious stones; printed material
Imports: $365.6 billion (2007 est.)
Imports Partners: China 46.3%; Japan 10%; Taiwan 7.1%; Singapore 6.8%; U.S. 4.9%; South Korea 4.2% (2009)
Imports Commodities: raw materials and semi-manufactures; consumer goods; capital goods; foodstuffs; fuel (most is re-exported)

SUGGESTED WEB SITES

www.cia.gov/library/publications/the-world-factbook/geos/hk.html
www.gov.hk/en/about/sitemap.htm
www.WorldBank.org
www.state.gov
www.civic-exchange.org

Hong Kong Report

Hong Kong, the "fragrant harbor" situated on the southeastern edge of China, was under British rule, characterized as the "pearl of the Orient," from 1842 until 1997. The British colonial administration supported a market economy in the context of a highly structured and tightly controlled political system. This allowed Hong Kong's dynamic and vibrant people to shape the colony into one of the world's great success stories. The history of Hong Kong's formation and development, its achievements, and its handling of the difficult issues emanating from a "one country, two systems" formula since it was returned to Chinese rule in 1997 provide one of the most fascinating stories of cultural, economic, and political transition in the world. Hong Kong's "borrowed time" has ended; but its efforts to shape itself into the "Manhattan of China" are in full swing.

HISTORY

In the 1830s, the British sale of opium to China was creating a nation of drug addicts. Alarmed by this development, the Chinese imperial government banned opium; but private British "country traders," sailing armed clipper ships, continued to sell opium to the Chinese by smuggling it (with the help of Chinese pirates) up China's coast and rivers. In an effort to enforce the ban, the Chinese imperial commissioner, Lin Zexu, detained the British in Canton (Guangzhou) and forced them to surrender their opium. Commissioner Lin took the more than 21,000 chests of opium and destroyed them in public.[1] The British, desperate to establish outposts for trade with an unwilling China, used this siege of British warehouses as an excuse to declare war on the Chinese. Later called the Opium War (1839–1842), the conflict ended with China's defeat and the Treaty of Nanjing.

Great Britain did not wage war against the Chinese in order to sell an addictive drug that was banned by the Chinese government. Rather, the Chinese government's attack on the British opium traders, whose status as British citizens suddenly proved convenient to the British government, provided the necessary excuse for Great Britain getting what it really wanted: free trade with a government that had restricted trade with the British to one port, Canton. It also allowed London to assert Great Britain's diplomatic equality with China, which considered itself the "Central Kingdom" and superior to all other countries. The Chinese imperial government's demand that all "barbarians," including the British, kowtow to the Chinese emperor, incensed the British and gave them further cause to set the record straight.

The China trade had been draining the British treasury of its gold and silver species; for the British purchased large quantities of Chinese porcelain, silk, tea, and spices, while the Chinese refused to purchase the products of Great Britain's nineteenth-century Industrial Revolution. Smug in their belief that their cultural and moral superiority was sufficient to withstand any military challenge from a "barbarian" country, the Chinese saw no need to develop a modern military or to industrialize. An amusing example of the thought process involved in "Sinocentrism"—the belief that China was the center of the world and superior to all other countries—was Imperial Commissioner Lin's letter to Queen Victoria. Here he noted "Britain's dependence on Chinese rhubarb, without which the English would die of constipation."[2] China's narrow world view blinded it to the growing power of the West and resulted in China's losing the opportunity to benefit from the Industrial Revolution at an early stage. The Opium War turned out to be only the first step in a century of humiliation for China—the step that led to a British foothold on the edge of China.

For their part, the British public did not generally see the sale of opium as a moral issue, or that large-scale addiction was a possible outcome for China. Opium was available for self-medication in Britain, was taken orally (not smoked as it was in China), was administered as a tranquilizer for infants by the working class, and was not considered toxic by the British medical community at that time.[3] Great Britain's colonial government in Hong Kong remained dependent on revenues from the sale of opium until Hong Kong was occupied by Japan during World War II.[4]

The Treaty of Nanjing gave the British the right to trade with the Chinese from five Chinese ports; and Hong Kong, a tiny island off the southern coast of China, was ceded to them "in perpetuity." In short, according to the practices of the colonizing powers of the nineteenth century, Hong Kong became a British colony forever. The Western imperialists were still in the acquisition phase of their history. They were not contemplating that one day the whole process of colonization might be reversed. As a result, Great Britain did not foresee that it might one day have to relinquish the colony of Hong Kong, either to independence or to Chinese rule.

Hong Kong Island's total population of Chinese villagers and people living on boats then numbered under 6,000. From 1842 onward, however, Hong Kong became the primary magnet for Chinese immigrants fleeing the poverty, chaos, and cruelty of China in favor of the relatively peaceful environment of Hong Kong under British rule. Then, in 1860, again as a result of a British victory in battle, the Chinese ceded to the British "in perpetuity" Stonecutters' Island and a small (3 1/2 square miles) but significant piece of land facing the island of Hong Kong: Kowloon Peninsula. Just a few minutes by ferry (and, since the 1970s, by tunnel) from Hong Kong Island, it became an important part of the residential, commercial, and business sectors of Hong Kong. The New Territories, the third and largest part (89 percent of the total area) of what is now known as "Hong Kong," were not granted "in perpetuity" but were merely leased to the British for 99 years under the second Anglo–Chinese Convention of Peking in 1898. The New Territories, which extended from Kowloon to the Chinese land border, comprised the major agricultural area supporting Hong Kong.

The distinction between those areas that became a British colony (Hong Kong Island and Kowloon) and the area "leased" for 99 years (the New Territories) is crucial to understanding why, by the early 1980s, the British felt compelled to negotiate with the Chinese about the future of "Hong Kong"; for although colonies are theoretically colonies "in perpetuity," the New Territories were merely leased, and would automatically revert to Chinese sovereignty in 1997. Without this large agricultural area, the rest of Hong Kong could not survive; the leased territories had, moreover, become tightly integrated into the life and business of Hong Kong Island and Kowloon.

Thus, with the exception of the period of Japanese occupation (1942–1945) during World War II, Hong Kong was administered as a British Crown colony from the nineteenth century onward. After the defeat of Japan in 1945, however, Britain almost did not regain control over Hong Kong because of the United States, which insisted that it did not fight World War II in order to return colonies to its allies. But Britain's leaders, both during and after World War II, were determined to hold on to Hong Kong because of its symbolic, economic, and strategic importance to the British Empire. India, Singapore, Malaya, Burma—all could be relinquished, but not Hong Kong.[5] Moreover, although during World War II, a U.S. presidential order had stated that at the end of the war, Japanese troops in Hong Kong were to surrender to Chiang Kai-shek, the leader of the Republic of China, it did not happen. Chiang, more worried about accepting surrender of Japanese troops in the rest of China before the Chinese Communist could, did not rush to Hong Kong. Meanwhile, a British fleet moved rapidly to Hong Kong to pre-empt Chiang occupying Hong Kong, even though Chiang averred he would not have stayed. The British doubted this, and argued that Hong Kong was still British sovereign territory and would itself accept the surrender of the Japanese.[6]

At the end of the civil war that raged in China from 1945 to 1949, the Communists' Red Army stopped its advance just short of Hong Kong. Beijing never offered an official explanation. Perhaps it did not want to get into a war with Britian in order to claim Hong Kong, or perhaps the Chinese Communists calculated that Hong Kong would be of more value to them if left in British hands. Indeed, at no time after their victory in 1949 did the Chinese Communists attempt to force Great Britain out of Hong Kong, even when Sino–British relations were greatly strained, as during China's "Cultural Revolution."[7]

This did not mean that Beijing accepted the legitimacy of British rule. It did not. After coming to power on the mainland in 1949, the Chinese Communist Party held that Hong Kong was a part of China stolen by British imperialists, and that it

THE SECOND ANGLO/CHINESE CONVENTION CEDES THE KOWLOON PENINSULA TO THE BRITISH

The second Anglo/Chinese Convention, signed in 1860, was the result of a string of incidents and hostilities among the Chinese, the British, and the French. Although the French were involved in the outbreak of war, they were not included in the treaty that resulted from conflict.

The catalyst for the war was that, during a truce, the Chinese seized the chief British negotiator and executed 20 of his men. In reprisal, the English destroyed nearly 200 buildings of the emperor's summer palace and forced the new treaty on the Chinese. This called for increased payments ("indemnities") by the Chinese to the English for war-inflicted damages as well as the cession of Kowloon Peninsula to the British.

was merely "occupied" by Great Britain. Hence the notion of Hong Kong as "a borrowed place living on borrowed time." The People's Republic of China insisted that Hong Kong not be treated like other colonies; for the process of decolonization has in practice meant sovereignty and freedom for a former colony's people.[8] China was not about to allow Hong Kong to become independent. After the People's Republic of China gained the China seat in the United Nations in 1971, it protested the listing of Hong Kong and Macau (a Portuguese colony) as colonies by the UN General Assembly's Special Committee on Colonialism. In a letter to the Committee, Beijing insisted they were merely

> part of Chinese territory occupied by the British and Portuguese authorities. The settlement of the questions of Hong Kong and Macao is entirely within China's sovereign right and does not at all fall under the ordinary category of colonial territories. Consequently they should not be included in the list of colonial territories covered by the declaration on the granting of independence to colonial countries and peoples. . . . The United Nations has no right to discuss these questions.[9]

China made it clear that, unlike other colonies, Hong Kong's colonial subjects did not have the option of declaring independence, for overthrowing British colonial rule would have led directly to the re-imposition of China's control. And although there is for the Hong Kong Chinese a cultural identity as Chinese, after 1949 few wanted to fall under the rule of China's Communist Party government. Furthermore, Beijing and London as a rule did not interfere in Hong Kong's affairs, leaving these in the capable hands of the colonial government in Hong Kong. Although the colonial government formally reported to the British Parliament, in practice it was left to handle its own affairs. Still, the colonial government did not in turn cede any significant political power to its colonial subjects.[10]

No doubt the Chinese Communists were ideologically uncomfortable after winning control of China in 1949 in proclaiming China's sovereign rights and spouting Communist principles while at the same time tolerating the continued existence of a capitalist and British-controlled Hong Kong on its very borders. China could have acquired control within 24 hours simply by shutting off Hong Kong's water supply from the mainland. But China profited from the British presence there and, except for occasional flareups, did little to challenge it.

By 1980, the Hong Kong and foreign business communities had grown increasingly concerned about the expiration of the British lease on the New Territories in 1997. The problem was that all land in the New Territories (which by then had moved from pure agriculture to becoming a major area for manufacturing plants, housing, and commercial buildings) was *leased* to businesses or individuals, and the British colonial government could not grant any land lease that expired after the lease on the New Territories expired. Thus, all land leases—regardless of which year they were granted—would expire three days in advance of the expiration of the main lease on the New Territories on July 1, 1997. As 1997 grew steadily closer, then, the British colonial government had to grant shorter and shorter leases. Investors found buying leases increasingly unattractive. The British colonial government felt compelled to do something to calm investors.[11]

For this reason, it was the British, not the Chinese, who took the initiative to press for an agreement on the future status of the colony and the rights of its people. Everyone recognized the inability of the island of Hong Kong and Kowloon to survive on their own, because of their dependence upon the leased New Territories for food, and because of the integrated nature of the economies of the colonial and leased parts of Hong Kong. Everyone (everyone, that is, except for British prime minister Margaret Thatcher) also knew that Hong Kong was militarily indefensible by the British and that the Chinese were unlikely to permit the continuation of British administrative rule over Hong Kong after it was returned to Chinese sovereignty.[12] So, a series of formal Sino–British negotiations over the future of Hong Kong began in 1982. By 1984, the two sides had reached an agreement to restore all three parts of Hong Kong to China on July 1, 1997.

The Negotiations over the Status of Hong Kong

Negotiations between the People's Republic of China and Great Britain over the future status of Hong Kong got off to a rocky start in 1982. Prime Minister Thatcher set a contentious tone for the talks when she claimed, after meeting with Chinese leaders in Beijing, that the three nineteenth century treaties that gave Great Britain control of Hong Kong were valid according to international law; and that China, like other nations, had an obligation to honor its treaty commitments. Thatcher's remarks infuriated China's leaders, who denounced the treaties that resulted from imperialist aggression as "unequal," and lacking legitimacy in the contemporary world.

Both sides realized that Chinese sovereignty over Hong Kong would be reestablished in 1997 when the New Territories lease expired, but they disagreed profoundly on what such sovereignty would mean in practice. The British claimed that they had a "moral commitment" to the people of Hong Kong to maintain the stability and prosperity of the colony. Both the British and the Hong Kong population hoped that Chinese sovereignty over Hong Kong might be more symbolic than substantive and that some arrangement could be worked out that would allow for continuing British participation in the administration of the area. The Chinese vehemently rejected what they termed "alien rule in Chinese territory" after 1997, as well as the argument that the economic value of a Hong Kong *not* under its administrative power might be greater.[13] Great Britain agreed to end its administration of Hong Kong in 1997, and together with China worked out a detailed and binding arrangement for how Hong Kong would be governed under Chinese sovereignty.

The people of Hong Kong itself did not formally participate in these negotiations over the colony's fate. Although the British and Chinese consulted various interested parties in the colony, they chose to ignore many of their viewpoints. China was particularly adamant that the people of Hong Kong were Chinese and that the government in Beijing represented *all Chinese* in talks with the British.

In September 1984, Great Britain and the People's Republic of China initialed the Joint Declaration on the Question of Hong Kong. It stated that, as of July 1, 1997, Hong Kong would become a "Special Administrative Region" (SAR) under the control of the central government of the People's Republic of China. The Chinese came up with the idea of "one country, two systems," whereby, apart from defense and foreign policy, the Hong Kong SAR would enjoy a high degree of autonomy. Hong Kong would maintain its current social, political, economic, and legal systems alongside China's systems; would remain an international financial center; and would retain its ability to establish independent economic (but not diplomatic) relations with other countries.

The Sino–British Joint Liaison Group was created to oversee the transition to Chinese rule. Any changes in Hong Kong's laws made during the transition period, if they were expected to continue after 1997, had to receive final approval from the Joint Liaison Group. If there were disagreement within the Liaison Group between the British and Chinese, they were obligated to talk until they reached agreement. This procedure gave China veto power over any proposed changes in Hong Kong's governance and laws proposed from 1984 to 1997.[14] When London's newly appointed governor, Christopher Patten, arrived in 1992 and attempted to change some of the laws that would govern Hong Kong after 1997, China had reason to use that veto power.

The Basic Law

The Basic Law is the crucial document that translates the *spirit* of the Sino–British Joint Declaration into a legal code. Often referred to as a "mini-constitution" for Hong Kong after it became a SAR on July 1, 1997, the Basic Law essentially defines where Hong Kong's autonomy ends and Beijing's governance over Hong Kong begins. The British had no role in formulating the Basic Law, as the Chinese considered it an internal, sovereign matter. In 1985, China established the Basic Law Drafting Committee, under the direction of the National People's Congress (NPC). The Committee had 59 members—36 from the mainland, 23 from Hong Kong. Of the latter, almost all were "prominent figures belonging to high and high-middle strata," with Hong Kong's economic elite at its core. In addition, China established a "Consultative Committee" in Hong Kong of 180 members. Its purpose was to function as a nonofficial representative organ of the people of Hong Kong from all walks of life, an organ that would channel their viewpoints to the Basic Law Drafting Committee. By so including Hong Kong's elite and a Hong Kong-wide civic representative organ in consultations about the Basic Law, China hoped to provide political legitimacy to the Basic Law.[15] Once the Basic Law was approved in April 1990 by China's NPC, the final draft was promulgated.

The Basic Law gave Hong Kong a high degree of autonomy after 1997, except in matters of foreign policy and defense, which fell under Beijing's direct control. The government was to be made up of local civil servants and a chief executive chosen by an "Election Committee" appointed by the Standing Committee of the National People's Congress.[16] The chief executive was given the right to appoint key officials of the Special Administrative Region (subject to Beijing's approval). Provisions were made to allow some British and other foreign nationals to serve in the administration of the SAR, if the Hong Kong government so desired. An elected Legislature was made responsible for formulating the laws.[17] The maintenance of law and order remained the responsibility of local authorities, but China took over from the British the right to station military forces in Hong Kong. The local judicial and legal system were to remain basically unchanged, but China's NPC reserved the right to approve all new laws written between 1990 and 1997.[18]

Thus, the Joint Declaration and Basic Law brought Hong Kong under China's rule, with the National People's Congress in Beijing accorded the right of the final interpretation of the meaning of the Basic Law in case of dispute; but the Basic Law allows Hong Kong considerable independence over its economy, finances, budgeting, and revenue until the year 2047. China is thus committed to preserving Hong Kong's "capitalist system and lifestyle" for 50 years and has promised not to impose the Communist political, legal, social, or economic system on Hong Kong. It also agreed to allow Hong Kong to remain a free port, with its own internationally convertible currency (the Hong Kong dollar), over which China would not exercise authority. The Basic Law states that all Hong Kong residents shall have freedom of speech, press, publication, association, assembly, procession, and demonstration, as well as the right to form and join trade unions, and to strike. Freedom of religion, marriage, choice of occupation, and the right to social welfare are also protected by law.[19]

Beijing agreed to continue to allow the free flow of capital into and out of Hong Kong. It also agreed to allow Hong Kong to enter into economic and cultural agreements with other nations and to participate in relevant international organizations as a separate member. Thus, Hong Kong was not held back from membership in the World Trade Organization (WTO) by China's earlier inability to meet WTO membership qualifications. Similarly, Hong Kong is a separate member of the World Bank, the Asian Development Bank, and the Asian-Pacific Economic Conference (APEC). Hong Kong is also allowed to continue issuing its own travel documents to Hong Kong's residents and to visitors.

When China promulgated the Basic Law in 1990, Hong Kong residents by the thousands took to the streets in protest, burning their copies of it. Some of Hong Kong's people saw Britain as having repeatedly capitulated to China's opposition to plans for political reform in Hong Kong before 1997, and as having traded off Hong Kong's interests in favor of Britain's own interests in further trade and investment in China. Hong Kong's business community, however, supported the Basic Law, believing that it would provide for a healthy political and economic environment for doing business. Other Hong Kong residents believed that it was Hong Kong's commercial value, not the Basic Law, that would protect it from a heavy-handed approach by the Chinese government.

The Joint Declaration of China and Great Britain (1984), and the Basic Law (1990) are critical to understanding China's anger in 1992 when Governor Patten proceeded to push for democratic reforms in Hong Kong without Beijing's agreement—particularly since Patten's predecessor, Governor David Wilson, always did consult Beijing and never pushed too hard. After numerous threats to tear up the Basic Law, Beijing simply stated in 1994 that, after the handover of Hong Kong to Chinese sovereignty in 1997, it would nullify any last-minute efforts by the colonial government to promote a political liberalization that went beyond the provisions in the Basic Law. And that is precisely what China did on July 1, 1997. As is noted later in this report, the changes that Patten advocated were largely last-ditch efforts to confer on Hong Kong's subjects democratic rights that they had never had in more than 150 years of British colonial rule. These rights related largely to how the Legislature was elected, the expansion of the electorate, and the elimination of such British colonial regulations as one requiring those who wanted to demonstrate publicly to first acquire a police permit.

The Chinese people were visibly euphoric about the return of Hong Kong "to the embrace of the Motherland." The large clock in Beijing's Tiananmen Square counted the years, months, days, hours, minutes, and even seconds until the return of Hong Kong, helping to focus the Chinese people on the topic. Education in the schools, special exhibits, the movie *The Opium War* (produced by China), and even T-shirts displaying pride in the return of Hong Kong to China's control reinforced a sense that a historical injustice was at last being corrected. On July 1, 1997, celebrations were held all over China, and the pleasure was genuinely and deeply felt by the Chinese people. In Hong Kong, amidst a drenching rain, celebrations were also held. At midnight on June 30, 1997, 4,000 guests watched as the Union Jack was lowered, and China's flag, together with the new Hong Kong Special Administrative Region flag, were raised. President Jiang Zemin, and Charles, the Prince of Wales, and Tony Blair, Prime Minister of Great Britain, represented their respective countries at the ceremony.[20]

THE SOCIETY AND ITS PEOPLE

Immigrant Population

In 1842, Hong Kong had a mere 6,000 inhabitants. Today, it has over 7 million people. What makes this population distinctive is its predominantly immigrant composition. Waves of immigrants have flooded Hong Kong ever since 1842. Even today, barely half of Hong Kong's people were actually born there. This has been a critical factor in the political development of Hong Kong; for instead of a foreign government imposing its rule on submissive natives, the situation has been just the reverse. Chinese people voluntarily emigrated to Hong Kong, even risking their lives to do so, to subject themselves to alien British colonial rule.

In recent history, the largest influxes of immigrants came as a result of the 1945–1949 Civil War in China, when 750,000 fled to Hong Kong; as a result of the "three bad years" (1959–1962) following the economic disaster of China's Great Leap Forward policy; and from 1966 to 1976, when more than 500,000 Chinese went to Hong Kong to escape the societal turmoil generated by the Cultural Revolution. After the Vietnam War ended in 1975, Hong Kong also received thousands of refugees from Vietnam as that country undertook a policy of expelling many of its ethnic Chinese citizens. Many Chinese from Vietnam risked their lives on small boats at sea to attain refugee status in Hong Kong.

Although China's improving economic and political conditions after 1979 greatly stemmed the flow of immigrants from the mainland, the absorption of refugees into Hong Kong's economy and society remained one of the colony's biggest problems. Injection of another distinct refugee group (the Chinese from Vietnam) generated tension and conflict among the Hong Kong population.

(Getty Images 52010120 NSEA)
Hong Kong handover ceremony, July 1, 1997.

Because of a severe housing shortage and strains on the provision of social services, the British colonial government first announced that it would confine all new refugees in camps and prohibit them from outside employment. It then adopted a policy of sending back almost all refugees who were caught before they reached Hong Kong Island and were unable to prove they had relatives in Hong Kong to care for them. Finally, the British reached an agreement with Vietnam's government to repatriate some of those Chinese immigrants from Vietnam who were believed to be economic rather than political refugees. The first few attempts at this reportedly "voluntary" repatriation raised such an international furor that the British were unable to systematize this policy. By the mid-1990s, however, better economic and political conditions in Vietnam made it easier for the British colonial government to once again repatriate Vietnamese refugees.

Before the July 1, 1997, handover, moreover, Beijing insisted that the British clear the camps of refugees. It was not a problem that China wanted to deal with. As it turned out, the British failed to clear the camps, leaving the job to the Chinese after the handover. The last one was closed in the summer of 2000. Today, China still maintains strict border controls, in an effort to protect Hong Kong from being flooded by Chinese from the mainland who are hoping to take advantage of the wealthy metropolis, or just wanting to look around and shop in Hong Kong.[21]

The fact that China's economy has grown rapidly for the last 30 years, especially in the area surrounding Hong Kong, has diminished the poverty that led so many Mainlanders to try illegally emigrating to Hong Kong. Nevertheless, overpopulation is still an important social issue because it stretches Hong Kong's limited resources and has contributed to the high levels of unemployment in recent years. Rulings that have greatly limited the right of Mainlanders with at least one Hong Kong parent (so-called "right-of-abode" seekers) to migrate to Hong Kong have eased concerns somewhat.

Today, the largest number of immigrants to Hong Kong are still Mainlanders; but they are more likely than in the past to come from distant provinces. Although Hong Kong is made up almost entirely of immigrants and their descendants, the older immigrants look down upon their non-Cantonese-speaking country cousins as "uncivilized." They are socially discriminated against and find it difficult to get the better-paying jobs in the economy. This is in striking contrast to past attitudes: From the 1960s to the 1980s, Hong Kong residents generally expressed deep sympathy with Mainlanders, building rooftop schools for them, and throwing food onto the trucks when the British colonial government transported Mainlanders back to China against their will.[22]

Language and Education

Ninety-five percent of Hong Kong's people are Chinese. The other 5 percent are primarily European, Vietnamese, Filipino, and Indonesian. Although a profusion of Chinese dialects are spoken, the two official languages, English and the Cantonese dialect of Chinese, predominate. Since the Chinese written language is in ideographs, and the same ideographs are usually used regardless of how they are pronounced in various dialects, all literate Hong Kong Chinese are able to read Chinese newspapers—and 95 percent of them do read at least one of the 16 daily newspapers available in Hong Kong.[23] Even before the handover, moreover, the people were intensively studying spoken Mandarin, the official language of China.

Since the handover in 1997, a source of bubbling discontent has been the decision of the Executive Council to require all children to be taught in Chinese. The government's rationale was that the students would learn more if they were taught in their own language. This decision caused an enormous furor. Many Hong Kong Chinese, especially from the middle classes, felt that if Hong Kong were going to remain a major international financial and trading center, its citizens must speak English. Many suspected that the real reason for insisting on Chinese was to respond to Beijing's wishes to bind the Hong Kong people to a deeper Chinese identity. In response to strong public pressures, the Hong Kong government finally relented and allowed 100 schools to continue to use English for instruction. Many of the other schools are, Hong Kong parents complain, suffering from a decline in the quality of education generally, and language skills in particular. Those who can afford it now try to get their children into the growing number of private schools.

Chinese cultural values of diligence, willingness to sacrifice for the future, commitment to family, and respect for education have contributed to the success of Hong Kong's inhabitants. (The colonial government guaranteed nine years of compulsory and free education for children through age 15, helping to support these cultural values.) As a result, the children of immigrants have received one of the most important tools for material success. Combined with Hong Kong's rapid post–World War II economic growth and government-funded social-welfare programs, education improved the lives of almost all Hong Kong residents, and allowed remarkable economic and social mobility. A poor, unskilled peasant who fled across China's border to Hong Kong to an urban life of grinding poverty—but opportunity—could usually be rewarded by a government-subsidized apartment, and by grandchildren who graduated from high school and moved on to white-collar jobs.

Since the 1997 handover, the Hong Kong SAR has continued to support compulsory and free education. But in this rapidly changing society, juvenile delinquency is on the rise, because parents are working long hours and spend little time with their children, and an increasing number of those who finish the basic nine years of school now leave the educational system. Criminal gangs *(triads)* recruit them to promote criminal activities.[24]

For those who wish to continue their education beyond high school, access to higher education is limited, so ambitious students work hard to be admitted to one of the best upper-middle schools, and then to one of the even fewer places available in Hong Kong's universities. Hong Kong's own universities are becoming some of the best in the world, and admission is highly competitive. An alternative chosen by many of Hong Kong's brightest students is to go abroad for a college education. This has been important in linking Hong Kong to the West.

(Copyright IMS Communication Ltd/Capstone Design RF (DAL027ASI))
A woman, from the Hakka minority in Hong Kong.

Living Conditions

Hong Kong has a large and growing middle class. By 1995, in fact, Hong Kong's per capita income had surpassed that of its colonial ruler, Great Britain.[25] Its people are generally well-dressed; and restaurants, buses, and even subways are full of people yelling into their cell phones. Enormous malls full of fashionable stores can be found throughout Hong Kong. McDonald's is so much a part of the cityscape that most residents do not realize it has American origins. After school, the many McDonald's are full of teenagers meeting with their friends, doing homework, and sharing a "snack" of the traditional McDonald's meal—hamburgers, fries, and Coca-Cola (served hot or cold). The character Ronald McDonald, affectionately referred to by his Cantonese name, Mak Dong Lou Suk Suk (Uncle McDonald), is recognized throughout Hong Kong.[26]

Nevertheless, Hong Kong's people suffer from extremes of wealth and poverty. The contrast in housing that dots the landscape of the colony dramatically illustrates this. The rich live in luxurious air-conditioned apartments and houses on some of the world's most expensive real estate. They may enjoy a social life that mixes such Chinese pleasures as mahjong, banqueting, and participation in traditional Chinese and religious rituals and festivals with British practices of horseracing, rugby, social clubs, yacht clubs, and athletic clubs for swimming and croquet.[27] (These practices have faded greatly with the exit of the British.) The Chinese have removed the name "Royal" from all of Hong Kong's old social clubs, including the Royal Jockey Club, which today is neither royal nor British. Times have changed: Hong Kong is no longer a colony, and exclusivity is a thing of the past. The mix of Hong Kong people and foreigners in business clubs reflects the fact that Hong Kong's business elite is now both British and Chinese. Of course, memberships in some clubs are still all-British, like cricket clubs, because the Chinese are not interested.[28]

The wealthy are taken care of by cooks, maids, gardeners, and chauffeurs, most of whom are Filipino. The Hong Kong people—and the government—greatly prefer hiring Filipinos to hiring mainlander Chinese, as the former are better trained, speak English, and are less likely to try to apply for permanent residency in Hong Kong. (The preference for hiring Filipinos adds to the sense the Chinese Mainlanders have of being discriminated against.) Virtually all members of the Filipino workforce (estimated to be about 200,000) have Sundays off, and they more or less take over the parks, and even the streets in Hong Kong Central, where they camp out for the day with their compatriots. The government dares not intervene to get them off the streets, lest it be accused of racism or violating their civil liberties. The Filipinos often complain about exploitation and abuse by Hong Kong employers.

There is a heavier concentration of Mercedes, Jaguars, Rolls Royces, and other luxury cars—not to mention cellular phones, car faxes, and fine French brandy—in Hong Kong than anywhere else in the world.[29] Some Hong Kong businessmen spend lavishly on travel, entertainment, and mistresses. Their mistresses, however, now tend to live in Shenzhen, the "Special Economic Zone" (SEZ) just 40 minutes by train from Hong Kong Central. Purchasing an apartment for them and paying for their upkeep are cheaper there, as are the karaoke bars, golf courses, bars, and massage parlors. Hong Kong is now for the wealthier and more sedate businessmen, while Shenzhen is more attractive to Hong Kong's young professionals.

The wealthiest business people have at least one bodyguard for each member of their families. Conspicuous consumption and what in the West might be considered a vulgar display of wealth are an ingrained part of the society. Some Hong Kong businesspeople, who have seemingly run out of other ways to spend their money, spend several hundred thousand dollars just to buy a lucky number for their car license plate. One Hong Kong businessman built a magnificent home in Beijing that replicates the style of the Qing Dynasty and features numerous brass dragons on the ceilings, decorated with three pounds of gold leaf.[30]

In stark contrast to the lifestyles of the wealthy business class, the vast majority of Hong Kong's people live in crowded high-rise apartment buildings, with several poor families sometimes occupying one apartment of a few small rooms, and having inadequate sanitation facilities.[31] Beginning in the mid-1950s, the British colonial government built extensive low-rent public housing, which today accommodates about half of the population. These government-subsidized housing projects easily become run down and are often plagued by crime. But without them, a not-insignificant percentage of the new immigrant population would have continued to live in squalor in squatter villages with no running water, sanitation, or electricity.

When he took office in 1997, Hong Kong's first chief executive, Tung Chee-hwa, made a significant commitment to provide more government-funded housing and social-welfare programs—some 85,000 new apartments were to be built each year. But in 2000, Tung suddenly announced that he had scrapped that policy two years earlier because of falling real-estate prices (owing to the Asian economic crisis)—without telling anyone. The result is a housing scarcity, and the rents for the apartments in the few new residential buildings are far beyond the means of the average Hong Kong resident.

The Economy

A large part of the allure of Hong Kong has been its combination of a dynamic economy with enlightened social welfare policies. The latter were possible not just because of the British colonial government's commitment to them but also because the flourishing Hong Kong economy provided the resources for them. Hong Kong had a larger percentage of the gross domestic product (GDP) available for social welfare than most governments, for two reasons. First, it had a low defense budget to support its approximately 12,000 British troops (including some of the famous Gurkha Rifles) stationed in the colony for external defense (only 0.4 percent of the GDP, or 4.2 percent of the total budget available). Second, the government was able to take in substantial revenues (18.3 percent of GDP) through the sale of land leases.[32]

Now, of course, China is in charge of Hong Kong's defense. Land leases came to an end in 1997 with the return of Hong Kong's leased territories to China, but the sale of land is still the primary source of government revenue. Continuing the system established by the British colonial government, the Hong Kong government makes money by selling land one piece at a time to its handful of real-estate developers, who then build on the land.[33] This is how Hong Kong can continue to finance governmental expenditures without any income tax, sales taxes, or capital gains tax; and with a low profit tax (16.5 percent) for corporations, and a flat 15 percent tax on salaries that kicks in at such a high level that a large percentage of those working in Hong Kong pay nothing at all.

The result is that Hong Kong is a great place to do business, but the cost of private housing is beyond the means of most of Hong Kong's middle class; and, having made its money by selling land to the land developers, the government must then turn around and use a substantial portion of that money to subsidize public housing so that rent is affordable. Half of Hong Kong's populace receives housing free or at minimal (subsidized) rent, but the middle class cannot afford to buy housing. Many have moved to more affordable housing in the Shenzhen Special Economic Zone across the border in China. This may explain in part why the Hong Kong population has fallen by at least 300,000 people in the last few years.

Hong Kong's real estate system has affected the people's viewpoint on housing. Unlike in the West, people do not view the purchase of an apartment as a place where they might want to reside for an indefinite period. Rather, they see it more like buying a stock, and they might buy and sell it within one year in order to make a profit. Before the Asian financial crisis that hit Hong Kong in 1997, this was a fairly sure bet; but housing prices have been much lower, and many people have lost their savings by speculating in housing.

Beijing's greatest concern before the handover in 1997 was that the colonial administration had dramatically increased welfare spending—65 percent in a mere five years.[34] From Beijing's perspective, the British appeared determined to empty Hong Kong's coffers, leaving little for Beijing to use elsewhere in China. The Chinese believed that the British were setting a pattern to justify Hong Kong's continuing expenditures in the next 50 years of protected autonomy. As it turned out, however, the British did not try to deplete Hong Kong assets, and China's companies were deeply involved even in the vastly expensive new airport that the British insisted on building before their departure. (It opened in July 1998, one year after the handover.)

No sooner had China taken back Hong Kong than, suddenly, the Asian economic financial crisis broke out, first in Thailand, and then throughout Asia. It wreaked havoc on Hong Kong's economy and challenged its financial and economic system. Beijing, instead of interfering with the decisions made by Hong Kong to address the

(Copyright IMS Communication Ltd/Capstone Design/FlatEarth Images RF (DAL022C14))
Bamboo scaffolding in Hong Kong.

crisis, took a hands-off approach, except to offer to support the Hong Kong dollar against currency speculators by using China's own substantial foreign-currency reserves of U.S. $150 billion to sustain the Hong Kong dollar's peg to the U.S. dollar. Furthermore, China did not take the easy route of devaluing the Chinese yuan, which would have sent Asian markets, and Hong Kong's in particular, into a further downward spiral.

The Asian financial and economic crisis (1997–2002) brought a severe downturn in living conditions in Hong Kong. Growing increasingly anxious about the situation, the Hong Kong population pressured their government to intervene and do more to ease the pain. They demanded that the government take a more activist role in providing social welfare, controlling environmental degradation, and regulating the economy. In response to such pressure, the government required taxis to stop using diesel fuel and enacted some of the toughest emissions standards in the world.[35] The government also committed more to social welfare than it had under British rule. In addition, it intervened in the financial markets, purchasing large amounts of Hong Kong dollars to foil attempts by speculators to make a profit from selling Hong Kong dollars. It also intervened in the Hong Kong stock market, using up some 25 percent of its foreign-currency reserves to purchase large numbers of shares in Hong Kong companies in a risky but ultimately successful effort to prevent a further slide of the stock market.

Hong Kong's woes were aggravated, however, by the world economic turndown that began in 2000, made worse by the September 11, 2001 terrorist attacks on the United States, and by the outbreak of the SARS (severe acute respiratory syndrome) epidemic in early 2003 in the next-door Chinese province of Guangdong. It quickly spread to Hong Kong, showing the vulnerability of Hong Kong's geographical position on the edge of China. The life-threatening illness (with close to a 5 percent mortality rate) and the initial reluctance of China to furnish information about the spread of the epidemic in China led to disaster for Hong Kong's tourist industry and economy in 2003. Meanwhile, pollution from the coal-fired plants in the flourishing economy next door in southern China led to rapidly deteriorating air quality in Hong Kong. A survey taken each year from 1992 to 2007 indicates that by the end of 2003, the satisfaction of the Hong Kong people with the government reached its nadir, with only 51 percent of those surveyed "satisfied" with their life in Hong Kong.[36]

Before the 1997 return of Hong Kong to China, many analysts predicted that Beijing would undercut Hong Kong's prosperity through various political decisions limiting political freedom, tampering with the legal system, and imposing economic regulations that would endanger growth. Instead, Beijing left Hong Kong in the hands of its hand-picked Chief Executive, Tung Chee-hwa. Tung, however, proved unable to save Hong Kong's economy from the Asian financial crisis and rapidly lost popularity. To wit, although the economy recovered, its growth rate has remained only half that of the China mainland. The local press severely criticized Tung for having too literally interpreted Beijing's promise to the rest of the world in 1997 that Hong Kong would not change. Indeed, Hong Kong has changed so little in the years since the handover that it seems to be losing its dominant position in Asia as a major financial and commercial center to Shanghai and Singapore. Just as worrisome, Taiwan has emerged as the Asian leader in computer technology, while Hong Kong has fallen further behind in the technology sweepstakes. Tung resigned in March 2005 due to "health reasons," just 3 years into his second term. Donald Tsang, a lifelong civil servant, became the new Chief Executive.

The erosion of Hong Kong's leadership in Asia is blamed not on Beijing, but on the government's lack of vision and its catering to the real-estate "tycoons"—the mere half dozen individuals who hold the vast majority of Hong Kong's wealth in their hands.[37] More than anything else, Hong Kong's loss of its international character and relative decline as an Asian city are blamed on its "official obsession with mainland relations."[38] Yet, regardless of how others view Hong Kong, its citizens seem generally pleased with their government's relationship with the mainland, especially in the last few years.[39]

The collapse of Hong Kong's real estate prices in the wake of the Asian economic crisis did, however, result in their turning to technology as a new source of wealth for Hong Kong, and themselves. The Internet is empowering small entrepreneurs. But Hong Kong's economy is plagued with other problems emanating from monopolistic and duopolistic control of many sectors in the economy.[40] A more cynical view is that Hong Kong businesspeople are sycophants to both the Hong Kong and Beijing governments: their customary operating procedure is to make money not through creativity but through their connections—a practice common on the mainland as well.

Hong Kong's real estate and stock markets have bounced back from the Asian financial crisis, but remain volatile. For the time being, Hong Kong's greatest protection against substantial turmoil and a prolonged recession is the stability and growth of China's economy, which continues to bubble along at an average of 9 percent per year. Hong Kong businesspeople have heavily invested in China. Were its growth to falter, so would Hong Kong's.

Hong Kong as a World Trade and Financial Center

From the moment Hong Kong became a colony, the British designated it as a free port. The result is that Hong Kong has never applied tariffs or other major trade restrictions on imports. Such appealing trade conditions, combined with Hong Kong's free-market economy, deepwater harbor, and location at the hub of all commercial activities in Asia, have made it an attractive place for doing business. Indeed, from the 1840s until the crippling Japanese occupation during World War II, Hong Kong served as a major center of China's trade with both Asia and the Western world.

The outbreak of the Korean War in 1950 and the subsequent United Nations embargo on exports of strategic goods to China, as well as a U.S.–led general embargo on the import of Chinese goods, forced Hong Kong to reorient its economy. To combat its diminished role as the middleman in trade with the mainland of China, Hong Kong turned to manufacturing. At first, it manufactured mainly textiles. Later, it diversified into other areas of light consumer goods and developed into a financial and tourist center.

Today, Hong Kong continues to serve as a major trade and financial hub, with thousands of companies located there for the purpose of doing business with China. Even Mainland China companies try to have headquarters in Hong Kong, for reasons related to foreign exchange, taxes, and stock listing issues. Hong Kong also remains a middleman in trade with China, with more than one-third of China's total trade still flowing in and out of China through Hong Kong. A full 80 percent of Hong Kong's container port traffic, in fact, goes to and from Southern China. So Hong Kong's economy is still closely tied to China's economy.

In addition, Hong Kong has actually shifted its own manufacturing base into China. Back in 1980, when almost half of Hong Kong's workforce labored in factories and small workshops, Hong Kong was on the verge of pricing itself out of world markets because of its increasingly well-paid labor and high-priced real estate. Just then, China, with its large and cheap labor supply, inexpensive land, and abundant resources, initiated major internal economic reforms that opened up the country for foreign investment. As a result, Hong Kong transferred its manufacturing base over the border to China, largely to the contiguous province of Guangdong. By the late 1990s, more than 75 percent of the Hong Kong workforce was in the service sector, with only 13 percent remaining in manufacturing. In short, the vast majority of Hong Kong workers in the manufacturing sector have been replaced by some 5 million Chinese laborers who now work in the tens of thousands of factories either owned or contracted out to Hong Kong businesses on the mainland.[41]

(Annie Reynolds/PhotoLink/Getty Images)

Airplane landing on a Hong Kong runway.

Hong Kong's many assets, including its hard-working, dynamic people, have made it into the world's tenth-largest trading economy. But it is facing much competition. In 2005, Singapore surpassed it to become the world's busiest container port, with Shanghai's and Shenzhen's ports hot on its heels. This does not mean that Hong Kong is not profiting from this shift, however, as it is Hong Kong investors who are helping the Mainland's ports grow. Similarly, as of 2007, Hong Kong had dropped from third to seventh as a place for foreign exchange trade. Still, it remains one of the world's largest banking centers, and has grown to become Asia's third-largest stock market, thanks to increased listings of companies from the China Mainland. Considering its tiny size and population, these are extraordinary achievements.

Hong Kong has, then, many competitors from other "dragons" in Asia; but given the rapid growth of the Asian economies over the last 30 years, there is no reason why other major financial and trade centers in Asia would necessarily harm Hong Kong's economy. Shanghai's or Shenzhen's growth need not come at Hong Kong's expense. In fact, the booming SEZ of Shenzhen is really an extension of Hong Kong's growth and power, as is most of Southern China. Perhaps Singapore presents the greatest challenge to Hong Kong's position as a financial center. And changes in the trade environment, such as South Korea establishing full diplomatic relations with China in 1992, have allowed it to deal directly with China, thereby bypassing Hong Kong as an entrepôt for trade and business with China. But Hong Kong's entrepreneurs and businesses remain efficient, flexible, and able to incorporate such changes into their business strategies successfully.

The Special Economic Zones (SEZs)

As part of its economic reform program and "open door" policy that began in 1979, China created Special Economic Zones in areas bordering or close to Hong Kong in order to attract foreign investment. SEZs, until recent years under far more liberal regulations than the rest of China, blossomed in the 1980s and 1990s. Various branches of China's government themselves invested heavily in the SEZs in hopes of making a profit. In the 1990s, even China's military developed an industrial area catering to foreign investors and joint ventures in one of China's SEZs, Shenzhen, as part of its effort to compensate for insufficient government funding for the military. It called its policy "one army, two systems"—that is, an army involved with both military and economic development.[42] Brushing aside its earlier preference for a puritanical society, China's military was as likely to invest in nightclubs, Western-style hotels, brothels, and health spas in the SEZs as it was in the manufacturing sector. In 1998, however, Beijing ordered the military to divest itself of its economic enterprises, so it no longer runs nonmilitary enterprises in Shenzhen.

The bulk of foreign investment in the SEZs comes from Hong Kong Chinese, either with their own money or acting as middlemen for investors from Taiwan, the United States, and others. Most direct foreign investment in China, in fact, comes *through* Hong Kong, either by setting up companies in Hong Kong, or using Hong Kong companies as intermediaries. In turn, China is the single largest investor in Hong Kong, and its state-owned enterprises and joint ventures also set up companies in Hong Kong. Thus, this integrated area of South China, encompassing Hong Kong, the SEZs, and the provinces of Guangdong, Fujian, and Hainan Island, has become a powerful new regional economy on a par with other Asian "little dragons."

Indeed, even before China took over Hong Kong in 1997, South China had already become an integral part of Hong Kong's empire, with profound political as well as economic implications. Many Hong Kong people who regularly cross into Shenzhen SEZ (and even own property there) pressure the Shenzhen government to be responsive to their interests. Hong Kong's media coverage of Shenzhen affairs also puts pressure on the SEZ's administrators to be more responsive to public concerns, such as the exploitation of Mainlanders working under contract in Hong Kong–owned firms.[43]

At the same time, Hong Kong's growing ties with the SEZs and cross-border trade generally is causing serious problems, including the restructuring of the workforce as jobs in the manufacturing sector move across the Hong Kong border for cheaper labor. The result is a downward pressure on wages in Hong Kong, although many economists believe this will make Hong Kong more competitive. In any event, the Hong Kong business community is moving to integrate the economy even more fully with the mainland to take advantage of China's future growth.

Sensitivity of the Economy to External Political Events

Hong Kong's economic strength rests on its own people's confidence in their future—a confidence that has fluctuated wildly over the years. When Beijing undertook economic-retrenchment policies, partially closed the "open door" to international trade and investment, engaged in political repression, or rattled its sabers over Taiwan, Hong Kong's stock market would gyrate, its property values declined, and foreign investment would go elsewhere. Not knowing what the transition to Chinese sovereignty would bring, Hong Kong's professional classes emigrated at the rate of about 60,000 people per year between 1990 and 1997. This drain of both talent and money out of the colony was as serious a concern for China as it was for Hong Kong.

London's refusal to allow Hong Kong citizens to emigrate to the United Kingdom contributed to a sense of panic among the middle and upper classes in Hong Kong—those most worried about their economic and political future under Communist rule. Other countries were, however, more than willing to accept these well-educated, wealthy immigrants, who came ready to make large deposits in their new host country's banks. Once emigrants gained a second passport (a guarantee of residency abroad in case conditions warrant flight), however, they tended to return to Hong Kong, where opportunities abound for entrepreneurs and those in the professions, such as doctors, architects, and engineers.

Beijing's verbal intimidation of Hong Kong dissidents who criticized China in the period following the Tiananmen crackdown in 1989, and again when Governor Christopher Patten began whipping up Hong Kong fervor for greater democratic reforms from 1992 to 1997, also aroused anxiety in the colony. Hong Kong's anxiety that Great Britain would trade the colony's democratic future for good relations with China was later counterbalanced by concern that Patten's efforts to inject Hong Kong with a heavy dose of democratization before its return to China's control would lead to China dealing harshly with Hong Kong's political freedoms after 1997. In the end, Patten's blunt refusal to abide by the terms agreed to in the Basic Law, namely, that China would have to agree to any changes made before 1997, merely led to a reversal of Patten's changes after the handover, and no more.

China's sovereignty over Hong Kong has not had a negative effect on its economy. Occasionally China's statements concerning Hong Kong's economy, judicial system, or politics have sent shock waves throughout the colony, causing the Hang Sang stock market to take a nose dive out of fears that China would ignore the principles in the Basic Law guaranteeing Hong Kong's 50 years of autonomy. Similarly, the 2003 SARS epidemic had a catastrophic impact on Hong Kong's economy. Such volatility demonstrates just how sensitive Hong Kong is to Beijing's policies and actions. Nevertheless, in the years since the handover, millions of Hong Kong citizens have relinquished their British passports for Hong Kong Special Administrative Region passports. Apart from pressures from China for them to do so, it indicates a belief that their future lies with Hong Kong and China, not Great Britain.

China is sensitive to the possibility of its policies or statements destabilizing Hong Kong. Indeed, in the ten years since the handover, Beijing has exercised unusual restraint so as to avoid being seen to interfere in Hong Kong's affairs. For example, Beijing no longer permits Chinese ministry officials to visit or oversee their counterparts in Hong Kong without clearance, lest it be interpreted as interference. Furthermore, it is the Hong Kong government, not the Ministry of Foreign Affairs office in Hong Kong, that deals with all of the foreign consulates in Hong Kong. And, unlike the British Commonwealth Office, which always sent copies of government documents to the Hong Kong government, China's Ministry of Foreign Affairs does not, again to avoid being accused of interference.[44]

Because Beijing has tread lightly in Hong Kong, most businesses already located there have remained. In fact, many foreign corporations rushed to establish themselves in Hong Kong before the handover in order to avoid the unpredictable, lengthy, and expensive bureaucratic hassle of trying to gain a foothold in the China mainland lying beyond Hong Kong. Even Taiwan's enterprises in Hong Kong have stood firm; for without direct trade and transport links between China and Taiwan, Hong Kong is still the major entrepôt for trade between the two places. Many Chinese mainland corporations also establish footholds in Hong Kong to ease the problem of foreign hard-currency transfer and to avoid a host of other difficulties that plague mainland businesses.

Nevertheless, the overall profile of the foreign business community has changed substantially since 1997. Many British went home, leaving their companies in the hands of capable Hong Kong Chinese managers, a reflection of the end of the colonial era in the business community as well as in the political system; but the percentage of Americans and Europeans doing business in Hong Kong has increased significantly.

Crime

Although Hong Kong is still characterized by a high level of social stability, a high crime rate continues to plague society. For more than a decade, ordinary criminality has been steadily augmented by crime under the control of competing Chinese triads. This is in part because the housing and community and mutual-aid groups of the 1970s and 1980s, which used to help the police track down criminals, disappeared. Their disappearance also contributed to increasing juvenile delinquency.[45] Opium, largely controlled by the triads, continues to be used widely by the Chinese. As a commentator once put it:

> Opium trails still lead to Hong Kong . . . and all our narcotic squads and all the Queen's men only serve to make the drug more costly and the profits more worthwhile. It comes in aeroplanes and fishing junks, in hollow pipes and bamboo poles and false decks and refrigerators and pickle jars and tooth paste tubes, in shoes and ships and sealing wax. And even cabbages.[46]

Today, Hong Kong remains one of the largest entrepôts for drugs, and the number of drug addicts is skyrocketing. This is in no small part because social and economic liberalization on the mainland has allowed its people to move about freely. Hong Kong triads work in collaboration with triads across the border. Young people cross the border to Shenzhen to buy drugs (originating in Myanmar (Burma), which they then smuggle back across Hong Kong's border. In turn, those recruited in Shenzhen provide a base in the mainland for Hong Kong triads to deal in drugs, prostitutes, and guns, and to set up underground banks and transport illegal immigrants across the border.[47]

Cooperation between Hong Kong and Chinese mainland drug investigators is complicated by Hong Kong's legal system, which still differs from the legal system of the rest of China. The critical difference is that Hong Kong does not have capital punishment. And China, committed to not changing Hong Kong's legal system for 50 years, has not pressured Hong Kong to change its law on the death penalty. Dozens of crimes such as drug dealing, punishable by execution in the rest of China, will result at most in a life sentence in Hong Kong.

Before the handover, when Hong Kong investigators asked the Chinese to turn over drug dealers to the Hong Kong authorities, the Chinese expended significant resources to find the criminals and turn them over. But when the Chinese asked the Hong Kong drug authorities to do the same, they went so far as to arrest the suspects, but refused to turn them over to China's public security office because of the fairly strong chance that a person convicted on charges of selling drugs in China would be executed. At first China wanted to copy the Singapore model of executing drug dealers, but it soon realized that would mean the execution of thousands. Now only the biggest drug dealers are executed. (The problems emanating from cross-border crime are discussed further in this report, under the topic "The Legal System and the Judiciary.")

Hong Kong's organized crime has long been powerful in the areas of real estate; extortion from massage parlors, bars, restaurants, and clubs; illegal gambling; smuggling; the sale of handguns (illegal for ordinary people to purchase); prostitution; and drugs. And, as is common in other Asian countries, gangs are often hired by corporations to deal with debtors and others who cause them difficulties. Triads have also expanded into kidnapping for ransom, and taken on some unexpected roles. As an example, when the British governor in Hong Kong, Christopher Patten, upset Beijing with his proposals for further democratization of Hong Kong before 1997, the Chinese Communist regime allegedly recruited triad members to begin harassing those within the Hong Kong government who were supporting Patten's proposals. (And, when Patten's dog disappeared one day in 1992 during the crisis stage of Sino–British relations, one rumor had it that the Chinese Communists had kidnapped the dog and were going to ransom it in exchange for halting political reform in Hong Kong. The other rumor was that Patten's pet had been flown into China to be served up for breakfast to Deng Xiaoping. Of course, neither rumor was true.)

POLITICS AND POLITICAL STRUCTURE

Politics and the political structure have changed greatly since the days of Hong Kong's colonial government, when the British monarch, acting on the advice of the prime minister, would appoint a governor, who presided over the Hong Kong government's colonial administration. Colonial rule in Hong Kong may be characterized as benevolent, consultative, and paternalistic, but it was nonetheless still colonial. Although local people were heavily involved in running the colony and the colonial government interfered very little in the business activities and daily lives of Hong Kong Chinese, the British still controlled the major levers of power and filled the top ranks in the government.

The colony's remarkable political stability until the handover in 1997 was, then, hardly due to any efforts by the British to transplant a form of Western-style democracy to Hong Kong. But, the colonial Hong Kong government did seek feedback from the people through the hundreds of consultative committees that it created within the civil service. Similarly, although the British ultimately controlled both the Legislative Council (LegCo) and Executive Council (ExCo), these governmental bodies allowed Hong Kong's socioeconomic elites to participate in the administration of the colony, even if they were unable to participate in the *formulation* of policy. Some 300 additional advisory groups as well as numerous partly elected bodies—such as the municipal councils (for Hong Kong Island and Kowloon), the rural committees (for the New Territories), and district boards—also had considerable autonomy in managing their own affairs. This institutionalized consultation among Chinese administrators and the colonial government resulted in the colony being governed by an elite informed by and sensitive to the needs of the Hong Kong people. As was common to British colonial administration elsewhere, the lower levels of government were filled with the local people. Rarely was political dissent expressed outside the government.[48]

The relatively high approval rating of British colonial rule helps explain why only a small portion of the mere 6 percent of registered voters actually voted. With the government assuring both political stability and strong economic growth, the people of Hong Kong spent most of their time and energy on economic pursuits, not politics. In any event, given the limited scope of democracy in Hong Kong, local people had little incentive to become politically involved. For this reason, as the handover came nearer, Hong Kong residents grew increasingly concerned that there were few competent and trustworthy leaders among the Hong Kong Chinese to take over.[49] They also worried that a government controlled by leaders and bureaucrats who held foreign passports or rights of residence abroad would not be committed to their welfare. Although Beijing withdrew its demand before the handover that all governmental civil servants swear an oath of allegiance to the government of China and turn in their British passports, many did so anyway (including the first chief executive, Tung Chee-hwa).

The colonial government remained stable, then, because it was perceived to be trustworthy, competent, consultative, and capable of addressing the needs of Hong Kong's people. Most Hong Kong citizens also believed that a strong political authority was indispensable to prosperity and stability, and they worried that the formation of multiple political parties could disrupt that strong authority. Thus, what is seen in the West as a critical aspect of democracy was viewed by the people in Hong Kong as potentially destabilizing.

Nevertheless, by the late 1980s, many Hong Kong Chinese began to demand that democratic political reforms be institutionalized before the Chinese Communists took over in 1997. The ability of the departing colonial government to deal with these increased pressures to democratize Hong Kong was, however, seriously constricted by the 1982 Joint Declaration and the Basic Law of 1990, which required Beijing's approval before the British could make any changes in the laws and policies governing Hong Kong. The people of Hong Kong awoke to the fact that their interests and those of the colonial government were no longer compatible. Britain's policy toward Hong Kong had become a mere appendage of British policy toward China, and the status quo was frozen. The Hong Kong colonial government had, essentially, lost its independence to Beijing and London.[50]

What was "handed over" on July 1, 1997, was sovereign control of Hong Kong. Hong Kong became a Special Administrative Region of China, with Beijing guaranteeing autonomy for 50 years in the political, legal, economic, and social realms. But Hong Kong would be governed by its new "constitution," the Basic Law, written by China's leaders. This document provided for certain changes to be made *after* the handover. Notably, Article 23 required the Hong Kong Legislative Council to outlaw treason, succession, subversion, and sedition, as well as other activities that could endanger China's national security. That is, Hong Kong was expected to outlaw, and punish, those individuals and organizations operating in Hong Kong who in the view of Beijing might pose a threat to China's security; but for five years, nothing was done to define how Article 23 could be implemented (see below).

Similarly, for five years after the 1997 handover, there was no real change in the government—except, of course, that Hong Kong's chief executive reported to Beijing, not London. The structure of the post-1997 government, outlined in great detail in the Basic Law, was to be, like Hong Kong's colonial government: structured on a separation of powers among the executive, legislative, and judicial branches of government, serving to check the arbitrary use of power by any single individual or institution of the government. This separation of powers, however, never did exist within the framework of a representative democracy. Thus, the government today remains similar in many respects to what it was under colonial rule, with power continuing to be centralized in the executive branch. LegCo cannot even initiate substantive legislation without the approval of the chief executive, or hold the chief executive accountable for his actions. In effect, Hong Kong under the Basic Law has retained the colonial model put in place by the British in the nineteenth century.

China has made some changes in Hong Kong's government to bring it into greater correspondence with its own government structure, even if in some cases this is merely a matter of changing names. In 2002, under a new "ministerial" system, Beijing changed the Basic Law to allow the chief executive to appoint all 14 policy secretaries in the cabinet of 20. (The others are leading politicians, including two heads of progovernment parties, and close personal advisers.) Previously, cabinet secretaries were senior civil servants who were, at least in theory, politically neutral. Now that they are appointed, they serve at the chief executive's discretion, which means that their political views weigh heavily in their selection. This has raised further questions concerning the accountability of the government to the Legislature.[51]

The Executive Council is run by Hong Kong's chief executive. Fortunately, continuity was maintained in the first years after the handover, when most of the cabinet heads under British colonial rule agreed to serve under the first chief executive. Sole decision-making authority remains vested in the chief executive, although ultimately Beijing must approve of any of ExCo's policies. So far, Beijing has chosen not to exercise this approval in a way that hampers the chief executive's policy-making authority.

Since 1982, the Hong Kong Transition Project has taken the pulse of Hong Kong each year through an extensive survey of public opinion on a variety of issues related to the impact of the transition of Chinese sovereignty. The level of satisfaction has been consistently much higher in the post-handover period than in the 15 years leading up to it.[52] Furthermore, the percentage of those satisfied with the performance of the PRC government in dealing with Hong Kong affairs remained unusually high. In fact, Hong Kong public opinion in Nov. 2003 continued to give a far higher approval rating to Beijing's leadership than to Hong Kong's chief executive: Those "satisfied" or "very satisfied" with Beijing's dealing with Hong Kong affairs were 72 percent of those polled; whereas for Chief Executive Tung, the approval rating in the same 2003 poll was a humiliating 21 percent. Moreover, satisfaction with the overall performance of the Hong Kong government fell from 66 percent approval in June 1997 (still under British rule), to 20 percent by Nov. 2003.[53]

Problems with becoming the first chief executive of Hong Kong were understandable: As a colony, Hong Kong had bred strong civil servants, but no real political leaders who had had opportunities to make political decisions. So Beijing had to pick someone whom they felt was safe, but who inevitably lacked real leadership experience. Tung Chee-hwa was a businessman, an elitist who opposed the welfare state and supported "Asian values" of a strong central leadership and obedient citizenry.[54] Tung Chee-hwa's popularity one year after he took office (49 percent were satisfied or very satisfied with his performance in July 1998) was quickly eroded by his inability to effectively address a number of problems, including fallout from the Asian financial and economic crisis that began within a few months of his entering office. Tung was blamed not only for plummeting land prices and the government's incompetence, but also for his reluctance to intervene in order to relieve the people's suffering that resulted from the crisis. The infrequency of his appearances in front of LegCo, and his refusal to consult with LegCo representatives, caused significant conflict and anger. Although it was the executive branch's role to deal with the financial crisis in Hong Kong, LegCo still believed that it should have been consulted.[55]

Although Beijing reappointed Tung for a second term in 2002, Beijing was not pleased with his performance. This was largely because of the growing number of demonstrations and protests directed at his government, and the dissatisfaction of the people of Hong Kong with his performance. By late 2003, some 58 percent of Hong Kong people polled said they would support China's President Hu Jintao and Premier Wen Jiabao if they were to dismiss Tung for his performance.[56]

Suddenly, in March 2005, Tung resigned, apparently under pressure from Beijing. Because two years remained in his term as Chief Executive, it led to a major constitutional crisis; for the Basic Law lacked procedures for dealing with a mid-term replacement. Beijing managed to engineer the choice of Donald Tsang, a member of the Executive Council, to fill out the term, that ended in 2007. Beijing had already ruled out any possibility that it would allow universal suffrage and full democracy for the elections of the next chief executive at that time (or of the Legislative Council of 2008—only 30 of its 60 legislators were directly elected at that time). Although China has pledged significant autonomy for Hong Kong, Beijing argued that nothing in the Basic Law obligates it to allow Hong Kong to fully democratize on its own timetable. The real reason for China's reluctance to let Hong Kong move ahead is, no doubt, because it will put Hong Kong far in front of the pace of democratization in some of China's major cities, such as Beijing and Shanghai.[57]

Efforts to get the economy back on track were, as noted above, stymied greatly by events beyond the chief executive's control—the Asian financial crisis, the global recession that followed the terrorist attacks on the United States on September 11, 2001; and the SARS epidemic that erupted in 2003. Nevertheless, as the public and the press saw it, the fault for failure to revive the economy lay at the feet of the Hong Kong government, and especially the Chief Executive, not external forces (or Beijing). If, they argued, the Chief Executive had appointed advisers based on competence instead of politics, he would have received better advice.

Hong Kong residents generally believe, then, that Hong Kong's problems arise not from Beijing's control but, rather, from the incompetence of their own government. So the critical question for the Chief Executive is no longer how autonomous he is of Beijing's control. Instead, it is whether he can maintain his legitimacy as the head of Hong Kong's government. In this respect, Donald Tsang has thus far done quite well. In a March 2006 survey concerning his performance since taking office, fully 76 percent were "satisfied" with Tsang as Chief Executive.[58] But most Hong Kong people would prefer direct elections for all LegCo members and district council members, as well as for the chief executive. Presently, the latter is appointed by the Beijing-controlled Election Committee.

(Courtesy of Bruce Argetsinger (Bargetsinger01))

Small businesses on a busy Hong Kong street.

Article 23 of the Basic Law

As noted earlier, Article 23 outlaws treason, secession, subversion, and sedition. It also prohibits the theft of "state secrets," political activities of foreign political organizations in Hong Kong, and the establishment of ties with foreign political organizations by political organizations in Hong Kong. It had been left up to the Hong Kong government to give teeth to Article 23 so that it could be implemented, but it had not done so.

The purpose of Article 23 is to protect the national security of China as a whole by defining the types of activities in Hong Kong that would be punishable by law. It was one thing to define terms such as "subversion" as it applied to the rest of mainland China; it was another thing to define these terms for residents of Hong Kong. This had to be done without jeopardizing the fundamental rights given to Hong Kong people in the Basic Law, some of which have not been given to the rest of the Chinese.

The terrorist attacks of 9/11 spurred the Hong Kong government's Security Bureau to submit a bill to amend Article 23. In early 2003, after a year-long period of public consultation, Chief Executive Tung submitted a draft bill to LegCo. This predictably caused a public uproar, as it brought back fears that Beijing might have pressured the government to limit the guaranteed rights of people in Hong Kong. The public was particularly worried about a constriction of freedom of the media to report news; and the rights to demonstrate and protest, to associate with foreign organizations or with organizations banned on the mainland, to access Internet sites, and to organize pressure groups and political groups. There was also concern that the new definitions of "treason," "subversion," and "sedition" might mean that journalists who managed to get hold of unpublished government documents, or refusal to reveal their anonymous sources, could be charged with "theft of state secrets."[59]

The concerns about the potential limits on individual freedom by Article 23 has spilled over into a concern for the rule of law in Hong Kong, which is the basis for protecting all other freedoms. The possibility that the Hong Kong government might resort to secret trials on cases relating to proscription of certain local organizations, and that "mere association between a local organization and a proscribed one on the Mainland" might be grounds for arrest, also caused anxiety. On July 1, 2003, 500,000 people demonstrated in Hong Kong against the draft bill. Finally, after further demonstrations, and Beijing's realization that the suggested amendments were destabilizing Hong Kong and causing hostility toward Beijing, the bill to amend Article 23 was withdrawn without any date for reconsideration. The impact of this is dramatically illustrated by polls that asked about the performance of China's government in dealing with Hong Kong affairs: In June 2003, before the July 1st demonstrations, there was a 57 percent satisfaction rate with Beijing; 6 months later, after the withdrawal of the legislation, the satisfaction rate shot up to 72 percent.[60] The fact that Beijing in this case catered to the will of the Hong Kong people seems to have won enormous good will for China.

(© IMS Communications, Ltd/Capstone Design/FlatEarth Images RF)
Flags of Hong Kong and China flying side by side.

Political Parties and Elections

On July 1, 1997, China immediately repealed Governor Patten's expansion of the franchise for the 1995 elections, which had lowered the voting age to 18 and extended the vote to all of Hong Kong's adult population (adding 2.5 million people to the voting rolls)—without consulting Beijing. As the Joint Declaration required that China had to agree to even a small increase in democracy in the Hong Kong colony, Beijing was within its rights when it repealed the changes and replaced the Legislature elected under those rules. A 400-member "Provisional Legislature," already selected by China's Preparatory Committee, replaced it on July 1, 1997.

Why did Beijing cancel the results of the 1995 elections? China's leaders viewed the last-ditch colonial government's efforts to install a representative government in Hong Kong as part of a conspiracy to use democracy to undercut China's rule in Hong Kong after 1997. They believed that the last-minute political reforms could, in fact, jeopardize Hong Kong's prosperity and stability by permitting special interests and political protest to flourish; and that Hong Kong's social problems—narcotics, violence, gangs, prostitution, an underground economy—required that Hong Kong be controlled, not given democracy. China therefore adopted a status quo approach to Hong Kong. After all, Hong Kong's political system under British colonial control had been imposed from the outside. This system worked well and kept Hong Kong stable and prosperous. Beijing merely wanted to replace a colonial ruler with a Chinese one.[61]

Nor did Beijing want a situation in which those citizens living in Hong Kong enjoyed substantially more rights than the rest of China's population. This could easily lead to pressure on Beijing to extend those rights to all Chinese. If Beijing is going to give greater political rights to its people, it would rather do so on its own timetable. Nevertheless, Beijing has kept its promises not to interfere in Hong Kong's political system after the handover.

In the 1995 elections, Hong Kong's Democratic Party, whose political platform called for major changes to the Basic Law in order to give Hong Kong people more democratic rights, won two-thirds of the vote. Although Beijing canceled the results of that election upon gaining control over Hong Kong in July 1997, Beijing immediately rescheduled the elections for May 1998; and when those elections led to the re-election of the same number of Democratic Party members as before, Beijing made no effort to remove them from the legislature.[62]

In the fall 2004 LegCo elections, the Democratic Party and their allies won 25 of the 60 seats, meaning that the pro-government/pro-Beijing parties still control the legislature after several rounds of elections. In that election, however, only 30 of the seats were chosen by direct election from the citizenry. The other 30 were chosen by what are called "functional constituencies," which tend to be more conservative, pro-business, and pro-Beijing. So, even though the democratic camp won the majority of the electoral vote, they still were not able to gain a majority of the seats.

The pan-democrats managed to win 24 of the 60 seats in the 2008 LegCo election. Many analysts attribute the electoral results to the preference of the electorate for better local governance over deadlines for democratization. The democracy camp lacks a meaningful platform of political, social, and economic policies that would improve conditions in Hong Kong and be attractive to the electorate. Analysts also point to the disarray within the democracy camp, which seems unable to agree on anything other than that more democracy would be a good thing, and that they should stand in opposition to government policies.

Beijing has warned the pro-democracy group not to move forward with any plan to hold a referendum on direct elections before the original timetable called for them. Combined with this warning, however, have been significant efforts by China to enhance economic integration between Hong Kong and the mainland provinces bordering it. China has also made it easier for Chinese tourists to go to Hong Kong—a potential boon to the flagging tourist industry in Hong Kong.[63] This may help explain why the pro-Beijing forces maintained control of LegCo.

Under British colonial rule, ordinary citizens were rarely involved in politics. Today, there is much greater involvement, but a lingering suspicion of and lack of enthusiasm for politics and political parties in Hong Kong. The lack of a coherent platform and strong leadership in the democracy camp, as well as its failure to bring about change in government policies, contributes to a feeling of political inefficacy—a sense that there is little ordinary citizens can do to shape policy. This view is also influenced by a belief that there is a not altogether healthy alliance between big business and government.

On the other hand, the government has the support of some remarkably strong pro-Beijing parties, especially the Democratic Alliance for the Betterment of Hong Kong (DAB). (It may not be appropriate to call the DAB pro-Beijing, but it does make

(United Nations Photo (UN74037))

Apartment complexes in Hong Kong.

more efforts to work with China for common goals and opposes it far less frequently than many in the democracy camp do.) The DAB has joined up with other "patriotic" (that is, pro-China/pro-government) parties in a loose coalition. The coalition works *with* the government to make policies focused on environmental, social, and economic issues. Many are suspicious of the DAB ties to the government, and concerned about the funding of the DAB by the Chinese government; but it and Hong Kong's "patriotic" organizations are highly effective at mobilizing people at the local level and responding to their concerns on issues important to their daily lives.

It is difficult for the Democratic Party to make inroads on the seats assigned to functional constituencies, which usually go to pro-government business people. This unfair, indeed undemocratic, system, whereby a mere 180,000 individuals could vote for the 30 functional constituency seats, while 3.5 million voted for the other 30 seats, is a residual system set up by the British colonial government.

The complicated electoral system virtually guarantees business groups dominance in the Legislature. This is one factor accounting for both Beijing and the Hong Kong government's tolerating the Democratic Party and its political protests, even when the protesters publicly denounce Beijing's leaders. That is, with rare exceptions (such as the protests led by the democracy camp against the bill to amend Article 23), their protests are without consequences. Besides, LegCo is a rather powerless political body. It is poorly funded, so legislators cannot afford the kind of research support that is essential to making effective policy proposals. And while in session, LegCo meets only once a week, at which time legislators give speeches written for them by their staffs.[64] This is hardly the stuff of impassioned legislative debate and policy making.

Beijing has tolerated the gadfly role of the Democratic Party, no doubt hoping that this hands-off policy will help bring Taiwan into negotiations for unification with the mainland sooner. The Hong Kong government has also kept channels open to Democratic Party leaders. The Democratic Party has, in turn, adopted a more moderate stance toward both Beijing and the government than it might otherwise have done—a position facilitated by the departure of several of the party's more radical leaders—and the public's concerns that it not stir up problems that could destabilize Hong Kong.[65]

The Democratic Party's minority position within the Legislature condemns it to the position of a critic and complainer.[66] In the post-transition period, the Hong Kong people believe that the party's demands for greater democracy have not been appropriate to the problems at hand. These problems include the "bird flu," which threatened public health and forced the government to kill the entire stock of chickens in Hong Kong; the red tide, which killed hundreds of thousands of fish; a disastrous opening of the new airport in 1998; the right-of-abode seekers in Hong Kong; the drop in real-estate values, increased unemployment, bankruptcies of retail stores, and a dramatic decline in tourism. These were not the sorts of problems for which the Democratic Party's call for changes in the Basic Law, a timetable for democratization, and opposition to Beijing have been relevant. Indeed, the vast majority of the public has wanted government intervention to address these problems. Thus, the many street demonstrations that occur tend not to be directed to broad demands for democracy but, rather, to pressing public policy issues. They reflect the frustration of the population in the legislature's ineffectiveness in shaping government policy.

The Legal System and the Judiciary

Under colonial rule, Hong Kong's judiciary was independent. After 10 years under the "one country, two systems" envisioned in the Basic Law, it has remained so. Judges are appointed and serve for life. English common law, partly adapted to accommodate Chinese custom, has been at the heart of the legal system. Much of the confidence in Hong Kong as a good place to live and do business has been based on the reputation of its independent judiciary for integrity and competence, the stability of the legal and constitutional system, and Hong Kong's adherence to the rule of law.

China has allowed Hong Kong's legal system to rely on such legal concepts as habeas corpus, legal precedent, and the tradition of common law, which do not exist in China. Beijing has not subjected Hong Kong's legal system to the Communist Party Politburo's guidelines for the rest of China. But on political matters such as legislation, human rights, civil liberties, and freedom of the press, the Basic Law offers inadequate protection. For example, the Basic Law provides for the Standing Committee of the National People's Congress (NPC) in Beijing, not the Hong Kong courts, to interpret the Basic Law and to determine whether future laws passed by the Hong Kong Legislature conflict with the Basic Law.

Furthermore, although Beijing had promised Hong Kong that the chief executive will be accountable to the Legislature, the Basic Law gives the chief executive the power to dissolve the Hong Kong Legislature and veto bills. The relationship of Hong Kong's Basic Law to China's own Constitution remains in limbo. The fundamental incompatibility between the British tradition (in which the state's actions must not be in conflict with the laws) and China's practice of using law as a tool of the state, as well as China's conferring and withdrawing rights at will, is at the heart of the concern about Chinese rule over Hong Kong.[67]

The first case to be tried by the Hong Kong courts after July 1, 1997, was that of an American streaker who ran nude in crowded downtown Hong Kong and was easily apprehended. In this rather amusing case, the court fined and released the defendant. Serious questions soon arose, however, about the handling of cross-border crime. As noted earlier, the source of the problem is that Hong Kong does not have capital punishment, and China does. In the past, when Hong Kong has arrested a P.R.C. citizen wanted for a criminal activity that is punishable in China by execution, it has refused to turn the person over to China's judicial authorities.

The position of the Hong Kong courts is that a Hong Kong citizen cannot be tried in China for *any* crimes committed in Hong Kong, even if caught in China, because of its judicial independence under the "one country, two systems" structure. But, according to Article 7 of China's Criminal Code, China has the right to prosecute a Chinese national who commits a crime *anywhere* in the world, providing the crime was either planned in, or had consequences in, China.[68] The assertion of such a broad jurisdiction over Chinese nationals inevitably puts it in conflict with Hong Kong's judicial system. Because so many criminals straddle the border between the Hong Kong SAR and mainland China, and because they have different criminal codes, there are many levels of conflict and ambiguity to resolve. In the 14 years since the handover, the increasingly porous borders between Hong Kong and the rest of China, the greater mobility of the Chinese people, and the open market economy have seemed like an open invitation to Hong Kong's triads not only to expand their crime rings within China, but also to take refuge there. At the same time, citizens of China who commit capital crimes in mainland China sometimes try to enter Hong Kong, where capital punishment is prohibited.

Polls over the years have indicated that residents (including those who really do not know anything about the Basic Law's provisions) are satisfied with the Basic Law and the rule of law in Hong Kong.[69] In addition, most residents think that Beijing has adhered to the Basic Law, but they have not been pleased that the Hong Kong government has several times dragged Beijing into Hong Kong affairs. One of the most notable examples was the Hong Kong government's request to China's National People's Congress in 1999 to rule on the "right of abode" for Mainlanders with at least one Hong Kong parent. (At the time of the NPC's ruling, 1.4 million Mainlanders were claiming eligibility.) The Hong Kong judiciary's decision was overturned, a result that came as a great relief to Hong Kong residents who were more afraid of overcrowding than they were of Beijing's interference. Nevertheless, as of 2008, the right to abode is still an uncomfortable issue; for although there is sympathy for these settlers' plight (many have lived and worked in Hong Kong for most of their lives), the settlers are usually the fruit of relationships between mainland Chinese women and married Hong Kong men who frequently go to the mainland for business, a point that does not sit well with many citizens, especially Hong Kong women.

Public Security

Under the "one country, two systems" model, Hong Kong continues to be responsible for its own public security. The British colonial military force of about 12,000 has been replaced by a smaller Chinese People's Liberation Army (PLA) force of about 9,000. The military installations used by the British forces were turned over to the PLA. In efforts to reassure the Hong Kong populace that the primary purpose of the military remains the protection of the border from smuggling and illegal entry, and for general purposes of national security, soldiers are mostly stationed across the Hong Kong border in Shenzhen. Some are also stationed, however, in heavily populated areas, to serve as a deterrent to social unrest and mass demonstrations against the Chinese government. No doubt the high cost of keeping soldiers in metropolitan Hong Kong was also a factor in the PLA decision to move most of its troops to the outskirts.

Apart from diminishing the interactions between the PLA soldiers and the Hong Kong people, China has done much to ensure that the troops do not become a source of tension. To the contrary, China wants them to serve as a force in developing a positive view of Chinese sovereignty over Hong Kong. As part of their "charm offensive," the soldiers must be tall (at least 5 feet 10 inches); have "regular features" (that is, be attractive); be well read (with all officers having college training, and all ordinary soldiers having a high school education); know the Basic Law; and be able to speak both the local dialect of Cantonese and simple English.[70] In short, the PLA, which has no such requirements for its regular soldiers and officers, has trained an elite corps of soldiers for Hong Kong. No doubt the hope is that a well-educated military will be less likely to provoke problems with Hong Kong residents. In fact, PLA troops are even permitted to date and marry local Hong Kong women!

Press, Civil Rights, and Religious Freedom

China's forceful crackdown on protesters in Beijing's Tiananmen Square in 1989 and subsequent repression traumatized the Hong Kong population. China warned the Hong Kong authorities that foreign agents might use such organizations as the Hong Kong Alliance in Support of the Patriotic Democratic Movement in China (a coalition of some 200 groups) to advance their intelligence activities on the mainland, and even accused that group of "playing a subversive role in supporting the pro-democracy movement."[71] China also announced that it would not permit politically motivated mass rallies or demonstrations against the central government of China after the handover. As it turns out, Beijing has done little to stop the rallies and protests by such diverse groups as doctors, students, property owners, social workers, and civil servants, Mainlanders demanding the right to abode in Hong Kong, or even protests against the bill to amend Article 23. Such protests occur regularly in Hong Kong. Indeed, even Jiang Zemin, during one of his final trips to Hong Kong as president of China, had to face the potential humiliation by thc thousands of Falun Gong demonstrators protesting Beijing's crackdown on the religious group in China. (Rather than prohibiting the group members from protesting, Hong Kong authorities took Jiang along a route where they were not so visible.) Beijing has, however, used subtle and not-so-subtle intimidation to discourage Hong Kong from supporting prodemocracy activities, suggesting that such acts would be "treasonous." In addition, those individuals who assisted the Tiananmen demonstrators in Beijing in 1989, and those leaders of the annual demonstrations in Hong Kong to protest China's use of force to crack down on Tiananmen demonstrators, are blacklisted in China. Were they to try to go to the mainland, they could be arrested for sedition. Only when the chief executive himself has intervened on their behalf have they been allowed to cross the Hong Kong border.

One freedom that Hong Kong subjects under British rule had was freedom of the press. Hong Kong had a dynamic press that represented all sides of the political spectrum. Indeed, it even tolerated the Chinese Communist Party's sponsorship of both its own pro-Communist, pro-Beijing newspaper, and its own news bureau (the New China News Agency) in Hong Kong, which, until the handover, also functioned as China's unofficial foreign office in Hong Kong.

Concerns remain that China may one day crack down on Hong Kong's press, but they pale in comparison with other concerns. Indeed, when asked in a 2006 survey about 14 different issues, those evoking the lowest levels of concern were freedom of the press (8 percent), freedom of assembly (7 percent), and freedom of speech (7 percent). (By contrast, 48 percent were "very worried" about air and water pollution.)[72] This is not surprising, as China so far has done little to challenge Hong Kong's press freedom. The exceptions are primarily concerning matters that China considers "internal affairs." The Hong Kong press has received warnings not to speak favorably about Taiwan independence; but in general, members of the Hong Kong press have mastered the art of knowing where to draw the line so as not to offend Beijing. Self-censorship is nothing new for Hong Kong's (or China's) press; and even under the British, it was necessary. How Beijing will respond to a Hong Kong press that openly challenges the Chinese Communist Party remains to be seen; but in China itself, political analyses are no longer confined to parroting the Communist Party's line, and investigative reportage is encouraged. Further, the Hong Kong press does not hesitate to engage in searing criticisms of the chief executive, even though he is appointed by Beijing. In short, a major rollback of press freedom in Hong Kong seems unlikely.

As for religious freedom, Beijing has stated that as long as a religious practice does not contravene the Basic Law, it will be permitted. Given the fact that China's tolerance of religious freedom on the mainland has expanded dramatically since it began liberalizing reforms in 1979, this does not seem to be a likely area of tension. Even in the case of Falun Gong, a sect that is banned on the mainland but has strong support in Hong Kong, thus far Beijing does not seem inclined to pressure the Hong Kong government to restrict its activities.

THE FUTURE

The future is, of course, unpredictable; but so far, there is a "business as usual" look about Hong Kong. Hong Kong still appears much as it did before its return to China's sovereign control. And although some analysts believe that power is imperceptibly being transferred to Beijing, Deng Xiaoping's promise that "dancing and horseracing would continue unabated" has been kept. "U.S. aircraft carriers still drop anchor and disgorge their crews into the Wanchai district's red-light bars. Anti-China demonstrators continue their almost weekly parades through the Central district. . . . Meanwhile, the People's Liberation Army has made a virtue of being invisible."[73] It seems that fears of Hong Kong becoming just another Chinese city were misplaced.

China's leaders have promised that for 50 years after 1997, the relationship between China and the Hong Kong Special Administrative Region will be "one country, two systems." Moreover, China's leaders clearly want their cities to look more like Hong Kong, not the other way around. In fact, with China's commercial banks starting to act much the same as banks in any capitalist economy, the phasing out of China's state-run economy in favor of a market economy, a budding stock market, billions of dollars in foreign investment, and entry into the World Trade Organization, China's economy is looking more like Hong Kong's and less like a centrally planned socialist economy. China's leaders have repeatedly stated the importance of foreign investment, greater openness, and experimentation; and they are doing everything possible to integrate the country into the global economy. The imposition of a socialist economy on Hong Kong is, at this point, unthinkable.

China's political arena is also changing so profoundly that the two systems, which just 20 years ago seemed so far apart, are now much closer. China has undergone some political liberalization, (increasing electoral rights at the local level, permitting freedom in individual lifestyles, mobility, and job selection, and according greater freedom to the mass media). The rapid growth of private property and business interests is also bringing significant social change to China, including the proliferation of interest groups and associations to promote their members' interests.

Southern China's extraordinary economic boom has made Hong Kong optimistic about the future. Many of its residents see a new "dragon" emerging, one that combines Hong Kong's technology and skills with China's labor and resources. Others, however, do not see Hong Kong happily working as one unit with China. Instead, they see Hong Kong as a rival competing with Shanghai and with the many new ports that China is building; and even with Shenzhen, which borders Hong Kong and also now has a deepwater container port. Shenzhen is developing a high-quality pool of labor that costs just one-tenth that of Hong Kong, so many corporations are moving their operations—and their wealth—across the border to China. Moreover, the P.R.C. no longer needs Hong Kong as an entrepôt to export its products. Not only can it now do so through its own ports, but it is also building plants in other countries, such as India, where it will manufacture and export goods that otherwise would have gone through Hong Kong. In addition, China's membership in the WTO as of 2001, as well as its increasingly direct contacts and trade with Taiwan, could easily lead to a partial eclipse of Hong Kong.

Some Hong Kong analysts believe that the greatest threat to Hong Kong's success is not political repression and centralized control from Beijing; rather, it would take the more insidious form of China's bureaucracy and corruption simply smothering Hong Kong's economic vitality. One concern is that mainland companies operating in Hong Kong may be allowed to stand above the law or to use their Hong Kong ties to exert inappropriate influence on behalf of pro-China business interests and corrupt the Hong Kong economy.[74] So far, Hong Kong has escaped this fate, and is still considered one of the least corrupt business environments in the world. Another concern is that China may bring the features of the Singapore political model to Hong Kong, whereby it would be ruled by pro-China business tycoons who are insensitive to the political, social, and economic concerns of other groups, and where democratic parties would find it impossible to gain a majority in a legislative system stacked against them. Political rights and freedom would also be restricted by moving toward a Singapore political model.[75] Yet another worry is that the dangers lie *within* Hong Kong, notably the minimalist efforts of the Hong Kong government to reshape it in a way that allows it to remain competitive.

Beijing has an important stake in its take over of Hong Kong not being perceived as disruptive to its political or economic system, and that Hong Kong's residents and the international business community believe in its future prosperity. Policies and events that threatened that confidence in the 1980s led to the loss of many of Hong Kong's most talented people, technological know-how, and investment. China does not want to risk losing still more.[76] Beijing also wants to maintain Hong Kong as a major free port and the regional center of trade, financing, shipping, and information—although it is also doing everything possible to turn Shanghai into a competitive center.[77]

William Overholt has labeled the major underlying sources of tension between Hong Kong and Beijing as the "Three Confusions": Beijing's "confusion of Hong Kong, where there is virtually no separatist sentiment, with Taiwan"; confusion due to a failure to distinguish between the types of lawful demonstrations that have traditionally taken place in Hong Kong on a regular basis "with disruptive demonstrations in the mainland"; and confusion because of Beijing's failure to distinguish between some of the older leaders of the democracy movement "with the moderate loyal sentiments of the overwhelming majority of the democratic movement."[78] To the degree that China can eliminate such confusion, it will be able to avoid many problems with Hong Kong.

Finally, regardless of official denials by the government in Taiwan, Beijing's successful management of "one country, two systems" in Hong Kong will profoundly affect how Taiwan feels about its own peaceful integration with the mainland. If Beijing wants to regain control of Taiwan by peaceful means, it is critical that it handle Hong Kong well.

Timeline: PAST

A.D. 1839–1842
The first Opium War between China and Great Britain. Ends with Treaty of Nanjing, which cedes Hong Kong to Britain

1860
The Chinese cede Kowloon and Stonecutter Island to Britain. They became part of Hong Kong

1898
England gains a 99-year lease on the New Territories, which also became part of Hong Kong

1911
A revolution ends the Manchu Dynasty; the Republic of China is established

1941
The Japanese attack Pearl Harbor and take Hong Kong; Hong Kong falls under Japanese control

1949
The Communist victory in China produces massive immigration into Hong Kong

1980s
Great Britain and China agree to the return of Hong Kong to China

1990s
China resumes control of Hong Kong on July 1, 1997

PRESENT

1997–2002
Hong Kong's efforts to recover from the Asian economic and financial crisis are hindered by a worldwide economic downturn

2003
Outbreak of SARS. After massive demonstrations, government withdraws bill to amend the Basic Law, Article 23, on sedition from further consideration

2005
Chief Executive Tung Chee-hwa abruptly resigns—Donald Tsang becomes the new Chief Executive

2007
Beijing says it will allow the people of Hong Kong to directly elect their own leader in 2017 and their legislators by 2020

2008
LegCo Election. Pro-democracy camp wins more than a third of seats, retaining a key veto over future bills

2009
Hong Kong hosts East Asian Games

2010
Tens of thousands people march in honor of eight locals killed in a bus hijacking in Manila, denouncing the Philippine government for botching the rescue operation

NOTES

1. R. G. Tiedemann, "Chasing the Dragon," *China Now,* No. 132 (February 1990), p. 21.
2. Jan S. Prybyla, "The Hong Kong Agreement and Its Impact on the World Economy," in Jurgen Domes and Yu-ming Shaw, eds., *Hong Kong: A Chinese and International Concern* (Boulder, CO: Westview Special Studies on East Asia, 1988), p. 177.
3. Tiedemann, p. 22.
4. Steven Tsang, *A Modern History of Hong Kong* (New York: I.B.Tauris, 2004), p. 271.
5. Siu-kai Lau, "The Hong Kong Policy of the People's Republic of China, 1949–1997," *Journal of Contemporary China* (March 2000), Vol. 9, No. 23, p. 81.
6. Tsang, pp. 133–135.
7. Robin McLaren, former British ambassador to China, seminar at Cambridge University, Centre for International Relations (February 28, 1996).
8. Ambrose Y. C. King, "The Hong Kong Talks and Hong Kong Politics," in Domes and Shaw, p. 49.
9. Hungdah Chiu, Y. C. Jao, and Yual-li Wu, *The Future of Hong Kong: Toward 1997 and Beyond* (New York: Quorum Books, 1987), pp. 5–6.
10. Siu-kai Lau, "Hong Kong's 'Ungovernability' in the Twilight of Colonial Rule," in Zhiling, Lin and Thomas W. Robinson, *The Chinese and Their Future: Beijing, Taipei, and Hong Kong* (Washington, D.C.: The American Enterprise Institute Press, 1994), pp. 288–290.
11. McClaren.
12. McLaren noted that it was not easy to convince Prime Minister Thatcher in her "post-Falklands mood" (referring to Great Britain's successful defense of the Falkland Islands, 9,000 miles away, from being returned to Argentine rule), that Hong Kong could not stay under British administrative rule even after 1997.
13. T. L. Tsim, "Introduction," in T. L. Tsim and Bernard H. K. Luk, *The Other Hong Kong Report* (Hong Kong: The Chinese University Press, 1989), p. xxv.
14. Norman J. Miners, "Constitution and Administration," in Tsim and Luk, p. 2.
15. King, pp. 54–55.
16. Annex I, Nos. 1 and 4 of *The Basic Law of the Hong Kong Special Administrative Region of the People's Republic of China* (hereafter cited as *The Basic Law*). Printed in *Beijing Review,* Vol. 33, No. 18 (April 30–May 6, 1990), supplement. This document was adopted by the 7th National People's Congress on April 4, 1990.
17. For specifics, see Annex II of *The Basic Law* (1990).
18. Article 14, *The Basic Law* (1990).
19. Articles 27, 32, 33, and 36, *The Basic Law* (1990).

20. Tsang, p. 271.
21. Chinese cities on the mainland are overrun by transients. On a daily basis, Shanghai alone has well over a million visitors, largely a transient population of country people. Hong Kong does not feel that it can handle such an increase in its transient population.
22. James Tang and Shiu-hing (Sonny) Lo, University of Hong Kong, interview in Hong Kong (June 2000).
23. Michael E. DeGolyer, director, *1982–2007 Hong Kong Transition Project: Accountability & Article 23* (Hong Kong: Hong Kong Baptist University, December 2002), pp. 43–44. These newspapers include several that are pro-Beijing.
24. Tang and Lo.
25. In 1995, Hong Kong's per capita GDP was U.S.$23,500. Wang Gungwu and Wong Siu-lun, eds., *Hong Kong in the Asian-Pacific Region: Rising to the New Challenge* (Hong Kong: University of Hong Kong, 1997), p. 2.
26. James L. Watson, "McDonald's in Hong Kong: Consumerism, Dietary Change, and the Rise of a Children's Culture," in James L. Watson, ed., *Golden Arches East: McDonald's in East Asia* (Palo Alto, CA: Stanford University Press, 1997), pp. 77–109.
27. Tsim, in Tsim and Luk, p. xx.
28. Howard Gorges (South China Holdings Corporation), interview in Hong Kong (June 2000).
29. Wang and Wong, p. 3.
30. Keith B. Richburg, "Uptight Hong Kong Countdown," *The Washington Post* (July 2, 1996), pp. A1, A12.
31. An average apartment measures 20 feet by 23 feet, or 460 square feet.
32. Tsim, in Tsim and Luk, p. xxi.
33. The purchaser of the property pays tax, develops the property, and then pays a 2 percent tax on every real-estate transaction (renting or selling) that occurs thereafter.
34. Keith B. Richburg, "Chinese Muscle-Flexing Puts Hong Kong Under Pessimistic Pall," *The Washington Post* (December 26, 1996), p. A31.
35. For information on Hong Kong's problems with pollution, including from car emissions, see www.epd.gov.hk/epd/english/environmentinhk/air/air_main content.html
36. The Hong Kong Transition Project, *Parties, Policies, and Political Reform in Hong Kong* (Hong Kong: Hong Kong Baptist University, 2006), p. 15.
37. The six largest real-estate companies are among the top 20 companies in the Hang Seng Index of stocks. Others are banks and technology companies. The largest is Cheong Kong Holdings, under Li Kashing's control. Companies linked to him and his son accounted for 26 percent of the total market capitalization of the Hang Seng Index! *Asiaweek* (May 26, 2000), pp. 33–36.
38. Philip Bowring, "Meanwhile: China Changes, Not Hong Kong," *International Herald Tribune* (Feb. 12, 2007).
39. In surveys done since 1995, the percentage "satisfied" with the relationship with the mainland has fluctuated widely. The lowest level of satisfaction (21 percent) was recorded in 1995, the highest (71 percent) in November 2005. The November 2006 survey, recorded satisfaction at 62 percent. See The Hong Kong Transition Project, *Parties, Policies, and Political Reform in Hong Kong* (Hong Kong: Hong Kong Baptist University (2006), p. 51.
40. Al Reyes, journalist *(Asiaweek),* interview in Hong Kong (June 16, 2000).
41. "Is Hong Kong Ripe for a Bit of Central Planning?" *The Economist* (April 12, 1997).
42. Tammy Tam, "Shenzhen Industrial Estate Developed to Boost Military Funds," *The Hong-Kong Standard* (September 5, 1989), p. 1.
43. Shiu-hing Lo, "Hong Kong's Political Influence on South China," *Problems of Post-Communism,* Vol. 46, No. 4 (July/August 1999), pp. 33–41.
44. Tang and Lo; Christine Loh, LegCo legislator and founder of the Citizens' Party, interview in Hong Kong (June 2000). Loh did not run for re-election in September 2000. She is a businessperson-turned-politician. She has been a self-described "armchair critic" of the British and Hong Kong SAR governments, and is generally supportive of the business community.
45. Tang and Lo.
46. John Gordon Davies, "Introduction," *Hong Kong Through the Looking Glass* (Hong Kong: Kelly & Walsh, 1969).
47. Tang and Lo.
48. King, in Domes and Shaw, pp. 45–46.
49. Prybyla, in Domes and Shaw, pp. 196–197; and Lau, in Lin and Robinson, p. 302.
50. Lau, in Lin and Robinson, pp. 293–294, 304–305.
51. Willy Wo-Lap Lam, "New Faces to Star in Hong Kong's New Cabinet," CNN Web site (June 24, 2002).
52. In June 1997, just before the July 1 hand over, only 45 percent were "satisfied," and 41 percent were "dissatisfied." DeGolyer, *Hong Kong Transition Project,* 1982–2007, Table 145, p. 103; www.hkbu.edu.hk/ hktp.
53. Hong Kong Transition Project, *Listening to the Wisdom of the Masses: Hong Kong People's Attitudes toward Constitutional Reforms* (Hong Kong: Civic Exchange and Hong Kong Transition Project, 2004), Tables 38, 135, 9, pp. 32, 68, 14.
54. Loh, interview.
55. "Silent Treatment: Hong Kong's Chief and Its Legislature Aren't Talking," *Far Eastern Economic Review* (September 17, 1998), p. 50. Hong Kong Transition Project, *Listening to the Wisdom,* Table 25, pp. 22–23.
56. Hong Kong Transition Project, *Listening to the Wisdom* (2004), Table 135, p. 68.
57. Frank Ching, "Beijing Loath to Cast the Fate of Elections in Hong Kong to the Wind," *The Japan Times,* August 2, 2006.
58. The Hong Kong Transition Project, *Parties, Policies, and Political Reform in Hong Kong,* (2006), p. 54.
59. Data based on surveys done in November 2002. The government submitted the proposed changes to Article 23 two months earlier. In almost all categories of rights, and across all occupations, ages, education, and so on, Hong Kong people showed an increase in concern about rights because of the potential amendments to Article 23. Hong Kong Transition Project, *Listening to the Wisdom,* (2004) pp. 36–39, 49–66.
60. Ibid., Table 38, p. 32.
61. King, in Domes and Shaw, pp. 51, 56, 57.
62. In the 1998 elections, 20 members of LegCo were for the first time elected directly; and of these, 13 seats went to the Democratic Party. Of the remaining 40 seats, which were indirectly elected, seven went to the Democratic Party.
63. Mark Magnier, "Hong Kong Warned to Drop Vote Idea," *LA Times,* Nov. 10, 2004.
64. Loh, interview (June 2000).
65. Alvin Y. So, "Hong Kong's Problematic Democratic Transition: Power Dependency or Business Hegemony? *The Journal of Asian Studies,* Vol. 59, No. 2 (May 2000), pp. 375–376.
66. According to the Hong Kong Transition Project 2000 polls, only 30 percent of the people believe that political parties wield significant influence on the government, whereas 74 percent think that Beijing officials do. Michael E. DeGolyer, director, *The Hong Kong Transition Project: 1982–2007* (Hong Kong: Hong Kong Baptist University, 2000), p. 25.
67. James L. Tyson, "Promises, Promises. . . ." *The Christian Science Monitor* (April 20, 1989), p. 2.
68. "Another Place, Another Crime: Mainland Trial of Alleged Gangster Puts 'One Country, Two Systems' to Test," *Far Eastern Economic Review* (November 5, 1998), pp. 26–27.
69. DeGolyer, The Hong Kong Transition (2000) 2000, pp. 3–8. Of students, 54 percent are satisfied, 28 percent neutral.
70. Kevin Murphy, "Troops for Hong Kong: China Puts Best Face on It," *The International Herald Tribune* (January 30, 1996), p. 4.
71. Miu-wah Ma, "China Warns Against Political Ties Abroad," *The Hong Kong Standard* (September 1, 1989), p. 4; and Viola Lee,

“China ‘Trying to Discourage HK People,’ ” *South China Morning Post* (August 21, 1989). The article, which originally appeared in an *RMRB* article in July, was elaborated upon in the August edition of *Outlook Weekly,* a mouthpiece of the CCP.

72. The Hong Kong Transition Project, *Parties, Policies, and Political Reform in Hong Kong* (2006), p. 19.
73. “Hong Kong: Now the Hard Part,” *Far Eastern Economic Review* (June 11, 1998), p. 13.
74. Michael C. Davis, “Constitutionalism and Hong Kong’s Future,” *Journal of Contemporary China* (July 1999), Vol. 8, No. 21, pp. 271.
75. *Ibid.,* pp. 269, 273.
76. For extensive surveys on which types of people might want to leave Hong Kong, and for what reasons, see The Hong Kong Transition Project, *Parties, Policies, and Political Reform in Hong Kong* (Hong Kong: Hong Kong Baptist University, 2006).
77. Kai-Yin Lo, “A Big Awakening for Chinese Rivals: Hong Kong and Shanghai Look Afar,” *International Herald Tribune,* January 20, 2005.
78. William Overholt, Testimony, “The Hong Kong Legislative Election of Sept 12, 2004: Assessment and Implications” (Testimony presented to the Congressional-Executive Commission on China on Sept. 23, 2004), (Santa Monica: RAND Corporation, 2004), p. 7.

Taiwan Map

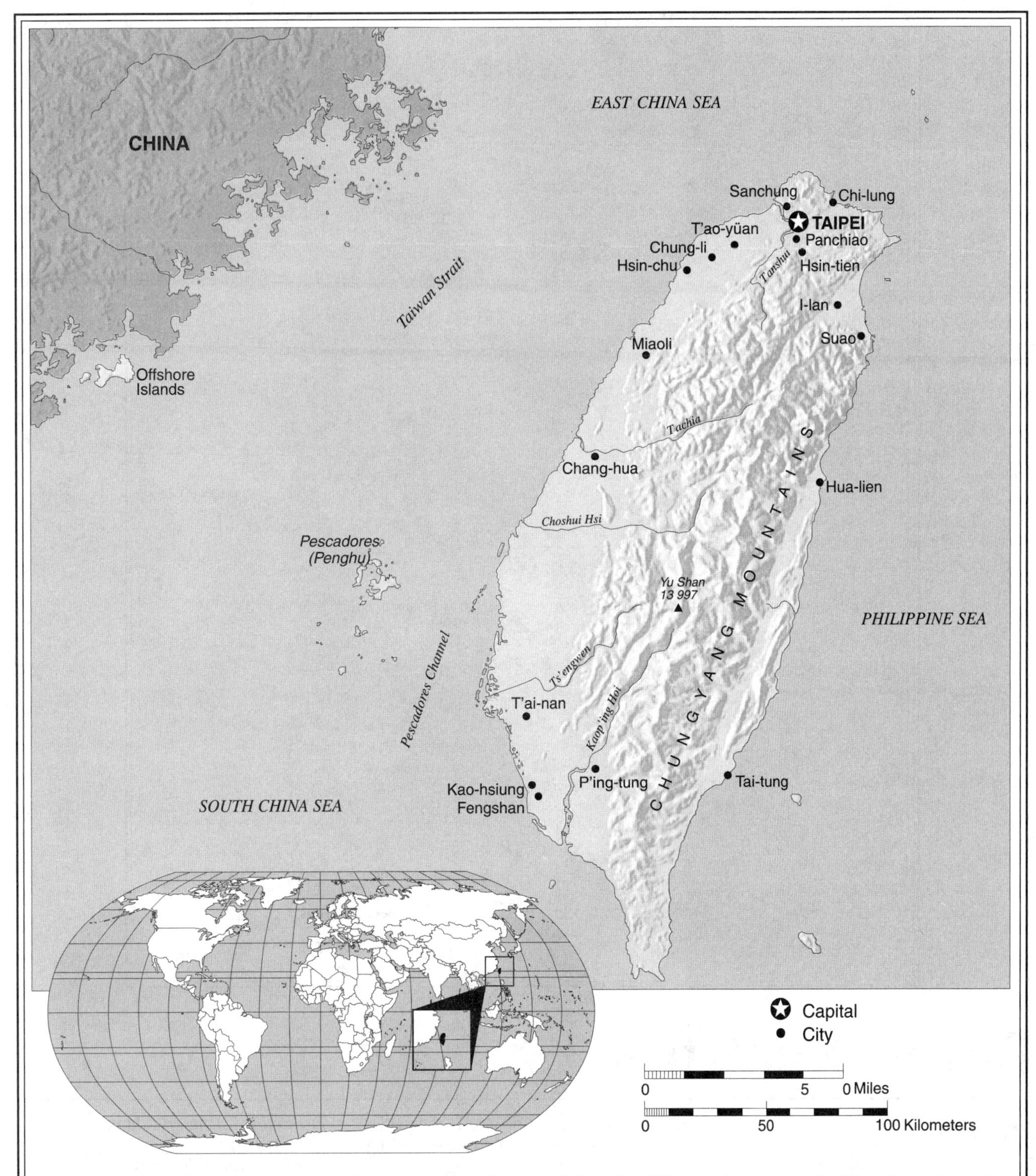

Taiwan was the center of the government of the Republic of China (Nationalist China) after 1949. According to international law, Taiwan is a part of China. Taiwan consists of the main island, 15 islands in the Offshore Islands group, and 64 islands in the Pescadores Archipelago. While the Pescadores are close to Taiwan, the Offshore Islands are only a few miles off the coast of mainland China.

Taiwan

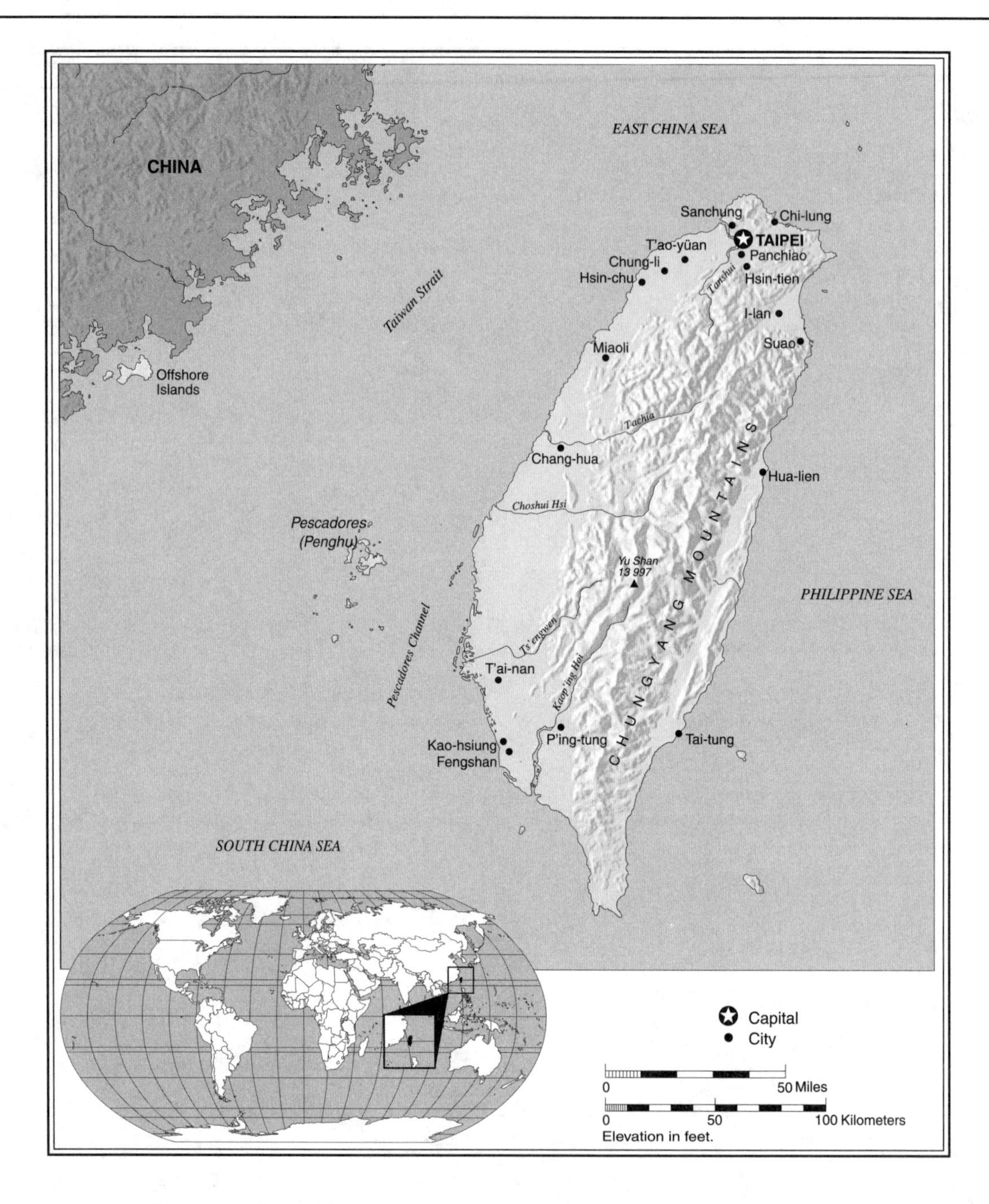

Taiwan Statistics

GEOGRAPHY

Area in Square Miles (Kilometers): 13,892 (35,980) (slightly smaller than Maryland and Delaware combined)
Capital (Population): Taipei (2,622,933) (September 2008)
Environmental Concerns: water and air pollution; contamination of drinking water; radioactive waste; trade in endangered species
Geographical Features: mostly rugged mountains in east; flat to gently rolling plains in west
Climate: tropical; marine

PEOPLE

Population

Total: 23,024,956 (July 2010 est.)
Population Growth Rate: 0.213% (2010 est.)
Sex Ratio: 1.02 male(s)/female (2010 est.)
Life Expectancy: Total Population: 78.15 years (*male:* 75.34 years; *female:* 81.2 years) (2010 est.)
Total Fertility Rate: 1.15 children born/woman (2010 est.)
Religions: mixture of Buddhist and Taoist 93%; Christian 4.5%; other 2.5%
Languages: Mandarin Chinese (official), Taiwanese, Hakka dialects

Education

Adult Literacy Rate: 96.1%

ECONOMY

Currency: New Taiwan dollar (TWD)
Exchange Rate: U.S.$1 = 30.415 TWD (December 2010)
GDP (Purchasing Power Parity): $734.3 billion (2009 est.)
GDP (Official Exchange Rate): $378.5 billion (2009 est.)
GDP Real Growth Rate: -19% (2009)
GDP Per Capita (Purchasing Power Parity): $32,000 (2009)
GDP Composition by Sector: agriculture: 1.4%; industry: 27.5%; services: 71.1% (2007 est.)
Unemployment Rate: 5.9% (2009 est.)
Inflation Rate (Consumer Prices): 1.8% (2009 est.)
Exports: $203.4 billion f.o.b. (2009 est.)
Exports Commodities: electronic and electrical products; metals; textiles; plastics; chemicals; auto parts
Exports Partners: China 32.4%; U.S. 12.9%; Hong Kong 8.5%; Japan 6.4%; Singapore 5% (2009)
Imports: $172.8 billion f.o.b. (2009 est.)
Imports Commodities: electronic and electrical products; machinery; petroleum; precision instruments; organic chemicals; metals (2002)
Imports Partners: Japan 22.1%; U.S. 10.49%; China 14.01%; South Korea 7.3%; Saudi Arabia 4.8%; Singapore 4.5% (2007)

COMMUNICATION

Telephones—Main Lines: 14.497 million (2008)
Telephones—Mobile Cellular: 25.412 million (2008)
Television Broadcast Stations: 76 (46 digital and 30 analog) (2007)
Internet Users: 15.143 million (2008)

TRANSPORTATION

Airports: 41 (2010)
Railways: 1,588 km (2010)
Roadways: 40,262 km; 38,171 km paved (includes 976 km of expressways) (2009)
Ports: Chilung (Keelung), Kaohsiung, Taichung

GOVERNMENT

Government Type: multiparty democracy
Major Parties: Kuomintang (Chinese Nationalist Party); Democratic Progressive Party; People First Party; Taiwan Solidarity Union; New Party
National Day: October 10 (Anniversary of the Chinese Revolution in 1911)
Head of State: Ma Ying-jeou (since May 2008)
Head of Government: Premier (President of the Executive Yuan) Wu Den-yih (since September 2009)
Suffrage: universal at 20 years of age

MILITARY

Military Branches: Army, Navy (includes Marine Corps), Air Force, Coast Guard Administration, Armed Forces Reserve Command, Combined Service Forces Command, Armed Forces Police Command
Military Expenditures: 2.2% of GDP (2006)

SUGGESTED WEB SITES

www.cia.gov/library/publications/the-world-factbook/geos/tw.html
www.gio.gov.tw
www.etaiwannews.com/Taiwan
www.mac.gov.tw/english/index1-e.htm

Taiwan Report

Taiwan, today a powerful economic player in Asia, was once an obscure island off the coast of China, just 90 miles away. Taiwan has also been known as Formosa, Free China, the Republic of China, Nationalist China and Republic of China on Taiwan. It was originally inhabited by aborigines from Southeast Asia. By the seventh century A.D., Chinese settlers had begun to arrive. The island was subsequently "discovered" by the Portuguese in 1590, and Dutch as well as Spanish settlers followed. Today, the aborigines' descendants, who have been pushed into the remote mountain areas by the Chinese settlers, number under 400,000, a small fraction of the 23 million people now living in Taiwan. Most of the population is descended from people who emigrated from the Chinese mainland's southern provinces before 1885, when Taiwan officially became a province of China. Although their ancestors originally came from China, they are known as *Taiwanese,* as distinct from the Chinese who fled the mainland from 1947 to 1949. The latter are called *Mainlanders* and represent less than 20 percent of the island's population. After 1949, the Mainlanders dominated Taiwan's political elite; but the "Taiwanization" of the political realm that began after Chiang Kai-shek's death in 1975 and the political liberalization since 1988 have allowed the native Taiwanese to take up their rightful place within the elite.

The Manchus, "barbarians" who came from the north, overthrew the Chinese rulers on the mainland in 1644 and established the Qing Dynasty. In 1683, they conquered Taiwan; but because Taiwan was an island 90 miles distant from the mainland and China's seafaring abilities were limited, the Manchus paid little attention to it and exercised minimal sovereignty over the Taiwanese people. With China's defeat in the Sino–Japanese War (1894–1895), the Qing was forced to cede Taiwan to the Japanese. The Taiwanese people refused to accept Japanese rule, however, and proclaimed Taiwan a republic. As a result, the Japanese had to use military force to gain actual control over Taiwan.

For the next 50 years, Taiwan remained under Japan's colonial administration. Japan helped to develop Taiwan's agricultural sector, a modern transportation network, and an economic structure favorable to later industrial development. Furthermore, the Japanese developed an educated workforce, which proved critical to Taiwan's economic growth.

With Japan's defeat at the end of World War II (in 1945), Taiwan reverted to China's sovereignty. In the meantime, the Chinese had overthrown the Manchu Dynasty (1911) and established a republican form of government on the mainland. Beginning in 1912, China was known as the Republic of China (R.O.C.). Thus, when Japan was defeated in 1945, it was Chiang Kai-shek who, as head of the R.O.C. government, accepted the return of the island province of Taiwan to R.O.C. rule. When some of Chiang Kai-shek's forces were dispatched to shore up control in Taiwan in 1947, tensions quickly arose between them and the native Taiwanese. The ragtag, undisciplined military forces of the KMT (the Kuomintang, or Nationalist Party) met with hatred and contempt from the local people. They had grown accustomed to the orderliness and professionalism of the Japanese occupation forces and were angered by the incompetence and corruption of KMT officials. Demonstrations against

MCCARTHYISM: ISOLATING AND CONTAINING COMMUNISM

The McCarthy period in the United States was an era of rabid anti-communism. McCarthyism was based in part on the belief that the United States was responsible for losing China to the Communists in 1949 and that the reason for this loss was the infiltration of the U.S. government by Communists. As a result, Senator Joseph McCarthy *(pictured here)* spearheaded a "witchhunt" to ferret out those who allegedly were selling out American interests to the Communists. McCarthyism took advantage of the national mood in the Cold War era that had begun in 1947. At that time, the world was seen as being divided into two opposing camps: Communists and capitalists.

The major strategy of the Cold War, as outlined by President Harry Truman in 1947, was the "containment" of communism. This strategy was based on the belief that if the United States attempted—as it had done with Adolf Hitler's aggression against Czechoslovakia (the first step toward World War II)—to appease communism, it would spread beyond its borders and threaten other free countries.

The purpose of the Cold War strategy was, then, to contain the Communists within their national boundaries and to isolate them by hindering their participation in the international economic system and in international organizations. Hence, in the case of China, there was an American-led boycott against all Chinese goods, and the United States refused to recognize the People's Republic of China as the legitimate representative of the Chinese people within international organizations.

(Wisconsin Historical Society (WHi8006))

Joseph McCarthy.

rule by Mainlanders occurred in February 1947. Relations were badly scarred when KMT troops killed thousands of Taiwanese opposed to mainland rule. Among those murdered were many members of the island's political and intellectual elite.

Meanwhile, the KMT's focus remained on the mainland, where, under the leadership of General Chiang Kai-shek, it continued to fight the Chinese Communists in a civil war that had ended their fragile truce during World War II. Civil war raged from 1945 to 1949 and diverted the KMT's attention away from Taiwan. As a result, Taiwan continued, as it had under the Qing Dynasty's rule, to function quite independently of Beijing. In 1949, when it became clear that the Chinese Communists would defeat the KMT, General Chiang and some 2 million members of his loyal military, political, and commercial elite fled to Taiwan, moving with them the R.O.C. government which they claim to be the true government of China. This declaration reflected Chiang's determination to regain control over the mainland, and his conviction that the more than 600 million people living on the mainland would welcome the return of the KMT to power.

During the McCarthy period of the "red scare" in the 1950s (during which Americans believed to be Communists or Communist sympathizers—"reds"—were persecuted by the U.S. government), the United States supported Chiang Kai-shek. In response to the Chinese Communists' entry into the Korean War in 1950, the United States applied its Cold War policies of support for any Asian government that was anti-Communist, regardless of how dictatorial and ruthless it might be, in order to "isolate and contain" the Chinese Communists. It was within this context that in 1950, the United States committed itself to the military defense of Taiwan and the offshore islands in the Taiwan Strait, by ordering the U.S. Seventh Fleet to the Strait and by giving large amounts of military and economic aid to Taiwan. General Chiang Kai-shek continued to lead the government of the Republic of China on Taiwan until his death in 1975, at which time his son, Chiang Ching-kuo, succeeded him.

One China, Two Governments

Taiwan's position in the international community and its relationship to the government in Beijing have been determined by perceptions and values as much as by actions. In 1949, when the R.O.C. government fled to Taiwan, the Chinese Communists renamed China the "People's Republic of China" (P.R.C.), and they proclaimed the R.O.C. government illegitimate. Later, Mao Zedong, the P.R.C.'s preeminent leader until his death in 1976, was to say that adopting the new name instead of keeping the name "Republic of China" was the biggest mistake he ever made, for it laid the groundwork for future claims of "two Chinas." Beijing claimed that the P.R.C. was the legitimate government of all of China, including Taiwan. Beijing's attempt to regain de facto control over Taiwan was, however, forestalled first by the outbreak of the Korean War and thereafter by the presence of the U.S. Seventh Fleet in the Taiwan Strait.

Beijing has always insisted that Taiwan is an "internal" Chinese affair, that international law is therefore irrelevant, and that other countries have no right to interfere. For its part, until 1995, the government of Taiwan agreed that there was only one China and that Taiwan was a province of China. But by 1995–1996, the political parties had begun debating the possibility of Taiwan declaring itself an independent state.

Although the Chinese Communists' control over the mainland was long evident to the world, the United States managed to keep the R.O.C. in the China seat at the United Nations by insisting that the issue of China's representation in the United Nations was an "important question." This meant that a two-thirds affirmative vote of the UN General Assembly, rather than a simple majority, was required. With support from its allies, the United States was able to block the P.R.C. from winning this two-thirds vote until 1971. Once Secretary of State Henry Kissinger announced his secret trip to Beijing in the summer of 1971 and that President Richard Nixon would be going to China in 1972, the writing was on the wall. Allies of the United States knew that U.S. recognition of Beijing as China's legitimate government would eventually occur, so there would no longer be pressure to block the P.R.C. from membership in the United Nations. They quickly jumped ship and voted for Beijing's representation.

CHIANG KAI-SHEK: DETERMINED TO RETAKE THE MAINLAND

Until his dying day, President Chiang Kai-shek, whose memorial hall is *pictured here,* maintained that the military, led by the KMT (Kuomintang, or Nationalist Party), would one day return to the mainland and, with the support of the Chinese people living there, defeat the Communist government. During the years of Chiang's presidency, banner headlines proclaimed daily that the Communist "bandits" would soon be turned out by internal rebellion and that the KMT would return to control on the mainland. In the last years of Chiang Kai-shek's life, when he was generally confined to his residence and incapable of directing the government, his son, Chiang Ching-kuo, always had two editions of the newspaper made that proclaimed such unlikely feats, so that his father would continue to believe these were the primary goals of the KMT government in Taiwan. In fact, a realistic appraisal of the situation had been made long before Chiang's death in 1975, and most of the members of the KMT only pretended to believe that recovery of the mainland was imminent.

Chiang Ching-kuo, although continuing to strengthen Taiwan's defenses, turned his efforts to building Taiwan into an economic showcase in Asia. Taiwan's remarkable growth and a certain degree of political liberalization were the hallmarks of his leadership. A man of the people, he shunned many of the elitist practices of his father and the KMT ruling elite, and he helped to bring about the Taiwanization of both the KMT party and the government. The "Chiang dynasty" in Taiwan came to an end with Chiang Ching-kuo's death in 1988. It was, in fact, Chiang Ching-kuo who made certain of this, by barring his own sons from succeeding him and by grooming his own successor, a native Taiwanese.

(Public Domain)

Chiang Kai-shek memorial.

As the Taiwanese have gained an increasing amount of power in the last 20 years, they have reasserted their Taiwanese identity. This has led to a fading out of the two Chiangs from Taiwan's history books and public images.

At this critical moment, when the R.O.C.'s right to represent "China" in the United Nations was withdrawn, the R.O.C. could have put forward the claim that Taiwan had the right to be recognized as an independent state, or at least to be granted observer status. Instead, the R.O.C. steadfastly maintained that there was but one China and that Taiwan was merely a province of China. As a result, today Taiwan has no representation in any major international organization under the name of "Republic of China;" and it has representation only as "Chinese Taipei" in organizations in which the P.R.C. is a member—if the P.R.C. allows it any representation at all. With few exceptions, however, Beijing has been adamant about not permitting Taiwan's representation regardless of what it is called.[1]

It is important to understand that at the time Taiwan lost its seat in the United Nations, it was still ruled by the KMT mainlanders, under the name of the Republic of China. The "Taiwanization" of the ruling KMT party-state did not begin until the mid 1970s. As a result, the near 85 percent of the population that was native Taiwanese lacked the right to express their preference for rule by native Taiwanese rather than KMT mainlanders, much less their desire for a declaration of independent statehood. Under martial law, those who had dared to demand independence were imprisoned, forcing the Taiwan independence movement to locate outside of Taiwan. In short, most of the decisions that have shaped Taiwan's international legal standing through treaties and diplomatic relations were made when the KMT mainlanders held political power. Only after the abandonment of martial law in 1987, and the KMT's decision to allow competing political parties, did the Taiwanese start to assert a different view of Taiwan's future. By then, many would agree, it was too late.

INTERNATIONAL RECOGNITION OF THE P.R.C., AT TAIWAN'S EXPENSE

The seating of the P.R.C. in the United Nations in 1971 thus led to the collapse of Taiwan's independent political standing as the R.O.C. in international affairs. Not wanting to anger China, which has a huge and growing economy and significant military power, the state members of international organizations have given in to Beijing's unrelenting pressure to exclude Taiwan. Furthermore, Beijing insists that in bilateral state-to-state relations, any state wishing to maintain diplomatic relations with it must accept China's "principled stand" on Taiwan—notably, that Taiwan is a province of China and that the People's Republic of China is its sole representative.

Commercial ventures, foreign investment in Taiwan, and Taiwan's investments abroad have suffered little as a result of countries' severing their *diplomatic* relations with Taipei. After being forced to close all but a handful of its embassies as one state after another switched recognition from the R.O.C. to the P.R.C., Taipei simply substituted offices that function as if they are embassies. They handle all commercial, cultural, and official business, including the issuance of visas to those traveling to Taiwan. Similarly, the states that severed relations with Taipei have closed down their embassies there and reopened them as business and cultural offices. The American Institute in Taiwan is a typical example of these efforts to retain ties without formal diplomatic recognition.

Still, Taiwan's government feels the sting of being almost completely shut out of the official world of international affairs. Its increasing frustration and sense of humiliation came to a head in early 1996. Under intense pressure to respond to demands from its people that Taiwan get the international recognition that it deserved for its remarkable accomplishments, Taiwan's president, Lee Teng-hui, engaged in a series of maneuvers to get the international community to confer de facto recognition of its statehood. Not the least of these bold forays was President

Lee's offer of U.S. $1 billion to the United Nations in return for a seat for Taiwan—an offer rejected by the United Nations' secretary-general.

Lee's campaign for election in the spring of 1996 proved to be the final straw for Beijing. Lee had as one of his central themes the demand for greater international recognition of Taiwan as an independent state. Beijing responded with a military buildup of some 200,000 troops in Fujian Province across the Taiwan Strait, and the "testing" of missiles in the waters around Taiwan. Under pressure from the United States not to provoke a war with the mainland, and a refusal on the part of the United States to say exactly what it would do if a war occurred, President Lee toned down his campaign rhetoric. A military conflict was averted. Since 1996, the pattern of Taiwan aggressively discussing independence, followed by China threatening the use of military force and the U.S. warning Taiwan to refrain from rhetoric about independence, has become well-established.

THE OFFSHORE ISLANDS

Crises of serious dimensions erupted between China and the United States in 1954–1955, 1958, 1960, and 1962 over the blockading of supplies to the Taiwan-controlled Offshore Islands in the Taiwan Strait. Thus, the perceived importance of these tiny islands grew out of all proportion to their intrinsic worth. The two major island groups, Quemoy (about two miles from the Chinese mainland) and Matsu (about eight miles from the mainland) are located about 90 miles from Taiwan. As a consequence, Taiwan's control of them made them strategically valuable for pursuing the government's professed goal of retaking the mainland—and valuable for psychologically linking Taiwan to the mainland.

In the first years after their victory on the mainland, the Chinese Communists shelled the Offshore Islands at regular intervals. When there was not a crisis, their shells were filled with pro-Communist propaganda materials, which littered the islands. When the Chinese Communists wanted to test the American commitment to the Nationalists in Taiwan and the Soviet commitment to their own objectives, they shelled the islands heavily and intercepted supplies to the islands. In the end, China always backed down; but in 1958 and 1962, it did so only after going to the brink of war with the United States. By 1979, most states had affirmed Beijing's claim that Taiwan was a province of China through diplomatic recognition of the P.R.C. Beijing's subsequent "peace initiatives" toward Taiwan moved the confrontation over the Offshore Islands to the level of an exchange of gifts by balloons and packages floated across the channel. As a 1986 commentary noted:

> The Nationalists load their balloons and seaborne packages with underwear, children's shoes, soap, toys, blankets, transistor radios and tape recorders, as well as cookies emblazoned with Chiang Ching-kuo's picture and audio tapes of Taiwan's top popular singer, Theresa Teng, a mainland favorite.
>
> The Communists send back beef jerky, tea, herbal medicines, mao-tai and cigarettes, as well as their own varieties of soap and toys.[2]

Because of the dramatic increase in contacts, tourism, trade, and smuggling since the early 1990s, however, such practices have ceased.

The lack of industry and manufacturing on the Offshore Islands led to a steady emigration of their natives to Southeast Asia for better jobs. The civilian population is about 50,000 (mostly farmers) in Quemoy and about 6,000 (mostly fishermen) in Matsu. Until the mid-1990s, the small civilian population in Quemoy was significantly augmented by an estimated 100,000 soldiers. The heavily fortified islands appeared to be somewhat deserted, since the soldiers lived mostly underground: hospitals, kitchens, sleeping quarters—everything was located in tunnels blasted out of granite, including two-lane highways that could accommodate trucks and tanks. Heavily camouflaged anti-artillery aircraft dotted the landscape, and all roads were reinforced to carry tanks.

Taiwan's military administration of Quemoy and Matsu ended in 1992.

Today, Taiwan's armed forces are streamlined, with only some 10,000 troops remaining on the Offshore Islands. Taiwan's Ministry of Defense believes, however, that these troops are better able than the former, larger forces to protect the Offshore Islands, because of improved weapons and technology. Many of the former military installations have become profit-making tourist attractions. Moreover, as of January 2001, the "three mini-links" policy was initiated, with both Quemoy and Matsu allowed to engage in direct trade, transportation, and postal exchange with the mainland. In truth, this change only legalized what had been going on for more than a decade—namely, a roaring trade in smuggled goods between the Offshore Islands and the mainland. Taiwan now seems far more interested in boosting the economy of these islands and increasing links with the mainland than it is concerned with the islands becoming hostages of the mainland. Indeed, Taiwan has moved forward with a plan to secure water for the islands from the mainland, to cope with an increased demand for water it assumes will be generated by growing tourism and business activities.[3] (The first PRC tourists went to Quemoy in Sept. 2004.)

Thus one fiction—that there was no direct trade between Taiwan's Offshore Islands and the mainland—has been abolished. Other fictions wait to be dismantled as Taiwan and the mainland move toward further integration. For example, there is still not supposed to be direct trade between the island of Taiwan and the mainland. All ships are supposed to transship their goods by way of another port, such as Hong Kong or the Ryuku Islands. But now, as soon as ships have left Taiwan, they simply process paperwork with port authorities in the Ryuku Islands by way of fax—that is, without actually ever going there—so they can go directly to the mainland. This eliminates the expensive formalities of transshipments via another port. In short, direct trade has actually been in existence for almost a decade.

CULTURE AND SOCIETY

Taiwan is a bundle of contradictions: "great tradition, small island; conservative state, drastic change; cultural imperialism, committed nationalism; localist sentiment, cosmopolitan sophistication."[4] Over time, Taiwan's culture has been shaped by various cultural elements—Japanese, Chinese, and American culture; localism, nationalism, cosmopolitanism, materialism; and even Chinese mainland culture (in the form of "mainland mania"). At any one time, several of these forces have coexisted and battled for dominance. As Taiwan has become increasingly affected by globalization, the power of the central government to control cultural development has diminished. This has unleashed not just global cultural forces but also local *Taiwanese* culture.[5]

The Taiwanese people were originally immigrants from the Chinese mainland; but their culture, which developed in isolation from that of the mainland, is not the same as the "Chinese" culture of the defeated "Mainlanders" who arrived from 1947 to 1949. Although the Nationalists saw Taiwan largely in terms of security and as a bastion from which to fight against and defeat the Chinese Communist regime on the mainland, "it also cultivated Taiwan as the last outpost of traditional Chinese high culture. Taiwanese folk arts, in particular opera and festivals, did thrive, but as low culture."[6]

The Taiwanese continue to speak their own dialect of Chinese (Taiwanese), distinct from the standard Chinese spoken by the Mainlanders, and almost all Taiwanese engage in local folk-religion practices.

However, until the mid-1990s, the Mainlander-controlled central government dictated a cultural policy that emphasized Chinese cultural values in education and the mass media. As a result, the distinctions institutionalized in a political system that discriminated against the Taiwanese were culturally reinforced.

The Taiwanese grew increasingly resistant to efforts by the KMT Mainlanders to "Sinify" them—to force them to speak standard Chinese and to adopt the values of the dominant Chinese Mainlander elite. By the 1990s, state-controlled television offered many programs in the Taiwanese dialect. Today, Taiwanese appear to have won the battle to maintain their cultural identity. Taiwanese legislators are refusing to use Chinese during the Legislature's proceedings, so now Mainlanders in the Legislative Yuan must learn Taiwanese. Indeed, the pendulum appears to have swung the other way, not just in language, but in terms of engineering an appropriate psycho-cultural milieu for an independent Taiwan. For example, former president Chen Shui-bian went through with plans to "de-Sinicise" Taiwan by ridding it of any symbolic connections with the Mainland. Thus, he has changed the names of government agencies and state-owned corporations that have "China" in them (for example, China Steel Corporation was renamed Taiwan Steel Corporation). "De-Chiangification," echoing the unofficial but popular "de-Maoification" campaign on the Mainland in the early 1980s, has led to changing the name of the Chiang Kai-shek Airport to Taiwan Taoyuan International Airport, and the gradual disappearance of Chiang memorabilia and removal of icons and statues of Chiang Kai-shek. Indeed the mayor of Kaoehsiung City personally oversaw the tearing down of a 24-foot-high statue at the Chiang Kai-shek Cultural Center, the second largest in Taiwan and then chopped it into two hundred pieces. Even Sun Yat-sen, the revolutionary who inspired the overthrow of the Manchu Dynasty in 1911, is fading from public memory. Sun Yat-sen established the Republic of China, and Chiang Kai-shek was his heir to political power on the Mainland, and brought the rule of Chinese mainlanders to Taiwan under the name of Republic of China. In short, the DPP was trying to cut off the Taiwanese from their Chinese heritage. Cynics say the real motivation was that Chen Shui-bian would like to divert the Taiwanese from all the corruption charges he and other members of the DPP were facing. After Ma Ying-jeou became the President in 2008, many of the names have been changed back (for example, Taiwan Post is now China Post again). In any event, such changes are merely a change in window dressing, given the billions of dollars of investment flowing from Taiwan into the mainland. Investments are doing far more to link Taiwan with the mainland than any mere name change can counter. And, ironically, in the convoluted cultural politics of Taiwan-Mainland relations, the Chinese on the Mainland are busily sprucing up sites for tourists that feature where Chiang Kai-shek was born, slept, or marched through, and are in the market for the statues of that defeated KMT Generalissimo!

Generally speaking, however, Taiwanese and Mainlander culture need not be viewed as two cultures in conflict, for they share many commonalities. Now that Taiwanese have moved into leadership positions in what used to be exclusively Mainlander institutions, and intermarriage between the two groups has become more common, an amalgamation of Taiwanese and traditional Chinese practices is becoming evident throughout the society. As is discussed later in this report, the real source of conflict is the Taiwanese insistence that their culture and political system not be controlled by Chinese from the mainland, whether they be Nationalists or Communists.

On the other hand, rampant materialism as well as the importation of foreign ideas and goods are eroding both Taiwanese *and* Chinese values. For example, there are more than 3,000 7-Eleven outlets in Taiwan, and customers in central Taipei rarely have to walk more than a few blocks to get to the next one. Starbucks coffee houses, often much larger than those in the United States, are usually so packed it is hard to find a seat. Most major American fast-food franchises, such as Kentucky Fried Chicken and McDonald's, are ubiquitous. They provide "snack" food for Taiwan's teenagers and children—before they settle down for "real" (Chinese) food. The Big Mac culture affects more than waistlines, contributing as it does to a more globalized culture in Taiwan. Although the government has at various times engaged in campaigns to reassert traditional values, the point seems lost in its larger message, which asks all to contribute to making Taiwan an Asian showplace. The government's emphasis on hard work and economic prosperity has seemingly undercut its focus on traditional Chinese values of politeness, the sanctity of the family, and the teaching of culturally based ethics (such as filial piety) throughout the school system. Materialism and an individualism that focuses on personal needs and pleasure seeking are slowly undermining collectively oriented values.[7] In many ways, this seems to parallel what is happening on the China Mainland. In the meantime, unlike their mainland cousins, who have had a 40-hour, five-day work week since 1996, the Taiwanese did not relinquish an exhausting six-day, 48-hour work week until 2001.While playing a part in Taiwan's economic boom of the past decade, the emphasis on materialism has contributed to a variety of problems, not the least of which are the alienation of youth, juvenile crime, the loosening of family ties, and the general decline of community values. The pervasive spread of illicit sexual activities through such phony fronts as dance halls, bars, saunas, "barber shops," and movies on-video/DVD and music-video/DVD establishments, as well as hotels and brothels, grew so scandalous and detrimental to social morals and social order that at one point the government even suggested cutting off their electricity. They nevertheless continue to thrive. Another major activity that goes virtually uncontrolled is gambling. Part of the problem in clamping down on either illicit sexual activities or gambling (both of which are often combined with drinking in clubs) is that organized crime is involved; and that in exchange for bribes, the police look the other way.[8]

THE ENVIRONMENT

The pursuit of individual material benefit without a concomitant concern for the public good has led to uncontrolled growth and a rapid deterioration in the quality of life, even as the people in Taiwan become richer. As Taiwan struggles to catch up with its own success, the infrastructure has faltered. During the hot, humid summers in its largest city, Taipei, air is so dense with pollution that eyes water, hair falls out, and many people suffer from respiratory illnesses. Inadequate recreational facilities leave urban residents with few options but to join the long parade of cars out of the city on weekends, often only to arrive at badly polluted beaches. Taiwan's citizens have begun forming interest groups to address such problems, and public protests about the government's neglect of quality-of-life issues have grown increasingly frequent. Environmental groups, addressing such issues as wildlife conservation, industrial pollution, and waste disposal, have burgeoned; but environmental campaigns and legislation have difficulty keeping pace with the rapid growth of Taiwan's material culture. According to Taiwan's Environmental Protection Agency, for example, from 1990 to 2000, the amount of garbage produced doubled, but without a doubling of capacity to dispose of it. Part of the problem is that in addition to being a very small island, Taiwan is mostly mountainous, so there is virtually nowhere to construct new landfills. Old landfills are filled to capacity and are leaching toxins into the soil. It is not uncommon for garbage simply to be dumped into a river, or left to rot in gullies or wherever it can be dumped unobserved.[9] Some is taken out to sea, and drifts back to the beaches. But new laws that require citizens to sort their garbage into three categories (regular waste, kitchen leftovers, and recyclable materials), and allow the government to fine citizens who do not sort properly, are intended to bring significant declines in the amount of garbage.[10]

Taiwan must continue to battle against the polluting effects of rapidly increasing wealth; for with almost all Taiwanese families owning a refrigerator, air conditioner, and a car or motorcycle (sometimes several), without a commensurate expansion of the island's roads, carbon emissions continue to grow, and air quality continues to deteriorate. Taipei's recent construction of a subway system has helped traffic flow and kept air pollution from worsening even more quickly. However, political roadblocks have confounded efforts to build a high-speed rail system, and efforts to build Taiwan's fourth nuclear-power plant, a critical project for providing clean electrical power, were plagued by political maneuvering. The DPP government, upon arrival in office, first cancelled the construction plan, and then, under pressure from the opposition and in consideration of the billions of dollars already invested in the plant, it overturned the cancellation. But in a sign of the strength of those environmentalists opposed to the building of the plant, the government made several concessions, including a pledge to a goal of a nuclear-free Taiwan in the future. In many respects, the government has shown good faith, and has supported the development of solar energy technology, in which Taiwan has become a leader.[11]

The public concerns about building nuclear-power plants are understandable. Taiwan sits astride an active earthquake zone, and the potential damage to the environment from a nuclear reactor hit by an earthquake is incalculable. Recent major earthquakes have measured as high as 6.7 on the Richter scale and caused significant damage to the island. The antinuclear movement is increasingly active, especially after an accident at one of Taiwan's nuclear-power plants and the discovery that there are more than 5,000 radioactive buildings in Taiwan, including more than 90 in the city of Taipei.[12] The antinuclear movement in Taiwan exemplifies the problem plaguing environmentalists elsewhere: those opposed to building nuclear plants out of fear of a nuclear accident are pitted against those who favor nuclear plants as a source of clean, carbon-free energy as the way to address the problem of carbon emissions and global warming gases. The debate goes on while, in the meantime, Taiwan's environment continues to be severely damaged and long-term environmental sustainability is brought into question.

(Courtesy of Suzanne Ogden (sogden02))
A busy Buddhist temple in Taipei. Local people make offerings of food and burning incense.

Taiwan is not alone in confronting the financial and political dilemmas caused by the need to create a sustainable environment in the context of a pro-growth economic policy, but its situation is perhaps more urgent because of its high population density and individual wealth. With a population of 1,600 persons per square mile, it is one of the most densely populated places in the world. When this is combined with Taiwan having "the highest density of factories and motor vehicles in the world," as well as being one of the highest per capita energy users in East Asia,[13] it is no surprise that the environment has become a serious mainstream issue. Decisions by tens of thousands of Taiwan's manufacturers to relocate abroad, especially to mainland China, is one way Taiwan is able to "export" its pollution.

RELIGION

A remarkable mixture of religions thrives in Taiwan. The people feel comfortable with placing Buddhist, Taoist, and local deities—and even occasionally a Christian saint—together in their family altars and local temples. Restaurants, motorcycle-repair shops, businesses small and large—almost all maintain altars. The major concern in prayers is for the good health and fortune of the family. The focus is on life in this world, not on one's own afterlife. People pray for prosperity, for luck in the stock market, and even more specifically for the winning lottery number. If the major deity in one temple fails to answer prayers, people will seek out another temple where other deities have brought better luck. Alternatively, they will demote the head deity of a temple and promote others to his or her place. The gods are thought about and organized in much the same way as is the Chinese bureaucracy. In fact, they are often given official clerical titles to indicate their rank within the deified bureaucracy.

Numerous Taiwanese religious festivals honor the more than 100 local city gods and deities. Offerings of food and wine are made to commemorate each one of their birthdays and deaths, and to ensure that the gods answer one's prayers. It is equally important to appease one's deceased relatives. The annual Tomb-Sweeping Festival (*qing ming*) in April is when the whole family cleans up their ancestral grave sites, makes offerings of food, money, and flowers, and burns incense to honor their ancestors.[14]

If they are neglected or offered inadequate amounts of food, money, and respect, they will cause endless problems for their living descendants by coming back to haunt them as ghosts. Even those having trouble with their cars or getting their computer programs to work will drop by the temple to pray to the gods and ancestors—just in case their problems have arisen from giving them inadequate respect. Whole businesses are dedicated to making facsimiles of useful or even luxury items out of paper, such as a car, a house, a computer, an airplane—and these days, paper Viagra as well—which are then brought to a temple and burned, thus sending them to their ancestors for their use.

The seventh month of the lunar calendar is designated "Ghost Month." For the entire month, most do whatever is necessary "to live in harmony with the omnipotent spirits that emerge to roam the world of the living." This includes "preparing doorway altars full of meat, rice, fruit, flowers and beverages as offerings to placate the anxious visitors. Temples [hang] out red lanterns to guide the way for the roving spirits. . . . Ghost money and miniature luxury items made of paper are burned ritualistically for ghosts to utilize along their desperate journey. . . ."[15]

In addition, the people heed a long list of taboos that can have an adverse effect on business during Ghost Month. The real estate industry is particularly hard hit, because people do not dare to move into new houses, out of fear that homeless ghosts might take up permanent residence. Many choose to wait until after Ghost Month to make major purchases, such as cars. Few choose to marry at this time as, according to folk belief, a man might discover that his new bride is actually a ghost! Pregnant women usually choose to undergo Caesarean sections rather than give birth during Ghost Month. And law suits decline because it is believed that ghosts dislike those who bring law suits.[16] (On the mainland of China, the Chinese Communist Party's emphasis on science and the eradication of superstition means that it is far less common there than in Taiwan for people to worry in such a systematic way about propitiating ghosts.)

Finally, there continues to be a preference for seeking medical cures from local temple priests, over either traditional Chinese or modern Western medicine. The concern of local religion is, then, a concern with this life, not with salvation in the afterlife. The attention to deceased ancestors, spirits, and ghosts is quite different from attention to one's own fate in the afterlife.

What is unusual in the case of Taiwanese religious practices is that as the island has become increasingly modern and wealthy, it has not become less religious. Technological modernization has seemingly not brought secularization with it. In fact, aspiring capitalists often build temples in hopes of getting rich. People bring offerings of food; burn incense and bundles of paper money to honor the temple gods; and burn expensive paper reproductions of houses, cars, and whatever other material possessions they think their ancestors might like to have in their ethereal state. They also pay real money to the owner of their preferred temple to make sure that the gods are well taken care of. Since money goes directly to the temple owner, not to a religious organization, the owner of a temple whose constituents prosper will become wealthy. Given the rapid growth in per capita income in Taiwan since 1980, then, temples to local deities have proliferated, as a builder of a temple was almost guaranteed to get rich if its constituents' wealth grew steadily.

Christianity is part of the melange of religions. About 4 percent of the population is Christian; but it is subject to local adaptations, such as setting off firecrackers inside a church during a wedding ceremony to ward off ghosts, and the display of flashing neon lights around the Virgin Mary. The Presbyterian Church in Taiwan, established in Taiwan by missionaries in 1865, was frequently harassed by the KMT because of its activist stance on social and human rights issues and because it generally supported the Taiwan-independence viewpoint.[17] There are more than 1,200 Presbyterian congregations in Taiwan; but in recent years, there has been a 10 percent drop in members—apparently due to the aging church leaders' inflexibility in responding to modernization and ideas from the outside.[18] The Catholic Church in Taiwan is likewise witnessing a decline in membership; and it is also suffering from the aging of its priests, most of whom had emigrated in the 1940s from the China mainland and are now dying off. There is an ever smaller number of young priests in training to replace them.

As for Confucianism, it is more a philosophy than a religion. Confucianism is about self-cultivation, proper relationships among people, ritual, and proper governance. Although Confucianism accepts ancestor worship as legitimate, it has never been concerned directly with gods, ghosts, or the afterlife. In imperial China, if drought brought famine, or if a woman gave birth to a cow, the problem was the lack of morality on the part of the emperor—not the lack of prayer—and required revolt.

When the Nationalists governed Taiwan, they tried to restore Chinese traditional values and to reinstitute the formal study of Confucianism in the schools. Students were apathetic, however, and would usually borrow Confucian texts only long enough to study for college-entrance exams. Unlike the system of getting ahead in imperial China through knowledge of the Confucian classics, students in present-day Taiwan need to excel in science and math. Yet, although efforts to engage students in the formal study of Confucianism have fallen on deaf ears, Confucian values suffuse the culture. The names of streets, restaurants, corporations, and stores are inspired by major Confucian virtues; advertisements appeal to Confucian values of loyalty, friendship, and family to sell everything from toothpaste to computers; children's stories focus on Confucian sages in history; and the vocabulary that the government and party officials use to conceptualize issues is the vocabulary of Confucianism—moral government, proper relationships between officials and the people, loyalty, harmony, and obedience.

EDUCATION

The Japanese are credited with establishing a modern school system in Taiwan in the first part of the twentieth century. After 1949, Taiwan's educational system developed steadily. Today, Taiwan offers nine years of free, compulsory education. Almost all school-age children are enrolled in elementary schools, and most go on to junior high schools. More than 70 percent continue on to senior high school. Illiteracy has been reduced to about 6 percent and is still declining. Night schools that cater to those students anxious to test well and make the cut for the best senior high schools and colleges flourish. Such extra efforts attest to the great desire of Taiwan's students to get ahead through education.

Taiwan has one of the best-educated populations in the world, a major factor in its impressive economic development. Its educational system is, however, criticized for its insistence on uniformity through a unified national curriculum, a lecture format that does not allow for student participation, the grueling high school and university entrance examinations, tracking, rote memorization, heavy homework assignments, and humiliating treatment of students by teachers. Its critics say that the system inhibits creativity.[19] There is also a gender bias in education, which results in women majoring in the humanities and social sciences, while men choose science and math majors. Reforms in recent years have tried to modify some of these practices; and Taiwan's burgeoning information-technology and high-tech sectors have added to pressures to train more women in science and technology.

The number of colleges and universities has more than doubled since martial law was lifted. But the 120 colleges and universities now in existence cannot meet the demand for spaces for all qualified students. As a result, many students go abroad for study. From 1950 to 1978, only 12 percent of some 50,000 students who studied abroad returned, a reflection of both the lack of opportunity in Taiwan and the oppressive nature of government in that period. Beginning in the late 1980s, as Taiwan grew more prosperous and the political system more open, this outward flood of human talent, or "brain drain," was stemmed. Nevertheless, the system of higher education has been unable to keep up with the demand for high-tech workers. This has led to a loosening of restrictions on importing high-tech workers from mainland China.[20]

As the economies of mainland China and Taiwan become more intertwined, the options that Taiwanese have in the educational arena are increasing. A growing number of Taiwan's high school graduates choose to go to a university on the mainland. This is in no small part because the vast majority of Taiwan's businesspeople are investing their money in the mainland economy. However, in spite of the fact that many students from Taiwan attend China's leading universities, which produce outstanding graduates whose degrees are readily accepted in the West, Taiwanese students find that their degrees are not properly honored once they return to Taiwan. Indeed, in a poll conducted in Taiwan in 2000, 40.5 percent of respondents believed that academic credentials earned by Taiwan's students in the mainland should have a stricter standard applied than to students graduating in Taiwan; and 6.3 percent believed that they should not be recognized at all.[21] Students have to take a set of exams upon returning from the mainland to validate the legitimacy of their degrees. It would appear, however, that the real reason is political—namely, to challenge the quality of any institutions under the control of the Chinese Communist rulers. The Ma Ying-jeou government is planning to recognize mainland degrees.

HEALTH CARE AND SOCIAL SECURITY

Although Taiwan's citizens have received excellent healthcare for many years, the healthcare system is now facing a crisis. Healthcare has been a major political issue for candidates running for office. But slower economic growth has produced lower governmental revenues for healthcare. The elderly, moreover, tend to visit their doctors on a weekly, if not a daily, schedule, not only because their visits are virtually free, but also because that's where all the other elderly people are! Lonely, with time on their hands and an obsession with longevity, the elderly often hang out at hospitals and health clinics. This contributes to the financial crisis of the "medicare" system, for doctors, who are paid by the number of patients they see, are more than happy to see dozens of patients each hour. Indeed, citizens complain that their visits usually last less than a minute!

Still, Taiwan's healthcare functions admirably. The one issue that continues to bedevil it is that the PRC continues to block its application for even a nonmember "observer" status in the World Health Organization (WHO). It does so on the grounds that Taiwan is simply a province of China, not a sovereign state. The 2003 SARS (severe acute respiratory syndrome) epidemic in Hong Kong, China, and Southeast Asian countries also broke out in Taiwan. More recently, bird flu has shown up in Taiwan. But because of the lack of either membership or observer status in WHO, Taiwan's ability to receive or contribute important health data was minimal. This situation potentially jeopardizes not only healthcare in Taiwan, but healthcare in the entire world, especially during epidemics.

The Labor Standards Law requires that Taiwanese enterprises provide a pension plan for their employees; but under the old labor standards regulations, 90 percent of employees did not qualify for a pension because they had not worked in the same company for a minimum of 25 years. In 2005, the Labor Pension Retirement Act came into effect, with the result that everyone is now covered. As a result, Taiwan's citizens are not so reliant on their families for support in old age.

WOMEN

The societal position of women in Taiwan reflects an important ingredient of Confucianism. Traditionally, Chinese women "were expected to obey their fathers before marriage, their husbands after, and their sons when widowed. Furthermore, women were expected to cultivate the "Four Virtues": morality, skills in handicrafts, feminine appearance, and appropriate language."[22] In Taiwan, as elsewhere throughout the world, women have received lower wages than men and have rarely made it into the top ranks of government and business—this in spite of the fact that it was women who, from their homes, managed the tens of thousands of small businesses and industries that fueled Taiwan's economic boom.

In the workplace outside the home, women have been treated differently than men. For example, women are not allowed to serve in the armed forces; but until the 1990s, "all female civil servants, regardless of rank, [were] expected to spend half a day each month making pants for soldiers, or to pay a substitute to do this."[23] Women, who make up 40 percent of Taiwan's civil service, find themselves "walking on glue" when they try to move from the lower ranks to the middle and senior ranks of the civil service. By the beginning of 2000, only 12 percent of the total senior level civil service, and less than a third of intermediate ranks, were made up of women. Those figures had not changed at all by the end of 2004.[24] There have, however, been greater opportunities for women in the last decade. Women are more visible in the media and politics than before. In 2000, for the first time, a woman was elected as vice-president (and was re-elected in 2004); but since 1949, not one of the five branches of government has been headed by a woman.

Because women may now receive the same education as men, and because employment in the civil service is now based on an examination system, women's social, political, and economic mobility has increased.[25] Better education of women has been both the cause and the result of greater advocacy by Taiwan's feminists of equal rights for women. It has also eroded the typical marriage pattern, in which a man is expected to marry a woman with an education inferior to his own.

THE ECONOMY

The rapid growth of Taiwan's economy and foreign trade has enriched Taiwan's population. A newly industrialized economy (NIE), Taiwan long ago shed its "Third World," underdeveloped image. Today, the World Bank classifies it as a "high-income" economy. With a gross domestic product per capita income that rose from U.S.$100 in 1951 to U.S.$16,590 (equivalent in purchasing power parity dollars to $30,100) in 2007, and a highly developed industrial infrastructure and service industry, Taiwan sits within the ranks of some of the most developed economies in the world. As with the leading industrial nations, however, the increasing labor costs for manufacturing and industry have contributed to a steady decline in the size of those sectors as companies relocate to countries with cheaper raw materials and lower wages. Taiwan's economic growth rate from 1953 to 1997 had averaged a phenomenal 8-percent-plus annually; but it has slowed down since the Asian financial crisis that began in 1997. In recent years, a growing percentage of Taiwan's economic growth is due to production on the mainland, where most of Taiwan's manufacturing base has moved.

The government elite initiated most of the reforms critical to the growth of Taiwan's economy, including land redistribution, currency controls, central banking, and the establishment of government corporations. Taiwan's strong growth and high per capita income does not, however, bring with them a lifestyle comparable to that in the most developed Western states. The government has had limited success in addressing many of the problems arising from its breathtakingly fast modernization. In spite of—and in some cases because of—Taiwan's astounding economic growth, the quality of life has deteriorated greatly. Taiwan's cities are crowded and badly polluted, and housing is too expensive for most urbanites to afford more than a small apartment. The overall infrastructure is inadequate to handle traffic, parking, electricity, and other services expected in a more economically advanced society.

Massive air and water pollution; an inadequate urban infrastructure for housing, transportation, electricity, and water; and the rapid acquisition of carbon-emitting consumer goods, especially air conditioners, motorcycles, and automobiles, have made the environment unbearable and transportation a nightmare. Complaints of oily rain, ignitable tap water, stunted crops due to polluted air and land, and increased cancer rates abound. "Garbage wars" over the "not-in-my-backyard" issue of sanitary landfill placement have led to huge quantities of uncollected garbage.[26] Numerous public-interest groups have emerged to pressure the government to take action. The government has tried to be responsive, but bitterly divisive politics and rampant corruption have complicated finding solutions to these woes, as everyone tries to get ahead in a now relatively open economy.

Taiwan's economic success thus far may also be attributed to a relatively open market economy in which businesspeople have developed international markets for their products and promoted an export-led economy. Taiwan's highly productive workers have tended to lack class consciousness, because they progress so rapidly from being members of the working-class "proletariat" to becoming capitalists and entrepreneurs. Even factory workers often become involved in small businesses.[27]

Now that Taiwan is privatizing those same government corporations that used to have complete control over many strategic materials as well as such sectors as transportation and telecommunications, workers are resisting the loss of their "iron rice bowl" of permanent employment in state enterprises and the civil service. Much like mainland China, the government is concerned that social instability may result if workers, instead of accepting the international trend toward privatization, resists it through street protests.[28] As an increasing number of industrial workers in Taiwan move into white-collar jobs and are replaced by relatively poorly organized immigrant laborers, who are often sent home when the jobs disappear, this problem seems to have been temporarily sidelined.

Sometimes called "Silicon Island," Taiwan has some 1.2 million small and medium-size enterprises, and only a handful of mega-giants. Most of these smaller enterprises are not internationally recognized names, but they provide the heart and even the backbone of technological products worldwide. They make components, or entire products (such as computer hardware), according to specifications set by other, often well-known, large firms, whose names go on the final product. Furthermore, because Taiwan's firms tend to be small and flexible, they can respond quickly to changes in technology. This is particularly true in the computer industry. Thanks to the many students who have gone to the United States to study and then stayed to work in the computer industry's Silicon Valley, there are strong ties with Taiwan's entrepreneurs.[29] Taiwan's development of the information-technology sector has benefited from governmental incentives and from Taiwan's educated labor force. In this sector, low-wage labor is not yet an issue, and Taiwan has become the leading Asian center for information technology (IT) and software.

A stable political environment facilitated Taiwan's rapid growth. So did Taiwan's protected market, which brought protests over Taiwan's unfair trade policies from those suffering from an imbalance in their trade with Taiwan. Since joining the World Trade Organization (WTO) in 2002, Taiwan has shed most of the regulations that have protected its industries from international trade competition and has come into compliance with international intellectual property rights legislation.

For the economy as a whole, there has been a dramatic turnaround since the 1997–2000 Asian financial crisis. By 2004, GDP growth has moved from negative numbers to a positive 3.2 percent. Taiwan came out of its economic downturn largely by increasing commercial links with China, where Taiwanese businesspeople can get better returns on their investment. There are more than 50,000 Taiwanese factories that have created more than 3 million jobs already in mainland China. A 10 percent of China's IT products that are exported are made in Taiwanese factories on the mainland, and Taiwanese firms control a quarter of China's export licenses. Some 10,000 Taiwanese businesspeople travel to the mainland each day. In 2001, Taiwan's government lifted the limits on investment in China by companies and enterprises from U.S.$5 million to U.S.$50 million. The new Ma Ying-jeou government has lifted more investment restrictions. In short, Taiwan's growth benefits enormously from its investments in China.

Indeed, most of Taiwan's businesspeople are more concerned about the survival of their businesses than about national security vis-à-vis the mainland. As a result, they have been sending delegations to China (without authorization from the government) to reassure its leaders that they will not let Taiwan declare independence and will continue to develop economic ties with the mainland. They believe (as does Beijing) that once the two economies are fully integrated, a declaration of independence of Taiwan will be highly unlikely.[30]

By 2007, Taiwanese had invested well over U.S.$150 billion in China. Because of restrictions on trade with the mainland, however, money must first move to Hong Kong or elsewhere, and only then to China. These sorts of maneuvers complicate business investments and irritate Taiwan's business community. Critics contend that former President Chen Shui-bian failed to fulfill his promises to liberalize current restrictions on travel to and from mainland China. As part of his 2004 election campaign, President Chen promised to open direct air passenger and cargo links, to ease restrictions on travel to and from the mainland, and to liberalize regulations prohibiting Taiwanese from raising capital for their China businesses on the Taiwan Stock Exchange.[31] But when Chen left office little progress had been made in these matters. Many in Taiwan's business community are, in fact, distressed that Chen is not doing more to protect and promote business ties with the mainland (which may explain why the business community generally supported the KMT in the 2004 elections). President Ma has been pushing for closer ties across the Taiwan Strait since he took office in May 2008.

Taiwanese enterprises run what amounts to a parallel economy on the mainland that is completely entangled with China's own. In fact, as of 2003, China had become Taiwan's number one trade partner. By 2006, over 22 percent of Taiwan's export trade was with China. Taiwan has a substantial trade surplus with the mainland. Much of China's trade surplus with the United States is, in fact, from Taiwan's enterprises doing business in China. It is estimated that up to one-third of exported consumer goods labeled 'Made in China' are actually made in firms owned by Taiwan's businesspeople in China. "Analysts attribute more than 70 percent of the growth in America's trade deficit with China to the exports of Taiwanese firms."[32] The result is that, as of December 2010, Taiwan held more than $381 billion in its central reserve bank—the fifth-largest holding of reserve currency and gold in the world.

Taipei has passed legislation so that free-trade zones can be established near Taiwan's international airports and harbors. The purpose of these tax-free areas is to encourage foreign businesspeople, including Chinese from the Mainland, to establish companies in Taiwan for the purpose of trade, processing, and manufacturing.[33] Clearly Taipei is trying both to stem the outward flow of Taiwan's investment moneys and to lure more investors to Taiwan by making its business conditions competitive with those of the mainland, Hong Kong, and Asian countries with free trade zones.

In 2010 Taiwan and mainland China signed a landmark free trade agreement—the Economic Cooperation Framework Agreement, or ECFA. This most significant agreement between the two sides since 1949 has further promoted trade and interactions across the Strait.

Agriculture and Natural Resources

After arriving in Taiwan, the KMT government carried out a sweeping land-reform program: The government bought out the landlords and sold the land to their tenant farmers. The result was equalization of land distribution, an important step in promoting income equalization among the farmers of Taiwan. Today, farmers are so productive that Taiwan is almost self-sufficient in agriculture—an impressive performance for a small island where only 25 percent of the land is arable.

The land-reform program was premised upon one of Sun Yat-sen's famous Three Principles, the "people's livelihood." One of the corollaries of this principle was that any profits from the increase in land value attributable to factors not related to the real value of farmland—such as through urbanization, which makes nearby agricultural land more valuable—would be turned over to the government when the land was sold. As a result, although the price of land has skyrocketed around Taiwan's major cities, and although many farmers are being squeezed by low prices for their produce, they would get virtually none of the increased value for their land if they sold it to developers. Many farmers have thus felt trapped in agriculture. In the meantime, the membership of both China and Taiwan in the World Trade Organization means that cheaper mainland produce flows quite freely into Taiwan. While this has undercut the profits of farmers in such areas as rice, their profits in fruits and other products have benefited by the lifting of trade barriers.

Natural resources, including land, are quite limited in Taiwan. Taiwan's rapid industrialization and urbanization have put a strain on what few resources exist.

(© Photodisc/Getty Images RF)

Taipei 101.

Taiwan's energy resources, such as coal, gas, and oil, are particularly limited. The result is that the government has had to invest in the building of four nuclear-power plants to provide sufficient energy to fuel Taiwan's rapidly modernizing society and economy. Taiwan has been able to postpone its energy and resource crisis by investing in the development of mainland China's vast natural resources. Taiwan's businesspeople have also moved their industries to other countries where resources, energy, land, and labor are cheaper. They now control a vast network of manufacturing and distribution facilities throughout the world.[34] Taiwan's economy has, in short, continued to grow while at the same time avoiding the problem of resource and energy scarcity through extensive participation in the global economy.

Internationalization of its economy is also part of Taiwan's strategy to thwart the PRC's efforts to cut off its relationships with the rest of the world. With Taiwan's important role in the international economy, it is virtually impossible for its trade, commercial, and financial partners to ignore it. This saves Taiwan from even greater international diplomatic isolation than it already faces in light of its current "non-state" status. In the meantime, its economy is becoming increasingly integrated with that of the Chinese mainland, to the mutual benefit of both economies.

Taiwan as a Model of Economic Development

Taiwan is often cited as a model for other developing economies seeking to lift themselves out of poverty. They could learn some useful lessons from certain aspects of Taiwan's experience, such as the encouragement of private investment and labor productivity, an emphasis on basic healthcare and welfare needs, and policies to limit gross extremes of inequality. But Taiwan's special advantages during its development have made it hard to emulate. These advantages include its small size, the benefits of improvements to the island's economic infrastructure and educational system made under the Japanese occupation, massive American financial and technical assistance, and a highly favorable international economic environment during Taiwan's early stages of growth.

What has made Taiwan extraordinary among the rapidly developing economies of the world is the government's ability—and commitment—to achieve and maintain a relatively high level of income equality. Although there are homeless people in Taiwan, their numbers are small. Government programs to help the disabled and an economy with moderate unemployment (3.9 percent in 2006) certainly help, as does a tight-knit family system that supports family members in difficult times. The government's commitment to Sun Yat-sen's principle of the "people's livelihood," or what in the West might be called a "welfare state," is still an important consideration in policy formation.

Like China's two stock markets, the regulatory regime of Taiwan's Stock Exchange Corporation (TSEC) is unreliable. Trading in Taiwan's stock market is highly speculative, and foreigners find it difficult to buy shares. The growth of Taiwan's stock market—a market often floating on the thin air of gossip and rumor—has created (and destroyed) substantial wealth almost overnight. Nevertheless, Taiwan's economic wealth remains fairly evenly distributed, contributing to a strongly cohesive social system.

At the same time, Taiwan's rapid economic growth rate, combined with a relatively low birth rate, has led businesses to import foreign laborers to do unskilled jobs that Taiwan's better-paid residents refuse to do. These foreign workers, largely from Thailand, the Philippines, and Indonesia, work at wages too low, with hours too long and conditions too dangerous for Taiwan's own citizens. By the end of 2001, there were more than 304,000 legal foreign workers in Taiwan, and a growing number of illegal foreign workers.

Today, one in five marriages is to someone from outside Taiwan, primarily from mainland China or Southeast Asian countries; and one in seven children is born in these marriages. Thus, the population of Taiwan, which used to be divided primarily

SUN YAT-SEN: THE FATHER OF THE CHINESE REVOLUTION

Sun Yat-sen (1866–1925) was a charismatic Chinese nationalist who, in the declining years of the Manchu Dynasty, played upon Chinese nationalist hostility both to foreign colonial powers and to the Manchu rulers themselves.

Sun (pictured at the right) drew his inspiration from a variety of sources, usually Western, and combined them to provide an appealing program for the Chinese. This program was called the Three People's Principles, which translates the phrase in the Gettysburg Address "of the people, by the people, and for the people" into "nationalism," "democracy," and the "people's livelihood."

This last principle, the "people's livelihood," is the source of dispute between the Chinese Communists and the Chinese Nationalists, both of whom claim Sun Yat-sen as their own. The Chinese Communists believe that the term means socialism, while the Nationalists in Taiwan prefer to interpret it to mean the people's welfare in a broader sense.

Sun Yat-sen is, in any event, considered by all Chinese to be the father of the Chinese Revolution of 1911, which overthrew the enfeebled Manchus. He thereupon declared China to be a republic and named himself president. However, he had to relinquish control immediately to the warlord Yuan Shih-K'ai, who was the only person in China powerful enough to maintain control over all other contending military warlords in China.

When Sun died in 1925, Chiang Kai-shek assumed the mantle of leadership of the Kuomintang, the Chinese Nationalist Party. After the defeat of the KMT in 1949, Sun's widow chose to remain in the People's Republic of China and held high honorary positions until her death in 1982.

(Library of Congress Prints and Photographs Division (LC-USZ62-5972))
Sun Yat-Sen: father of Chinese revolution (1866–1925).

along Mainlander-Taiwanese lines, has a fairly substantial immigrant component. Immigrants are still often looked down upon because they are seen as marrying Taiwanese to get ahead economically, and because they tend to come from relatively impoverished backgrounds. Some marry Taiwanese men who are from lower economic and social classes and may even be unemployed, or too sick or old to work. But these immigrant spouses, over 90 percent of whom are women, are essential to a society suffering from the same problems as Mainland China (from which so many of the brides come): a surplus of men of marriageable age, and a rapidly graying population because of a low birthrate.[35]

THE POLITICAL SYSTEM

From 1949 to 1988, the KMT justified the unusual nature of Taiwan's political system with three extraordinary propositions. First, the government of the Republic of China, formerly located on the mainland of China, was merely relocated temporarily on China's island province of Taiwan. Second, the KMT was the legitimate government not just for Taiwan but also for the hundreds of millions of people living on the Chinese mainland under the control of the Chinese Communist Party.[36] Third, the people living under the control of the Communist "bandits" would rush to support the KMT if it invaded the mainland to overthrow the Chinese Communist Party regime. Taiwan's political and legal institutions flowed from these three unrealistic propositions, which reflected hopes, not reality. Underlying all of them was the KMT's acceptance, in common with the Chinese Communist Party, that there was only one China and that Taiwan was a province of that one China. Indeed, until the early 1990s, it was a *crime* in Taiwan to advocate independence.

The Constitution

In 1946, while the KMT was still the ruling party on the mainland, it promulgated a "Constitution for the Republic of China." This Constitution took as its foundation the same political philosophy as the newly founded Republic of China adopted in 1911 when it overthrew China's Manchu rulers on the mainland: Sun Yat-sen's "Three People's Principles" ("nationalism," "democracy," and the "people's livelihood"). Democracy was, however, to be instituted only after an initial period of "party tutelage." During this period, the KMT would exercise virtually dictatorial control while in theory preparing China's population for democratic political participation.

The Constitution provided for the election of a National Assembly (a sort of "electoral college"); a Legislative Yuan ("branch") to pass new laws, decide on budgetary matters, declare war, and conclude treaties; a Judicial Yuan to interpret the Constitution and all other laws and to settle lawsuits; an Executive Yuan to run the economy and manage the country generally; a Control Yuan, the highest supervisory organ, to supervise officials through its powers of censure, impeachment, and auditing; and an Examination Yuan (a sort of personnel office) to conduct civil-service examinations. The Examination Yuan and Control Yuan were holdovers from Chinese imperial traditions dating back thousands of years.

Because this Constitution went into effect in 1947 while the KMT still held power on the mainland, it was meant to be applicable to all of China, including Taiwan. The KMT government called nationwide elections to select delegates for the National Assembly. Then, in 1948, it held elections for representatives to the Legislative Yuan and indirect elections for members of the Control Yuan. Later in 1948, as the Civil War between Communists and Nationalists on the mainland raged on, the KMT government amended the Constitution to allow for the declaration of martial law and a suspension of regular elections; for by that time, the Communists were taking control of vast geographical areas of China. Soon afterward, the Nationalist government under Chiang Kai-shek fled to Taiwan. With emergency powers in hand, it was able to suspend elections and all other democratic rights afforded by the Constitution.

By October 1949, the Communists had taken control of the Chinese mainland. As a result, the KMT, living in what it thought was only temporary exile in Taiwan, could not hold truly "national" elections for the National Assembly or for the Legislative and Control Yuans as mandated by the 1946 Constitution. But to foster its claim to be the legitimate government of all of China, the KMT retained the 1946 Constitution and governmental structure, as if the KMT alone could indeed represent all of China. With "national" elections suspended, those individuals elected in 1947 from all of China's mainland provinces (534 out of a total 760 elected had fled with General Chiang to Taiwan) continued to hold their seats in the National Assembly, the Legislative Yuan, and the Control Yuan—usually until death—without standing for re-election. Thus began some 40 years of a charade in which the "National" Assembly and Legislative Yuan in Taiwan pretended to represent all of China. In turn, the government of the island of Taiwan pretended to be a mere provincial government under the "national" government run by the KMT. At no time did the KMT government suggest that it would like Taiwan to declare independence as a state.

Although the commitment to retaking the China mainland was quietly abandoned by the KMT government even before Chiang Kai-shek's death, the 1946 Constitution and governmental structure remained in force. This was in spite of the fact that over those many years, numerous members of the three elected bodies died. Special elections were held just to fill their vacant seats. The continuation of this atavistic system raised serious questions about the government's legitimacy. The Taiwanese, who comprised more than 80 percent of the population but were not allowed to run for election to the Legislative Yuan or National Assembly, accused the KMT Mainlanders of keeping a stranglehold on the political system and pressured them for greater representation. Because the holdovers from the pre-1949 period were of advanced age and often too feeble to attend meetings of the Legislative Yuan (and some of them no longer even lived in Taiwan), it was virtually impossible to muster a quorum. Thus, in 1982, the KMT was forced to "reinterpret" parliamentary rules in order to allow the Legislative Yuan to get on with its work.

Chiang Ching-kuo, Chiang Kai-shek's son and successor, decided to concede reality and began bringing Taiwanese into the KMT. By the time a Taiwanese, Lee Teng-hui, succeeded Chiang Ching-kuo in 1988 as the new Nationalist Party leader and president of the "Republic of China," 70 percent of the party's membership was Taiwanese. Pressures therefore built for party and governmental reforms that would diminish the power of the old KMT Mainlanders. In July 1988, behind the scenes at the 13th Nationalist Party Congress, the leadership requested the "voluntary" resignation of the remaining pre-1949 holdovers. Allegedly as much as U.S.$1 million was offered to certain hangers-on if they would resign, but few accepted. Finally, the Supreme Court forced all those Chinese mainland legislators who had gained their seats in the 1948 elections to resign by the end of 1991.

Under the Constitution, the Legislative, Judicial, Control, and Examination branches hold certain specific powers. Theoretically, this should result in a separation of powers, preventing any one person or institution from the arbitrary abuse of power. In fact, however, none of these branches of government exercised much, if any, power independent of the KMT or the president who was chosen by the KMT until after the first completely democratic legislative elections of December 1992. In short, the KMT and the government were merged, in much the same way as the CCP was merged with the government on the mainland. Indeed, property and state enterprises owned by the government were claimed by the KMT as their own property when they lost control of the presidency in 2000.

Thanks to changes made by President Lee, however, Taiwan now has a far greater separation of powers as well as a two-headed government: The president is primarily responsible for Taiwan's security, and the prime minister (premier) is responsible for the economy, local government, and other broad policy matters. But problems remain because the president may appoint the prime minister without approval by the Legislature—yet the Legislature has the power to cast a no confidence vote and require new elections if it finds the government's actions unacceptable.

A final consequence of the three propositions upon which political institutions in Taiwan were created was that, after the KMT arrived in 1947, Taiwan maintained two levels of government. One was the so-called national government of the "Republic of China," which ruled Taiwan as just one province of all of China. The other was the actual provincial government of Taiwan, which essentially duplicated the functions of the "national" government, and reported to the "national" government, which for so many years pretended to represent all of China. In this provincial-level government, however, native Taiwanese always had considerable control over the actual functioning of the province of Taiwan in all matters not directly related to the Republic of China's relationship with Beijing. Taiwan's provincial government thus became the training ground for native Taiwanese to ascend the political ladder once the KMT reformed the political system. The Taiwan provincial government was officially "frozen" and essentially abolished in 1998.

Martial Law

The imposition of martial law in Taiwan from 1948 to 1987 is critical to understanding the dynamics of Taiwan's politics. Concerned with the security of Taiwan against subversion, or an invasion by the Chinese Communists, the KMT government had imposed martial law on Taiwan. This allowed the government to suspend civil liberties and limit political activity, such as organizing political parties or mass demonstrations. Thus it was a convenient weapon for the KMT Mainlanders to control potential Taiwanese resistance and to quash any efforts to organize a "Taiwan independence" movement. Police powers were widely abused, press freedoms were sharply restricted, and dissidents were jailed. As a result, the Taiwan Independence Movement was forced to organize abroad, mostly in Japan and the United States. Taiwan was run as a one-party dictatorship supported by the secret police.

Non-KMT candidates were eventually permitted to run for office under the informal banner of *tangwai* (literally, "outside the party"); but they could not advocate independence for Taiwan; and they had to run as individuals, not as members of opposition political parties, which were forbidden until 1989. The combination of international pressures for democratization, the growing confidence of the KMT, a more stable situation on the China mainland, and diminished threats from Beijing led the KMT to lift martial law in July 1987. Thus ended the state of "Emergency" under which almost any governmental use of coercion against Taiwan's citizens had been justified.

Civil Rights

Until the late 1980s, the rights of citizens in Taiwan did not receive much more protection than they did on the Chinese mainland. The R.O.C. Constitution has a "bill of rights," but most of these civil rights never really existed until martial law was lifted in 1987. Civil rights were repeatedly suspended when their invocation by the citizenry challenged KMT power or policies; and opposition political parties were not allowed to organize. Because the "Emergency" regulations provided the rationale for the restriction of civil liberties, the KMT used military courts (which do not accord basic protection of defendants' civil rights) to try what were actually civil cases,[37] arrested political dissidents, and used police repression, such as in the brutal confrontation during the 1980 Kaohsiung Incident.[38]

The 2002 survey by the Taipei-based Chinese Association for Human Rights, as well as the 2004 U.S. Department of State report on Taiwan's human rights indicated that the protection of judicial rights had improved somewhat. There was some improvement in the interrogation of suspects, but protection of suspects in police custody remained inadequate. Some cases of physical abuse of persons in police custody were reported, usually when lawyers were not present for the interrogation, or when the interrogation was not audio- and videotaped, as required by law. In 2003, a law was enacted to limit police powers. Search warrants and police raids of businesses suspected of illegal activities now require stricter proof of "probable cause." Furthermore, if police have failed to follow due process and therefore infringed on a suspect's rights, the suspect can immediately file an administrative appeal. In recent years, no reports of politically motivated disappearances or deaths have been made; and claims of unlawful dentention and arrest are rare.[39] Political rights (civil rights and freedom, equality, democratic consolidation, and media independence and objectivity) have progressed, if modestly. Improvement in the protection of women's rights has been attributed to better implementation of the Domestic Violence Prevention Law, and the promulgation of the Gender Equality Labor Law in 2002. The latter is aimed at eradicating sexual harassment and discrimination, requires employers to offer up to two years' paid maternity leave, and embodies principles of equal pay for equal work, and equal rights in places of employment. Taiwan also made progress in rights for children, elderly, and the handicapped.[40]

Political Reform in Taiwan

The Kuomintang maintained its political dominance until the turn of the century in part by opening up its membership to a broader segment of the population. It thereby allowed social diversity and political pluralism to be expressed *within* the KMT. The "Taiwanization" of both the KMT and governmental institutions after Chiang Kai-shek's death in 1975 actually permitted the KMT to ignore the demands of the Taiwanese for an independent opposition party until the late 1980s. The KMT also hijacked the most appealing issues in the platform of the Democratic Progressive Party—notably, the DPP's demand for more flexibility in relations with the P.R.C., environmental issues, and greater freedom of the press. By the time of the 1996 elections, the dominant wing of the KMT had even co-opted the DPP's platform for a more independent Taiwan. In short, as Taiwan became more socially and politically diverse, the KMT relinquished many of its authoritarian methods and adopted persuasion, conciliation, and open debate as the best means to maintain control.[41]

External pressures played a significant role in the democratization of Taiwan's institutions. American aid from the 1950s to the 1970s was accompanied by considerable American pressure for liberalizing Taiwan's economy, but the United States did little to force a change in Taiwan's political institutions during the Cold War, when its primary concern was to maintain a defense alliance with Taiwan against the Chinese Communists. Taiwan's efforts to bolster its integration into the international economy has allowed it to reap the benefits of internationalization. The government has also, after much prodding, responded positively to demands from its citizens for greater economic and cultural contact with China,[42] and for reform of the party and government. As a result, the KMT was able to claim responsibility for Taiwan's prosperity as well as for its eventual political liberalization; but as noted elsewhere, it must also take responsibility for the many problems it left on the platter for the incoming DPP administration in 2000.

The KMT's success in laying claim to key elements of the most popular opposition-party policies forced the opposition to struggle to provide a clear alternative to the KMT. Apart from the issue of Taiwan independence, when the DPP was acting as the opposition party it demanded more rapid political reforms and harshly criticized the KMT's corrupt practices. The DPP's exposure of the KMT's corruption, and of the infiltration into the political system by criminal organizations, brought public outrage. Over the decades of its rule, the KMT used money and other resources to help tie local politicians and factions to itself. The KMT's power at the grassroots level was lubricated through patronage, vote buying, and providing services to constituents.[43] The public demanded that the interweaving of political corruption, gangsterism, and business (referred to as "black and gold politics" in Taiwan) be brought under control, and that the KMT divest itself of corporate holdings that involve conflicts of interest and permit it to engage in corrupt money politics. (When it also controlled the government, the KMT *as a political party* possessed an estimated U.S.$2.6 billion of business holdings, a sizable percentage of corporate wealth in Taiwan.) According to statistics published in the *Taipei Times,* "two-thirds of gangs in Taiwan have lawmakers running on their behalf in the legislature, while one-quarter of elected public representatives have criminal records. . . . [T]hey are invulnerable because, as the KMT's legislative majorities have slowly declined, the ruling party needs the support of independent lawmakers—including those with organized crime backgrounds—to pass its legislative agenda."[44]

By the time of the first democratic elections for the president of Taiwan in 1996, much had changed in the platforms of both the KMT and the DPP. The KMT, which had by then developed a powerful faction within it demanding greater international recognition of Taiwan as an independent state, adopted what amounted to an "independent Taiwan" position. Although President Lee Teng-hui stated this was a "misinterpretation" by Beijing and the international community, and that Taiwan merely wanted more international "breathing space," his offer of U.S.$1 billion to the United Nations if it would give Taiwan a seat was hardly open to interpretation.

Angered by Lee's efforts to gain greater recognition for Taiwan as an independent state, China began missile "tests" in the waters close to Taiwan in the weeks leading up to the 1996 elections. Fortunately, none of the missiles accidentally hit Taiwan. As a result of Beijing's saber-rattling and pressures from the United States, the KMT had, by the time of the elections, retreated from its efforts to gain greater recognition as an independent state. Many members of the KMT regretted Lee's pushing for an independent Taiwan. By the 2000 election, moreover, many of those who did favor independence left the KMT to form a new party. In the 2004 legislative elections, the KMT ran on a platform rejecting independent statehood and supporting a positive and stable relationship with China to protect their business interests. It managed to keep the DPP and other parties sympathetic to declaring independence from winning a majority in the Legislative Yuan.

Taiwan's large middle class, with its diverse and complex social and economic interests arising from business interests and ownership of private property, has been a catalyst for political liberalization. Moreover, Taiwan has not suffered from vast economic disparities that breed economic and social discontent. The government's success in developing the economy meant at the least that economic issues did not provide fuel for political grievances. Thus, when martial law ended in 1987, the KMT could undertake political reform with some confidence. Its gradual introduction of democratic processes and values undercut much of its former authoritarian style of rule without its losing political power for more than a decade. Reform did generate tensions, but by the 1990s the KMT realized that street demonstrations would not bring down the government and that suppressing the opposition with harsh measures was unnecessary. This being the case, the KMT liberalized the political realm still further. Today, Taiwan's political system functions in most respects as a democracy; but as the KMT is the first to admit, it was so busy democratizing Taiwan that it neglected to democratize itself. Once it lost the presidency in the 2000 elections, and only narrowly remained in control of the Legislative Yuan, it rethought its political platform and tried to rid itself of the serious corruption and elitism that has alienated Taiwan's voters.[45] This may partially explain why in the 2004 legislative elections, the KMT, together with its political allies (the "Pan-Blue" alliance), was able to retain a narrow margin of control in the legislature.

Political Parties

Only in 1989 did the KMT pass laws legalizing opposition political parties. The Democratic Progressive Party, a largely Taiwanese-based opposition party, was for the first time recognized as a legal party. As with other political reforms, this decision was made in the context of a growing resistance to the KMT's continued restriction of democratic rights. Even after 1989, however, the KMT continued to regulate political parties strictly, in the name of maintaining political and social stability.

Angry factional disputes within both the DPP and the KMT have marred the ability of each party to project a unified electoral strategy. The KMT was particularly hurt by vitriolic disputes between pro-unification and pro-independence factions, and between progressive reformers who pushed for further liberalization of the economic and political systems, and conservative elements who resisted reform. These internal conflicts explain the KMT's inability to move forward on reform and its defeat in the presidential elections of 2000 and 2004.

((AP Photo/Wally Santana) (10070108333))
Ma Ying-jeou, ROC president since May 2008 and KMT chairman since October 2009.

In their first years in the Legislature, DPP legislators, no doubt frustrated by their role as a minority party that could not get through any of its own policies, sometimes engaged in physical brawls on the floor of the Legislature, ripped out microphones, and threw furniture. As the DPP steadily gained more power and influence over the legislative agenda, its behavior became more subdued, but its effectiveness was undermined by internal factionalism.[46] Poor performance, corruption, and internal power struggle contributed to the loss of the DPP in the 2008 presidential election.

Infighting became so serious in both parties that it led to the formation of three new breakaway parties, which competed for the first time in the 1998 elections. The Taiwan Independence Party, whose members had comprised the more radical faction within the DPP, refused to be intimidated by Beijing's possible military response to a public declaration of independence. Today, it has been renamed the Taiwan Solidarity Union. Within the KMT, members of the faction angered with the KMT leadership's slow-footedness in bringing about reunification with the mainland of China and its liberalizing reforms of the KMT, broke off to form the New Party, which first ran candidates in the 1994 elections. Its bitter disputes with former colleagues in the KMT made legislative consensus difficult. By the 2000 presidential elections, angry debates over who should lead the KMT led to the expulsion of James Soong, who then formed the People's First Party. Soong's effective campaign, which split the KMT vote, contributed to the victory of the DPP candidate, with Soong himself placing a close second and the KMT's candidate a distant third.[47] Today, the KMT, the New Party, and the People's First Party are in the "Pan-Blue" alliance in the Legislature, but are deeply divided on many issues.

Since the 1996 election, the most divisive issue has been whether or not to press for an independent Taiwan. Polls have indicated that although the Taiwanese people would like Taiwan to be independent, only a small minority has been willing to incur the risks of Beijing using force against Taiwan if the government were to endorse independence as a stated policy goal. The preference is for candidates who have promised a continuation of the status quo—namely, not openly challenging Beijing's stance that Taiwan is a province of China—but who favor Taiwan continuing to act as if it is an independent sovereign state.

The preconditions for reunification that the KMT set when it was in power are not likely to be met easily: democratization of the mainland to a (unspecified) level acceptable to Taiwan, and an (unspecified) level of economic development that would move the mainland closer to Taiwan's level of development. Symbolically, the most important step Chen Shui-bian made in advancing "the three links" (trade, postal, and transportation links) between mainland China and Taiwan was allowing direct flights to accommodate citizens at both ends to visit the other over the Chinese New Year's period. Although direct flights were already permitted in 2003, there has been a huge leap in the details: now commercial airliners from *both* sides (not just Taiwan) may take passengers across the Taiwan Strait; passengers may board in multiple cities on both sides; and China's planes only had to fly *through the air space,* not land in, either Hong Kong or Macao on their way to Taiwan, saving considerable time. Extremists condemned Chen even for this modest concession as selling out Taiwan's interests and taking further steps toward unification; but members of the KMT have taken far bolder steps to improve relations and economic integration with the Mainland. In an effort to promote cross-Strait relations, KMT chairman Lien Chan and People First Party chairman James Soong visited Chinese mainland in 2005. Both were received by CCP general secretary Hu Jintao.

THE U.S. SEVENTH FLEET HALTS INVASION

In 1950, in response to China's involvement in the Korean War, the United States sent its Seventh Fleet to the Taiwan Strait to protect Taiwan and the Offshore Islands of Quemoy and Matsu from an invasion by China. Because of improved Sino–American relations in the 1970s, the enhanced Chinese Nationalist capabilities to defend Taiwan and the Offshore Islands, and problems in the Middle East, the Seventh Fleet was eventually moved out of the area. The U.S.S. Ingersoll, a part of the Seventh Fleet, is shown at right. In 1996, however, part of the Seventh Fleet briefly returned to the Taiwan Strait when China threatened to use military force against Taiwan if its leaders sought independent statehood.

(Courtesy of U.S. Navy (USN_DD-652))

USS Ingersoll DD 652.

Money politics, vote-buying practices, and general dishonesty have plagued all of Taiwan's parties. In fact, they have burgeoned over the years, in part because of the growing importance of the legislature; for now that the Legislative Yuan is no longer a body of officials who fled the mainland in 1949 but a genuinely elected legislature with real power to affect Taiwan's policies, who wins really matters. As a result, candidates throw lavish feasts, make deals with business people, and spread money around in order to get the vote. Equally disturbing has been election-related violence, and the influence of the "underworld" on the elections.[48] The Chen administration has been plagued with corruption scandals in its second term, starting with what many commentators allege was a staged assassination attempt on the eve of the 2004 president election, to allegations throughout Chen's second term of family members engaging in insider trading and diverting public funds for private use. In 2006, there were two recall votes in the Legislative Yuan. Both failed to gain the two-thirds majority vote required to force Chen out of the presidency, but some opinion polls suggested that the majority of Taiwan's citizens favored him stepping down from the presidency before the end of his term in 2008.[49]

The positions of the KMT-led "pan-blue alliance," and the DPP-led "pan-green alliance" are far apart on the issue of Taiwan independence. The pan-blues do not support any policy that would risk a military confrontation with the PRC, and they are actively promoting "engagement" with China's officials. By contrast, the pan-greens have been highly confrontational, although there is considerable dissent about this strategy within its ranks. In part, this is because the people of Taiwan, regardless of how much they would like Taiwan to become an independent state, don't want to risk war. On social and economic issues, there is less difference between the two major parties. Both are also committed to democracy, advocate capitalism, and support an equitable distribution of wealth.

Finally, although Chinese Mainlanders are almost all within the KMT or the New Party, the KMT has become so thoroughly "Taiwanized" that the earlier clear divide between the DPP and KMT based on Mainlander or Taiwanese identity has eroded considerably. Even those who strongly favor reunification with the mainland have for many years identified themselves not as Mainlanders but as "new Taiwanese." They are, in short, "born-again Taiwanese."[50]

Interest Groups

As Taiwan has become more socially, economically, and politically complex, alternative sources of power have developed that are independent of the government. Economic interest groups comprised largely of Taiwanese, whose power is based on wealth, are the most important; but there are also public interest groups that challenge the government's policies in areas such as civil rights, the environment, women's rights, consumer protection, agricultural policy, aborigine rights, and nuclear power. Even before the lifting of martial law in 1987, these and other groups were organizing hundreds of demonstrations each year to protest government policy.

On average, every adult in Taiwan today belongs to at least one of the thousands of interest groups. They have been spawned by political liberalization and economic growth, and in turn add to the social pluralism in Taiwan. They are also important instruments for democratic change. Taiwan's government, has then, successfully harnessed dissent since the 1990s, in part by allowing an outlet for dissent through the formation of interest groups—and opposition parties.

Mass Media

With the official end to martial law, the police powers of the state were radically curtailed. The media abandoned former taboos and grew more willing to openly address social and political problems, including corruption and the abuse of power. A free press, strongly critical of the government and *all* the political parties, now flourishes. Taiwan, with 23 million people, boasts close to 4,000 magazines; about 100 newspapers, with a total daily circulation of 5 million; 150 news agencies, several with overseas branches; 29 television broadcasting stations, which are challenged by 75 cable stations that offer some 120 satellite channels; and more than 551 radio stations, which now include foreign broadcasts such as CNN, NHK (from Japan), and the BBC. More than 3,000 radio and 3,000 television production companies are registered in Taiwan. Although television and radio are still controlled by the government, they have become far more independent since 1988. About 75 percent of households buy a basic monthly package that gives them access to more than 70 satellite channels, many of which are operated by foreign companies. Programs from all over the world expose people to alternative ideas, values, and lifestyles, and contribute to social pluralism. Political magazines, which are privately financed and therefore not constrained by governmental financial controls, have played an important role in undercutting state censorship of the media and developing alternative perspectives on issues of public concern. New technology that defies national boundaries (including satellite broadcasts from Japan and mainland China), cable television, the Internet, and VCRs are diminishing the relevance of the state monopoly of television.[51]

THE TAIWAN–P.R.C.–U.S. TRIANGLE

From 1949 until the 1960s, Taiwan received significant economic, political, and military support from the United States. Even after it became abundantly clear that the Communists effectively controlled the China mainland and had legitimacy in the eyes of the people, the United States never wavered in its support of President Chiang Kai-shek's position that the R.O.C. was the legitimate government of all of China. U.S. secretary of state Henry Kissinger's secret trip to China in 1971, followed by President Richard M. Nixon's historic visit in 1972, led to an abrupt change in the American position and to the final collapse of the R.O.C.'s diplomatic status as the government representing "China."

Allies of the United States, most of whom had loyally supported its diplomatic stance on China, soon severed diplomatic ties with Taipei, a necessary step before they could, in turn, establish diplomatic relations with Beijing. Only one government could claim to represent the Chinese people; and with the KMT in complete agreement with the Chinese Communist regime that there was no such thing as "two Chinas" or "one Taiwan and one China," the diplomatic community had to make a choice between the two contending governments. Given the reality of the Chinese Communist Party's control over one billion people and the vast continent of China, and, more cynically, given the desire of the business community throughout the world to have ties to China, Taipei has found itself increasingly isolated in the world of diplomacy. It should be noted, however, that even when the R.O.C. had represented "China" in the United Nations, and at the height of the diplomatic isolation of the People's Republic of China from 1950 to 1971, Taipei was never officially recognized by any of its neighbors in Southeast Asia (unless they were in a defense alliance with the United States)—even though they distrusted China. That this was the case even at the height of China's unpopularity in the region was a bad omen for Taipei's dream of obtaining international legitimacy."[52]

Eventually the United States made the painful decision to desert its loyal Cold War ally, a bastion against communism in Asia, if not exactly an oasis of democracy. The United States had, moreover, heavily invested in Taiwan's economy. But on

(Republic of China (ROC) Taiwan Navy)

The ROC's naval exercise.

January 1, 1979, President Jimmy Carter announced the severing of diplomatic relations with Taipei and the establishment of full diplomatic relations with Beijing.

Taiwan's disappointment and anger at the time cannot be overstated, in spite of the fact that an officially "unofficial" relationship took its place. American interests in Taiwan are overseen by a huge, quasi-official "American Institute in Taiwan"; while Taiwan is represented in the United States by multiple branches of the "Taipei Economic and Cultural Office." In fact, the staffs in these offices continue to be treated in most respects as if they are diplomatic personnel. Except for the 23 countries that officially recognize the R.O.C. (primarily because they receive large amounts of aid from Taiwan), Taiwan's commercial, cultural, and political interests are represented abroad by these unofficial offices.

The United States's acceptance of the Chinese Communists' "principled stand" that Taiwan is a province of China and that the People's Republic of China is the sole legal government of all of China, made it impossible to continue to maintain a military alliance with one of China's provinces. Recognition of Beijing, therefore, required the United States to terminate its mutual defense treaty with the R.O.C. In the Taiwan Relations Act of 1979, the United States stated its concern for the island's future security, its hope for a peaceful resolution of the conflict between Taiwan's government and Beijing, and its decision to put a moratorium on the sale of new weapons to Taiwan.

Renewal of Arms Sales

In January 2010, the United States sold $6.4 billion worth of weapons to Taiwan, based on the Taiwan Relations Act. The Taiwan Relations Act was, however, largely ignored by the administration of President Ronald Reagan. Almost immediately upon taking office, it announced its intention to resume U.S. arms sales to Taiwan. The administration argued that Taiwan needed its weapons upgraded in order to defend itself. Irate, Beijing demanded that, in accordance with American agreements and implicit promises to China, the United States phase out the sale of military arms over a specified period. The U.S. has never actually agreed to this, but because of the gridlock in Taiwan's Legislative Yuan, in effect, U.S. arms sales have come to a halt (see below). Nevertheless, the fact that the U.S. *wants* to sell arms to Taiwan is a constant source of tensions with China. Similarly, the U.S. Congress's proposed Taiwan Security Enhancement Act in 2000—which, had it passed, would have amounted to a military alliance with Taiwan—and the possibility that Congress may authorize "theater missile defense" for the island, have been major irritants to the U.S.–China relationship.

When it took office in January 2001, the George W. Bush administration immediately stated its intention to go forward with deepening military ties with Taiwan. Tensions with Beijing generated by the U.S. sales of military equipment to Taiwan are aggravated by China's own sales of military equipment, such as medium-range missiles to Saudi Arabia, Silkworm missiles to Iran (used against American ships), nuclear technology to Pakistan,[53] and massive sales of semiautomatic assault weapons to the United States (one of which was used to attack the White House in 1994). These sales have undercut the P.R.C.'s standing on the moral high ground to protest American sales of military equipment to Taiwan. But views even within Taiwan concerning the purchase of U.S. military weapons and equipment are complicated. The DPP has argued that they are necessary to protect Taiwan against an attack by China. The KMT is, however, firmly opposed to further purchases, considering them a waste of money; for in the event of an attack, its defense would depend almost entirely on the United States—assuming that the Americans decided to come to its defense. Even worse, the KMT argues, additional arming of Taiwan would accelerate the arms race with China, and further destabilize the Taiwan Straits. The KMT has even accused President Chen of wanting greater military power as part of an effort to lay the groundwork for declaring Taiwan's independence.

This is a bizarre twist to the issue of Taiwan's defense; for in the past, the United States and other countries have repeatedly backed out of proposed arms deals in the face of Beijing's threatened punitive measures. Now it is Taiwan that is backing out. U.S. policymakers are frustrated with Taiwan's falling military expenditures and the perception that Taiwan's defense readiness has declined as a result.[54] But most analysts agree that, with or without arms purchases from the U.S., Taiwan knows that its own defenses would be overwhelmed by a military onslaught from the mainland. As to the U.S. position on what it would do if Taiwan were attacked by China, it remains one of "strategic ambiguity"; for even though President Bush in 2001 stated the U.S. would come to Taiwan's aid if it were attacked, since that time, the administration has made it clear to Taiwan's president that the U.S. will not be pulled into a war with China to defend Taiwan if it declares independence. The U.S. also has to consider that, with more than 120,000 troops in Iraq, with others tied down in Afghanistan, and with no end in sight, the U.S. simply does not have the resources to participate in a conflict with China. This does not stop the U.S. military from portraying the Chinese buildup across the Taiwan Strait as a grave threat and, on this basis, requesting more of the budget for "defense."

The United States wants Beijing to agree to the "peaceful resolution of the Taiwan issue." Meanwhile, the PRC claims it will peacefully resolve the Taiwan issue under the "one country two systems" policy, but it will not rule out the use of force in case Taiwan declares formal independence. China insists that Taiwan is an "internal" affair, not an international matter over which other states might have some authority. From China's perspective, then, it has the right as a sovereign state to choose to use force to settle the Taiwan issue. There is general recognition, however, that the purpose of China's military buildup across from Taiwan is not as much to attack Taiwan as to prevent it from declaring independence as a sovereign state.[55] Indeed, apart from a mild statement from Japan protesting China's "testing" of missiles over Taiwan in 1996, no Asian country has questioned China's right to display force when Taiwan pressed for independent statehood. Still, Beijing knows that American involvement may be critical to getting Taiwan to agree to unification with the mainland. One thing is certain: the day Beijing no longer threatens to use force against Taiwan if it declares independence, Taipei *will* declare independence. So we can expect Beijing to continue its bluster about using military force until Taiwan is reunified with the mainland.

Minimally, Taiwan wants the United States to insist that any solution to the China unification issue be *acceptable to the people of Taiwan.* One possible solution would be a confederation of Taiwan with the mainland: Taiwan would keep its "sovereignty" (that is to say, govern itself and formulate its own foreign policy), but China could say there was only one China. An interim solution bandied about by the KMT in 2006–2007 is that Taiwan would promise not to declare independence for 50 years, and Beijing would promise not to use force to gain control of Taiwan for 50 years. Then, in that context, both sides would go about their business of furthering the integration of the two economies and deepening cultural ties. Neither this proffered solution, or one that promotes the idea of a Chinese commonwealth for the relationship of Taiwan to China, will likely be adopted any time soon.[56]

CHINA'S "PEACE OFFENSIVE"

Since the early 1980s, Beijing has combined its threats and warnings about Taiwan's seeking independence with a "peace offensive." This strategy aims to draw Taiwan's leaders into negotiations about the future reunification of Taiwan with the mainland. Beijing has invited the people of Taiwan to visit their friends and relatives on the mainland and to witness the progress made under Communist rule. Many Taiwanese have traveled to the mainland. In turn, a mere trickle of Chinese from the mainland has been permitted by Taipei to visit Taiwan. The government did allow mainland Chinese students studying abroad to come for "study tours" of Taiwan. They have treated them as if they were visiting dignitaries, and the students usually returned to their universities full of praise for Taiwan. They also eventually loosened visa restrictions for those Mainland Chinese who had become residents of Hong Kong, Macao, or another country. By 2005, the number of Chinese visitors had increased to 5,000 per month. Taipei has said it is willing to increase that number to 30,000 per month, a decision motivated, it appears, by Taiwan's economy, which could use an infusion of tourist dollars to get it moving.[57] Economic issues have, at long last, trumped Taipei's worries about tourists overstaying their visas to find work, as well as political and security concerns. More recent mainland measures include giving two pandas to Taiwan as gifts, recognizing Taiwan's college degrees, allowing Taiwanese medical professionals to practice on the mainland, and giving more preferential policies to southern Taiwan farmers to export their produce to the mainland.

China's "peace offensive" is based on a nine-point proposal originally made in 1981. Its major points include Beijing's willingness to negotiate a mutually agreeable re-integration of Taiwan under the mainland's government; encouragement of trade, cultural exchanges, travel, and communications between Taiwan and the mainland; the offer to give Taiwan "a high degree of autonomy as a special administrative region" within China after reunification (the status it offered to Hong Kong when it came under Beijing's rule in 1997); and promises that Taiwan could keep its own armed forces, continue its socioeconomic systems, maintain political control over local affairs, and allow its leaders to participate in the national leadership of a unified China. This far exceeds what China offered Hong Kong.

The KMT's original official response to the Beijing "peace offensive" was negative. The KMT's bitter history of war with the Chinese Communists, and what the KMT saw as a pattern of Communist duplicity, explained much of the government's hesitation. So did Beijing's refusal to treat Taiwan as an equal in negotiations. Nevertheless, since 1992, Taiwan has engaged in unofficial "track 2" and "track 3" discussions on topics of mutual interest, such as the protection of Taiwan's investments in the mainland, tourism, cross-Strait communication and transportation links, and

(Republic of China (ROC) Taiwan Navy)

Chen Yunlin, president of the mainland-based Association for Relations Across the Taiwan Strait (ARATS) and Chiang Pin-kung, chairman of the Taiwan-based Strait Exchange Foundation (SEF). The two sides signed the historic ECFA (Economic Cooperation Framework Agreement) in June 2010.

the dumping of Taiwan's nuclear waste on the mainland.

Taipei remains sensitive, however, to the Taiwanese people's concern about the unification of Taiwan with the mainland. The Taiwanese have asserted that they will never accede to rule by yet another mainland Chinese government, especially a Communist one. When the DPP was in power, Taiwanese had fewer fears that the leadership would strike a deal with Beijing at their expense; for, as the long-time advocate of the interests of the Taiwanese people and an independent Taiwan, the DPP is trusted not to sell out their interests. To speak of the "Taiwanese" as a united whole is, however, misleading; for it must be remembered that the overwhelming majority of members in the KMT, which favors stronger integration with the mainland, are Taiwanese, and that Taiwanese who do business with the mainland are eager to see integration—if not necessarily political unification at this point in history—progress much more rapidly. In addition, the political influence of those age cohorts most opposed to improving cross-Strait ties is waning. For younger Taiwanese, "loving Taiwan does not mean hating China. If the PRC refrains from acting in ways that provoke negative reactions from young Taiwanese, current trends suggest that Taiwan's public will demand better relations between the two sides in the future."[58] An increasing Taiwanese identity does not, then, necessarily mean greater support for a pro-independence policy. Indeed, public opinion polls indicate that support for declaring independence, in spite of the fact that 80 percent of the population today was born on Taiwan, has rarely exceeded 10 percent of voters.

With only a handful of countries recognizing the R.O.C., and with Beijing blocking membership for the R.O.C. in most international organizations, Taipei is under pressure to achieve some positive results in its evolving relationship with Beijing. The introduction of direct commercial flights from Taiwan to Shanghai over the 2003 Chinese New Year was one of the first steps toward direct air links.[59] On the other hand, "indirect" trade between mainland China and Taiwan by way of Hong Kong has continued to soar, although the opening of Taiwan's Offshore Islands (Quemoy and Matsu) to direct trade with the mainland in 2001 has somewhat undercut the need for the Hong Kong connection. On December 15, 2008, the "three links" were officially launched between mainland China and Taiwan, with direct flight, shipping and postal services across the Taiwan Strait after a hiatus of almost 60 years. Now thousands of Mainland tourists are allowed to visit Taiwan per day, contributing to Taiwan's economy.

Although China has become Taiwan's largest export market and the single largest recipient of investment from Taiwan, Taipei still hesitates to move forward on many issues that would further bind Taiwan with the mainland. Its policy on mainland spouses is particularly stringent and clearly indicates a perception that China's citizens are potential enemies, not compatriots. In the 1990s, relatively few Taiwanese residents who married individuals from the mainland were permitted to live with them in Taiwan. By contrast, Beijing's policy was to welcome Taiwan spouses to come live on the mainland. Taiwan's government argued that the mainland spouses could be spies and that internal security forces were inadequate to follow them around. Over time, however, this policy has relaxed, and as of 2007, there were 240,000 Chinese spouses of Taiwan citizens. This number, expected to grow by the time of the 2012 election to 300,000 to 400,000 eligible to vote (though they may only vote after 8 years of residency in Taiwan), is causing concern to the ever-paranoid Taiwanese that they might vote as a block for political unity with the mainland.[60] As Mainlanders are smuggled into Taiwan in ever-larger numbers (primarily to satisfy the needs of entrepreneurs in Taiwan for cheap labor), and the number of fake marriages (marriages of convenience) with Taiwan's citizens grows (more than 16,000 had been discovered as of 2005),[61] the issue of surveillance has become a growing concern. At the same time, Taiwan spying on the mainland has grown steadily as its contacts with the mainland have increased. The Military Intelligence Bureau recruits from its approximately one million citizens who work, live, and tour on the mainland. They in turn form Taiwan spy networks that recruit local Chinese with sex, money, and "democratic justice" (an appeal to their sense of injustice at the hands of the Chinese government).[62]

In the meantime, China continues to deepen and widen harbors to receive ships from Taiwan; wine and dine influential Taiwanese; give preferential treatment to Taiwan's entrepreneurs in trade and investment on the mainland; open direct telephone links between Taiwan and the mainland; rebuild some of the most important temples to local deities in Fujian Province where Taiwanese like to visit; establish special tourist organizations to care solely for people from Taiwan; and refurbish the birthplace of Chiang Kai-shek, the greatest enemy of the Chinese Communists in their history.

Taiwan's businesspeople and scholars are eager for Taiwan's relationship with the mainland to move forward. They seek direct trade and personal contacts, and try to separate political concerns from economic interests and international scientific exchanges. The manufacturing and software sectors are particularly concerned with penetrating and, if possible, controlling the China market. Otherwise, they argue, businesspeople from other countries will do so.

The business community, faced with Taiwan's ever-higher labor cost and Taiwan's lack of cheap raw materials, has flocked to China. Whether they move outdated labor-intensive factories and machinery to the mainland, or invest in cutting-edge technology, Taiwan's businesspeople benefit from China's cheap, hard-working labor force.

Others are concerned, however, that, with more than 15 percent of its total foreign investment in the mainland, Taiwan could become "hostage" to Beijing. That is, if China were to refuse to release Taiwan's assets or to reimburse investors for their assets on the mainland in case of a political or military conflict between Taipei and Beijing, Taiwan's enterprises would form a pressure point that would give the advantage to Beijing. Any military defensive capabilities either Taiwan or the U.S. could offer are useless to counter possible Mainland economic warfare, such as an economic blockade, when most of Taiwan's trade is with the mainland itself and 60 percent of its investments are in the mainland.[63] Furthermore, without diplomatic recognition in China, Taiwan's businesses on the mainland are vulnerable in case of a conflict with local businesses or the government. High-level members of the government have even denounced those who invest in the mainland as "traitors." And members of the pro-independence press have attempted to whip up fear among its citizens that with Beijing's new anti-secession law, Taiwanese visiting or living in China will be vulnerable to "shakedown artists."[64] So far, affairs have turned out quite the opposite: China has actually *favored* Taiwan's businesses over all others; and Taiwan's investors have tended to turn a quick profit, and to construct many safeguards, so that any seizure of assets would result in negligible losses. To wit, Beijing learned that its heavy-handed approach to Taiwan in passing an anti-secession law was counter-productive, and now relies more on "soft power" in its relations with Taipei.

PROSPECTS FOR THE REUNIFICATION OF TAIWAN WITH THE MAINLAND

The December 2004 legislative elections kept an anti-independence majority in the legislature, and the 2008 presidential election returned the KMT to power, but most Taiwanese remain opposed to reunification. Most agree that, given China's threats to use force if Taiwan were to declare independence, it would be foolish to do so; and that the government's policy toward China should be progressive, assertive, and forward-looking. Lacking a long-term plan and simply reacting to Beijing's initiatives puts the real power to determine the future relationship in China's hands.

For the last 20 years, reform-minded individuals in Taiwan's political parties have insisted that the government needs to actively structure how the cross-Straits relationship evolves. One of the most widely discussed policies favored by those who opposed a declaration of independence has been a "one country, two regions" model that would approximate the "one country, two systems" model China has with Hong Kong. The problem is determining who would govern in that "one China" after reunification with the mainland. To call that country the People's Republic of China would probably never find acceptance in Taiwan—a point understood by Beijing. As a result, it now frequently drops the "people's republic" part of the name in its pronouncements. Symbolically, this eliminates the issue of two different governments claiming to represent China, and whether that China would be called a "people's republic" or a "republic." It would be called neither. But, would Beijing be in charge of the new unified government? Would the government of the "Republic of China," even were free and democratic elections for all of the mainland and Taiwan to be held, be the winner? There is certainly no evidence to suggest that most people on the mainland would welcome being governed by Taiwan's leaders.

In spite of the negative rhetoric, changes in Taipei's policies toward the P.R.C. have been critical to improving cross-Strait ties. For example, Taiwan's government ended its 40-year-old policy of stamping "communist bandit" on all printed materials from China and prohibiting ordinary people from reading them; Taiwan's citizens, even if they are government officials, are now permitted to visit China; scholars from Taiwan may now attend international conferences in the P.R.C., and Taipei now permits a few P.R.C. scholars to attend conferences in Taiwan; and KMT retired veterans, who fought against the Communists and retreated to Taiwan in 1949, are actually encouraged to return to the mainland to live out their lives because their limited KMT government pensions would buy them a better life there! Certainly some members of Taiwan's upper class are acting as if the relationship will eventually be a harmonious one when they buy apartments for their mistresses and purchase large mansions in the former international sector of Shanghai and elsewhere. (It is estimated that there are more than one million individuals from Taiwan living in China.) One survey indicated, in fact, that after the United States and Canada, mainland China was the preferred place to emigrate for Taiwan's citizens![65] And, from the perspective of the Democratic Progressive Party, things are getting worse: hundreds of thousands more Taiwanese are relocating to the mainland, and they are taking with them venture capital estimated at well over U.S.$100 billion—money that might otherwise be pumped into Taiwan's economy.

Some Taiwanese share with Beijing a common interest in establishing a Chinese trading zone in East Asia. A "Chinese common market" would incorporate China, Taiwan, Hong Kong, Macau, and perhaps other places with large ethnic-Chinese communities such as Singapore and Malaysia. Economically integrating these Chinese areas through common policies on taxes, trade, and currencies would strengthen them vis-à-vis the Japanese economic powerhouse, which remains larger than all the other Far Eastern economies put together.

Yet, with an ever-smaller number of first-generation Mainlanders in top positions in the KMT and Taiwan's government, few are keen to push for reunification. Indeed, the majority of people in Taiwan are not interested in reunification under current conditions. They are particularly concerned about two issues. The first is the gap in living standards. Taiwan is fully aware of the high price that West Germany paid to reunify with East Germany. Obviously the price tag to close the wealth gap with mammoth China would be prohibitive for tiny Taiwan; and any plan that would allow Beijing to heavily tax Taiwan's citizens would be unacceptable. Yet China is developing so quickly, at least along the densely populated eastern coast that faces Taiwan, that this issue should disappear. In developed coastal areas, such as Shanghai and Shenzhen, living standards are comparable to that of Taipei.

The second issue is democracy. Fears that they might lose certain political freedoms and control over their own institutions have made the Taiwanese wary of reunification. Finally, whether or not "one country, two systems" succeeds in protecting Hong Kong from Beijing's intervention, Taiwanese reject a parallel being drawn between Hong Kong and Taiwan. Hong Kong was, they argue, a British colony, whereas Taiwan, even if only in the last dozen years, is a fledgling democracy. The KMT stated that once China had attained a certain (unspecified) level of development and democracy, China and Taiwan would be reunified; but the DPP has made no such statement. In any event, with so much left open to its own arbitrary interpretation as to what is sufficient development and democracy, even the KMT seems to have made no real commitment to reunification. For its part, the DPP-led government was pressured into accepting greater interdependency with China, yet had tried desperately to move toward the formal declaration of an independent state, to no avail.

China's significant progress in developing the economy and increasing the rights of the Chinese people in the last 30 years should make reunification more palatable to Taiwan. Greater contacts and exchanges between the two sides should in themselves help lay the basis for mutual trust and understanding.

Many members of Taiwan's political and intellectual elite, including the DPP and KMT party leaders, think tanks, and the Ministry of Foreign Affairs, seem to spend the better part of each day pondering the meaning of "one China." In general, they would like China to return to the agreement that they reached with Beijing in 1992: namely, that each side would keep its own interpretation of the "one China" concept: Beijing would continue to see "one China" as the government of the P.R.C. and would continue to deny that the R.O.C. government existed. Taipei would continue to hold that Taiwan and the mainland are separate but politically equivalent parts of one China. Beijing now also supports the so-called "1992 Consensus."

Today, it is estimated that China has more than 900 (non-nuclear) missiles pointed at Taiwan; but unless Taiwan again pushes seriously for recognition as an independent state, an attack is unlikely; for the mainland continues to benefit from Taiwan's trade and investments, remittances and tourism; and it is committed to rapid economic development, which could be seriously jeopardized by even a brief war—a war that it might not win if the United States were to intervene on behalf of Taiwan. In spite of the fact that the U.S. has repeatedly made it clear that it does not support Taiwan independence, Beijing could well worry that the U.S. might suddenly be persuaded to come to Taiwan's support.

It is in Taiwan's best interests for the relationship with China to develop in a careful and controlled manner, and to avoid public statements on the issue of reunification versus an independent Taiwan, even as that issue haunts every hour of the day in Taiwan. It is also in Taiwan's interest to wait and see how well China integrates Hong Kong under its formula of "one country, two systems." In the meantime, Taiwan's international strategy—agreeing that it is part of China, while acting like an independent state and conducting business and diplomacy with other states as usual—has proved remarkably successful. It has allowed Taiwan to get on with its own economic development without the diversion of a crippling amount of revenue to military security. A continuation of the status quo is clearly the preferred alternative for Taiwan, the United States, and mainland China; for it avoids any possibility of a military conflict, which none would welcome; and it does not interrupt the preferred strategy of both the DPP and the KMT—"closer economic ties, lower tensions, and more communication with the mainland."[66]

At the same time, this strategy allows time for the mainland to become increasingly democratic and developed, in a way that one day might make reunification palatable to Taiwan. Meanwhile, Beijing's leadership knows that Taiwan acts as a de facto independent state, but it is willing to turn a blind eye as long as Taipei does not push too openly for recognition. Thus far, the reality has mattered less to Beijing than recognition of the symbolism of Taiwan being a province of China. Beijing's leadership is far more interested in putting resources into China's economic development than fighting a war with no known outcome. In part for this reason, and in part because of continuing efforts of the U.S. to sell armaments to Taiwan, Beijing has made it clear to Washington that China's own buildup of missiles targeted at Taiwan would be linked to American sales and efforts to install a theatre missile defense system around Taiwan.

Because China's sovereignty over Taiwan has been an emotional, historical, and nationalistic issue for the Chinese people, however, Beijing does not make a "rational" cost-benefit analysis of the use of force against a rebellious Taiwan. Taiwan does not like Beijing's militant rhetoric, but some mainland Chinese analysts believe that China's leadership is forced to sound more militant than it feels, thanks to the militant nationalism of ordinary Chinese people and the Chinese military. Indeed, some go so far as to say that, were China an electoral democracy, the people would have voted out

Timeline: PAST

A.D. 1544
Portuguese sailors are the first Europeans to visit Taiwan

1700s
Taiwan becomes part of the Chinese Empire

1895
The Sino-Japanese War ends; China cedes Taiwan to Japan

1945
Japan is forced to return the colony of Taiwan to China when Japanese forces are defeated in World War II

1947–49
Nationalists, under Chiang Kai-shek, retreat to Taiwan

1950s
A de facto separation of Taiwan from China; The Chinese Communist Party is unable to bring its civil war with the KMT to an end because the U.S. interposes its 7th Fleet between the mainland and Taiwan in the Taiwan Straits

1971
The People's Republic of China replaces the Republic of China (Taiwan) in the United Nations as the legitimate government of "China"

1975
Chiang Kai-shek dies and is succeeded by his son, Chiang Ching-kuo

1980s
40 years of martial law end and opposition parties are permitted to campaign for office

1990s
Relations with China improve; the United States sells F-16 jets to Taiwan; China conducts military exercises to intimidate Taiwanese voters. Democratization develops. The Asian financial crisis deals a setback to Taiwan's economic growth.

PRESENT

2000
Chen Shui-bian is elected President, ending the KMT rule

2004
Chen Shui-bian is re-elected in a controversial election

2005
China passes an anti-secession law that says the country can use force if Taiwan moves toward independence
Lien Chan, head of the opposition KMT, travels to Beijing and meets with Hu Jintao, general secretary of the CCP. It is the first meeting between KMT and CCP leaders since 1949

2008
Ma Ying-jeou is elected president, returning the KMT to power
Chen Shui-bian under corruption investigation
Chen Yunlin, chairman of the Association for Relations Across the Taiwan Straits, visits Taiwan

2009
China drops longstanding objections to Taiwan's participation in World Health Organization
President Ma Ying-jeou is elected head of the ruling KMT
Typhoon Morakot hits southern Taiwan leaving hundreds dead in floods and mudslides

2010
The United States approves the sale of air defense missiles to Taiwan, as part of a proposed $6.4bn arms sale package to the island
Taiwan and China sign a landmark free trade pact (ECFA) seen as most significant agreement in 60 years of separation.

the CCP leadership because it has done little to regain sovereignty over Taiwan.

As Taiwan's relationship with China deepens and broadens, it is possible that more arrangements could be made for the representation of both Taiwan and China in international organizations, without Beijing putting up countless roadblocks. Indeed, Beijing welcomed Taiwan's membership in the WTO, if for no other reason than that it allows it to pry open Taiwan's market. Further, Taiwan's WTO membership has led to even more investments in the mainland and further economic integration.

Taiwan eagerly embraced membership in the WTO, not because it would necessarily benefit from WTO trade rules, but so that it could become a player in a major international organization. But, the very trade practices that led to Taiwan's multibillion-dollar trade surplus will have to be abolished to gain compliance with WTO regulations. Still, had Taiwan not joined the WTO, it would have eventually lost its competitiveness in agriculture and the automobile industry anyway.

To conclude, the benefits of Taiwan declaring independence would be virtually nil. Already Beijing refuses to have diplomatic relations with any country that officially recognizes the Republic of China. Those countries that do recognize the R.O.C., or Taiwan, as a sovereign state also have difficulty trading with the P.R.C. Given the size of the China market, this is an unacceptable price for most countries to pay. Beijing would no doubt use this trump card to punish those that would dare to recognize Taiwan as an independent and sovereign state, just as it does now.

Taiwan is looking for a place for itself in the international system, and it can't seem to find it. But its government realizes that the island is a small place, and that if Taiwan ever were to stop demanding international status and attention, it might well discover that it had suddenly become, de facto, a province of China while the international community looked the other way. If only for this reason, it is in Taiwan's interest to continue to press its case for greater international recognition, and to continue to engage in pragmatic unofficial diplomacy and trade with states throughout the world. It may not buy Taiwan statehood, but it may well buy the government's continued independence of Beijing.

NOTES

1. Before it won its bid to host the 2008 Olympics, Beijing had said that it would not allow Taipei to co-host them unless it first accepted the "one China" principle. Taipei did not accept it, so Beijing is the sole host of the 2008 Olympics.
2. John F. Burns, "Quemoy (Remember?) Bristles with Readiness," *The New York Times* (April 5, 1986), p. 2.
3. Mainland Affairs Council, "Report on the Preliminary Impact Study of the 'Three Mini-links' Between the Two Sides of the Taiwan Strait" (October 2, 2000); and discussions at The National Security Council and Ministry of National Defense in Taiwan (January 2001).
4. Edwin A. Winckler, "Cultural Policy on Postwar Taiwan," in Steven Harrell and Chun-chieh Huang, eds., *Cultural Change in Postwar Taiwan* (Boulder, CO: Westview Press, 1994), p. 22.
5. *Ibid.*, p. 29.
6. Thomas B. Gold, "Civil Society and Taiwan's Quest for Identity," in Harrell and Huang, p. 60.
7. Thomas A. Shaw, "Are the Taiwanese Becoming More Individualistic as They Become More Modern?" Taiwan Studies Workshop, *Fairbank Center Working Papers,* No. 7 (August 1994), pp. 1–25.
8. "Premier Hau Bristling about Crime in Taiwan," *The Free China Journal* (September 13, 1990), p. 1; and Winckler, p. 41.
9. Paul Li, "Trash Transfigurations," *Taipei Review* (October 2000), pp. 46–53.
10. Central News Agency, Taipei, "Nearly 90% support new garbage classification policy, survey finds," *Taiwan News Online* (December 27, 2004). www.etaiwannews.com/Taiwan/Politics/2004/12/27/1104112881.htm.
11. U.S. Energy Information Administration. Report on Taiwan, available at www.eia.doe.gov/emeu/cabs/taiwanenv.html
12. A dormitory for employees of Tai Power is one of these buildings. Ninety percent of Taiwan's nuclear waste is stored on Orchid Island, where one of Taiwan's 9 aboriginal tribes lives. Christian Aspalter, *Understanding Modern Taiwan: Essays in Economics, Politics, and Social Policy* (Burlington, Vt.: Ashgate Publishers, 2001), pp. 103–107.
13. U.S. Energy Information Administration. Report on Taiwan, op.cit.
14. www.settlement.org/cp/english/taiwan/holidays.html
15. Lee Fan-fang, "Ghosts' Arrival Bad for Business," *The Free China Journal* (August 7, 1992), p. 4.
16. *Ibid.*
17. Marc J. Cohen, *Taiwan at the Crossroads* (Washington, D.C.: Asian Resource Center, 1988), pp. 186–190. For further detail, see his chapter on "Religion and Religious Freedom," pp. 185–215. Also, see Gold, in Harrell and Huang, p. 53.
18. "Presbyterian Church in Taiwan Calls for Reform," (February 18, 2004). Posted on www.christiantoday.com/news/asip/42.htm
19. See *Free China Review,* Vol. 44, No. 9 (September 1994), which ran a series of articles on educational reform, pp. 1–37.
20. Brian Cheng, "Foreign Workers Seen as a Mixed Blessing," *Taipei Journal* (October 20, 2000), p. 7.
21. Election Study Center, National Chengchi University, *Taipei: Face-to-Face Surveys* (February 2000). Funded by the Mainland Affairs Council.
22. Cohen, p. 107.
23. *Ibid.,* p. 108. For more on women, see the chapter on "Women and Indigenous People," pp. 106–126.
24. Jim Hwang, "The Civil Service: Walking on Glue," *Taipei Review* (October 2000), pp. 22–29; and "Taiwan's Civil Service Makes Headway on Gender Equality," *Taiwan Update,* Vol. 5, No. 3 (March 30, 2004), p. 7.
25. Cher-jean Lee, "Political Participation by Women of Taiwan," *Taiwan Journal* (August 20, 2004), p. 7.
26. Robert P. Weller, "Environmental Protest in Taiwan: A Preliminary Sketch," Taiwan Studies Workshop, *Fairbank Center Working Papers,* No. 2 (1993), pp. 1, 4.
27. Taiwan's workers could not get higher wages through strikes, which were forbidden under martial law. The alternative was to try starting up one's own business. Gold, in Harrell and Huang, pp. 50, 53.
28. Kelly Her, "Not-So-Iron Rice Bowl," *Free China Review* (October 1998), pp. 28–35.
29. "Taiwan: In Praise of Paranoia," *The Economist* (November 7, 1998) pp. 8–15.
30. Since 2005, a steady stream of high level KMT political and economic leaders have gone to China to assure China that they do not support Taiwan independence and want to expand their economic relations with China.
31. Peter Morris, "Taiwan business in China supports opposition," *Asia Times Online* (Feb. 4, 2004). www.atimes.com/atimes/China/FB04Ad04.html
32. "Taiwan: In Praise of Paranoia," p. 17. For 2004, Taiwan had an overall trade surplus (with all countries) of well over US$7 billion.
33. Francis Li, "Taiwan to Set Up Free-Trade Zones," *Taipei Journal* (October 11, 2002), p. 3.
34. "Taiwan: In Praise of Paranoia," p. 16.
35. Editor, "Standard-bearers for the Future," *Taiwan Review* (February 2007), p. 1; and Zoe Cheng, "The Biggest Leap," *Taiwan Review* (February 2007), pp. 4–11.
36. Until the DPP came into power in 2000 and began to promote the concept of Taiwan as an independent state, Taiwan was always shown on the map as a part of China, as were Tibet, Inner Mongolia, and even Outer Mongolia, which is an independent state.
37. From 1950 to 1986, military courts tried more than 10,000 cases involving civilians. These were in violation of the

Constitution's provision (Article 9) that prohibited civilians from being tried in a military court. Hung-mao Tien, *The Great Transition: Political and Social Change in the Republic of China* (Palo Alto, CA: Hoover Institution, Stanford University, 1989), p. 111.

38. The Kaohsiung rally, which was followed by street confrontations between the demonstrators and the police, is an instance of KMT repression of *dangwai* activities. These activities were seen as a challenge to the KMT's absolute power. The KMT interpreted the Kaohsiung Incident as "an illegal challenge to public security." For this reason, those arrested were given only semi-open hearings in a military, not civil, tribunal; and torture may have been used to extract confessions from the defendants. Tien, p. 97.
39. Bureau of Democracy, Human Rights, and Labor, U.S. Department of State, *China (Taiwan only): Country Report on Human Rights Practices (2003),* (Washington D.C. 2004). Available at: www.state.gov/g/drl/rls/hrrpt/2003/27767.htm; and Lin Fang-yan, "Rights Group Reports on ROC Progress," *Taipei Journal* (January 3, 2003), p. 1.
40. Lin Fangyan, *Ibid.*
41. Tien, p. 72.
42. A February 2000 poll conducted in Taiwan indicated that only 6.1 percent of the respondents opposed conditional or unconditional opening up of direct transportation links with the mainland. Election Study Center, National Chengchi University, Taipei. Face-to-face surveys. Funded by the Mainland Affairs Council, Executive Yuan (February 2000).
43. Shelley Rigger, "Taiwan: Finding Opportunity in Crisis," *Current History* (September 1999), p. 290.
44. Shelley Rigger, "Taiwan Rides the Democratic Dragon," *The Washington Quarterly* (Spring 2000), pp. 112–113. Reference is to the editor, *Taipei Times* (January 4, 2000).
45. Discussions with Shaw Yu-ming, deputy secretary-general of the KMT, at KMT headquarters, Taipei, 2001.
46. Myra Lu and Frank Chang, "Election Trends Indicate Future of Taiwan Politics," *Free China Journal* (November 27, 1998), p. 7.
47. According to Shaw Yu-ming, the KMT was defeated not because of its policies but because it was the KMT leadership who chose the candidates to run in the election. The implication was that if instead, the KMT membership had chosen candidates, they would have picked candidates who had a better chance of winning.
48. Cal Clark, "Taiwan in the 1990s: Moving Ahead or Back to the Future?" in William Joseph, *China Briefing: The Contradictions of Change* (Armonk, NY: M. E. Sharpe, 1997), p. 206; Myra Lu, "Crack-down on Vote-Buying Continues," *Free China Journal* (November 27, 1998), p. 2.
49. One poll, conducted by *The China Times* (June 18, 2006) indicated that 53 percent of respondents thought Chen should resign. Referenced in Kerry Dumbaugh: *China-U.S. Relations: Current Issues and Implications for U.S. Policy* (Washington: Congressional Research Service Report for Congress (February 2007), p. 9. Available at http://fpc.state.gov/documents/organization/81340.pdf.
50. Lee Chang-kuei, "High-Speed Social Dynamics," *Free China Review* (October 1998), p. 6.
51. *Taiwan Yearbook, 2007,* (Government Information Office, Taipei, 2007). Available at http://english.www.gov.tw/Yearbook/index.jsp?categid=28&recordid=52736; and Chin-chuan Lee, "Sparking a Fire: The Press and the Ferment of Democratic Change in Taiwan," in Chin-chuan Lee, ed., *China's Media, Media China* (Boulder, CO: Westview Press, 1994), pp. 188–192.
52. Chen Jie, *Foreign Policy of the New Taiwan: Pragmatic Diplomacy in Southeast Asia* (Northampton, MA: Edward Elgar Publishing, 2002), pp. 63–64.
53. In 2000, however, China agreed to stop selling nuclear and missile technology to Pakistan.
54. Kerry Dumbaugh: *China-U.S. Relations: Current Issues and Implications for U.S. Policy* pp. 2, 9.
55. Thomas J. Christensen, "The Contemporary Security Dilemma: Deterring a Taiwan Conflict," *The Washington Quarterly* (Autumn 2002), pp. 7–21.
56. Patrick L. Smith, "For Many in Taiwan, Status Quo with China Sounds Fine," *International Herald Tribute,* December 11, 2006.
57. Jimmy Chuang, "A 'foreigner's point of view' can boost tourist numbers: [Premier] Su," *Taipei Times* (January 4, 2007), p. 4; Kathrin Hille, "Taiwan May Allow More Mainland Chinese Visitors," *Financial Times,* October 21, 2005.
58. Shelley Rigger, *Taiwan's Rising Rationalism: Generations, Politics, and "Taiwanese Nationalism,"* Policy Studies 26 (Washington D.C.: East-West Center Washington, 2006), p. 84.
59. Taiwan's own airlines are now permitted to continue on to Shanghai.
60. China News Agency (Taipei), "Chinese Spouses of Taiwan Citizens Wont Sway Elections: Official," *Taiwan Headlines* (January 5, 2007). Available at www.taiwanheadlines.gov.tw/ctasp?xItem=57629&CtNode=5
61. Yulin County Government, "Over 16,000 PRC Citizens Found in Fake Marriages with Taiwanese" (2005). Available at http://en.yunlin.gov.tw/index3/en/03Bulletin/03Bulletin_01_01.asp?id=686. However, "fake marriages" are more likely to be related to human trafficking and prostitution than spying. See Zoe Cheng (Feb. 2007), p. 9.
62. Wendell Minnick, "The Men in Black: How Taiwan Spies on China," *Asia Times Online Co.,* (www.atimes.com), 2004.
63. Robert S. Ross, "Explaining Taiwan's Revisionist Diplomacy," *Journal of Contemporary China,* Vol. 15, no. 48 (August 2006), pp. 446–447, 450. Ross (2006), pp. 452–454. See the charts based on public opinion polls by the Mainland Affairs Council, done 3 to 4 times per year from May 2000 through December 2006, in which the highest percentage of those polled who wanted to declare independence immediately only reached more than 10 percent once (10.3 percent in November 2005). Charts available at www.mac.gov.tw/english/index1-e.htm.
64. Editorial, "The Chinese Gulag Beckons," *Taipei Times,* (January 10, 2005). www.taipeitimes.com/News/edit/archives/2005/01/10/2003218825
65. Chien-min Chao, "Introduction: The DPP in Power," *Journal of Contemporary China,* vol. 11, no. 33 (November 2002), p. 606.
66. "Taiwan Stands Up," *The Economist* (March 25, 2000), p. 24.

Article 1

Think Again: China

— In addition, search information on the internet that shows the severity of air pollution and water pollution in china.

It's often said that China is walking a tightrope: Its economy depends on foreign money, its leadership is set in its ways, and its military expansion threatens the world. But the Middle Kingdom's immediate dangers run deeper than you realize.

HARRY HARDING

"China's Biggest Risks Are Economic"

No. In fact, China's most severe risks are ecological—particularly its environmental problems and its vulnerability to communicable disease. Of course, this is not to say that China has no economic problems. No country is immune from the normal business cycle, and China today is subject to both inflationary and recessionary risks. But Beijing is developing the fiscal and monetary tools to regulate the economy so as to prevent these problems from becoming catastrophic once they emerge.

In contrast, China's ecological and health risks are far more serious than people realize. Air pollution in China is affecting the quality of life in cities like Beijing, Hong Kong, and Shanghai, among others. The risk of water shortages, both in agricultural areas and major cities, is high and growing; only 1 percent of the surface water available to Shanghai is safe to drink. In one harbinger of things to come, an explosion at a chemical plant in northeast China in November 2005 sent a benzene slick cascading down the Songhua River. Millions of people in the large, industrial city of Harbin were without water for a week. The probability of more acute environmental crises resulting from chemical spills or toxic emissions is high. The Chinese government is already warning that the country's emission of carbon dioxide and greenhouse gases will significantly damage China's agricultural production.

China is also experiencing epidemics of chronic disease. Reported cases of HIV increased by 30 percent to roughly 650,000 in 2006, and the United Nations projects that 10 million Chinese will be infected by 2010. Hepatitis infects 10 percent of the country's population. The probability of an outbreak of an acute communicable disease, such as the avian flu, remains high. The main issue is how virulent the virus will be, and whether its spread can be contained. The risk is exacerbated by the decay of the rural public health system due to lack of funding and by the reluctance of local officials to report new occurrences of the disease, making it more likely that an outbreak will become a deadly epidemic.

"A Second Tiananmen Crisis Is Inevitable"

Hardly. The Tiananmen Square crisis of 1989 involved mass protests in scores of cities across China—and the demonstrations in Beijing were so large that the government was able to suppress them only through the use of brutal military force. Though not inconceivable, another dramatic uprising on that scale is unlikely.

It is true, however, that China has many problems that are producing widespread popular discontent. These include environmental problems; gaps in the country's social safety net, particularly with regard to health insurance and old-age pensions; controversies over land and water rights; and chronic corruption among officials. These grievances have caused a sharp increase in grassroots protests. The Chinese government itself reported some 80,000 such incidents in 2005, some of which were quite large and even violent. In the most notorious uprising, in late 2005, riot police fired at protesting farmers in a rural Guangdong Province village. Witnesses claim as many as 20 villagers were killed.

But Chinese leaders are adopting policies to address the causes of rural grievance, such as increasing spending on rural projects, abolishing onerous agricultural taxes, and cracking down on local officials who squeeze villagers. When protests do occur, they arrest the leaders but often try to remedy the particular issues that caused the unrest. Six months after the fatal confrontation in Guangdong, a similar protest nearby ended with official promises to review the terms of the land confiscation that had provoked it. Above all, by controlling the media and suppressing independent political organization, Beijing is trying to ensure that protests remain localized. Moreover, in many quarters, particularly China's growing urban middle class, political support for the government appears to be quite high.

The real concern is whether bigger issues could foil these efforts. The emergence of serious and widespread economic problems (especially inflation and unemployment) or the government being blamed for a major domestic or international crisis (such as an environmental catastrophe or an incident during the upcoming 2008 Olympics) could lead to nationwide discontent. It would be particularly dangerous if the dissatisfaction were so widespread that it overwhelmed the party's control over the media and the Internet, or produced a divided leadership unable to respond effectively. In such a circumstance, there could be large-scale protests in several major cities that might be difficult to control, as was the case in 1989.

"Chinese Elite Politics Are Stable"

Yes, but less than you might think. Chinese politics has become increasingly institutionalized, the elite are more pragmatic, and top leaders want to avoid a perception of internecine feuding. But President Hu Jintao has had to tread more softly than his predecessors. Although he has been able to secure the dismissals of a few central and provincial leaders by charging them with corruption, he has not yet been able to replace them with his own protégés.

Hu is nearing the end of his first five-year term. Past practice suggests that one or two potential successors should have been appointed to the Politburo by now. The president will have to identify his heirs apparent by the 17th Communist Party Congress this fall so they have enough time to win broader support before Hu retires in 2012. If his choices are not widely accepted, the result could be a decline in the party's legitimacy. Indeed, the uncertainties surrounding succession now constitute the biggest political risk facing China this year.

What is even more worrying is that the grid-locked succession may reflect a lack of agreement on China's policy decisions. True, Hu recently secured formal endorsement of his stance that China needs to address its most serious domestic problems and spread the benefits of economic growth in order to tamp down social conflict. But that doesn't preclude a debate over how to achieve that goal. For one thing, the party has defined one of its primary goals as creating a "harmonious socialist society," which implies that any policy option that isn't "socialist"—for example, protecting private property rights or moving toward pluralistic democracy—should be taken off the table. For another, although Hu talks about the need for sustainable development, the party leaders still assign priority to rapid economic growth. Those decisions give easy rhetorical openings to any of Hu's rivals who may want to challenge his political agenda going forward.

"China's Banks Will Collapse"

Doubtful. Until very recently, China's banking system was in big trouble. It was the main mechanism for financing the country's high levels of investment, which made up 45 percent of gross domestic product (GDP) in 2005. The banks faced considerable pressure to lend money to inefficient state-owned enterprises. As a result, the volume of the country's nonperforming loans rose to alarming levels. But the banks survived, largely because depositors had few other outlets for their savings. The banking system was not that solvent, but it remained liquid.

In more recent years, though, China's corporate sector has become a less risky customer for the banks. Investment is increasingly being financed by means other than bank loans, such as bond issues, corporate profits, or stock offerings. The latter have yielded recordbreaking sums, with the Industrial and Commercial Bank of China leading the way with a $22 billion initial public offering last fall. And a gradual process of mergers, acquisitions, and privatization is increasing the profitability of formerly state-owned enterprises.

At the same time, the solvency of the banks has also improved. China has been recapitalizing banks, transferring nonperforming loans to management companies, and inviting partial foreign ownership of major banks. In addition, the banks' portfolios are being broadened through greater reliance on home mortgages and fee-generating services.

China's financial system isn't entirely out of the woods. The banks' lending decisions are still subject to political pressures, because the party still chooses senior bank executives from its ranks. The health of smaller local banks, various investment brokerages, and insurance companies is not ideal. And with a growing range of investment opportunities—including the stock market, real estate, and even foreign mutual funds—Chinese banks now have to worry that financial insolvency could more easily generate liquidity problems.

Even so, China's low level of foreign indebtedness gives the government the tools to contain the economic consequences of a financial crisis.

"China Is Too Dependent on Foreign Money"

Not really. China is certainly highly integrated into today's international economy. By abandoning the economic autarchy of the Maoist period, it has become a major trading nation. It exports large volumes of textiles, machinery, and electronic equipment. In turn, it imports advanced technology, petroleum, and other natural resources. It is also a favored destination for foreign direct investment (FDI), not only because of its attractiveness as a manufacturing platform for exports, but because of the size and dynamism of its own domestic market. China now attracts twice the FDI it did 10 years ago.

The relatively large share of exports to China's GDP and the volumes of incoming FDI have generated concern that China is too dependent on the international economy and is acutely vulnerable to a slowdown. But these concerns are overblown. For one thing, with $1 trillion in foreign exchange reserves, and an extraordinarily high domestic savings rate of roughly 47 percent, China is hardly dependent on foreign capital. It has relied on the technology and marketing networks that accompany foreign investment to promote its exports, but it would most likely survive a reduction in new investment fairly easily.

The same is true with trade. China is a large continental economy, and at 64 percent of GDP, its trade dependence is far lower than of places such as Hong Kong or Singapore. Moreover, much of the value of Chinese exports is provided by imported components and raw materials, with local elements providing relatively less value. Computers bearing the tag

"Made in China" may be assembled there, but their screens and microprocessors likely come from Taiwan or South Korea. Processing and assembly accounted for 55 percent of China's total exports in 2006. That means that the net contribution of trade to the Chinese economy is less than the gross figures imply. Sure, China would take a hit if there were a severe global recession or a terrorist attack that crippled international trade flows. But its economy could weather that challenge far better than most.

"Chinese Nationalism Is on the Rise"

Yes, but don't exaggerate its implications. Popular nationalism has been a part of China's fabric since the middle of the 19th century. It emerged as a reaction to invasion by nations more technologically advanced than China—first from Europe, then Japan. More recently, the Chinese Communist Party, whose ideological appeal began to erode in the 1980s, has been encouraging nationalism as a source of legitimacy.

But the party recognizes that nationalism is a double-edged sword. Although it can be a source of domestic legitimacy, it can also generate apprehension and mistrust abroad. That lesson was borne out by several anti-foreign protests, including those against the United States for the accidental bombing of the Chinese Embassy in Belgrade in 1999 and for the collision between an American reconnaissance aircraft and a Chinese fighter in 2001. Although these episodes did no lasting damage to China's relations with the United States, Beijing was alarmed by the fervor of the protests and the time it took to bring them to an end. China's leaders understand that nationalism can generate public criticism of leaders who "capitulate" to foreign governments just as easily as it generates support for those who are perceived as upholding Chinese interests.

Accordingly, the promotion of nationalism now plays a smaller role in the party's search for legitimacy. It has been replaced to a degree by the quest for a "harmonious socialist society." The media now repeatedly emphasize that China's rise should be peaceful, and officials try to keep nationalist sentiment in check.

The problem is that popular nationalism can have its own momentum, independent of the wishes of the party's leadership. But without further democratization, nationalistic public opinion isn't powerful enough to determine Chinese foreign policy. At the margins, however, it reduces the flexibility of Chinese foreign policymakers. It could be a source of political instability if the Chinese government were accused of failing to uphold the national interest in the event of an international crisis.

"China's Rise Will Lead to Military Conflict"

Highly unlikely, at least for the foreseeable future. Yes, China is modernizing its military, seeking not only a stronger nuclear deterrent but also a greater ability to project conventional force. And like any powerful country, China will use force if it believes its vital national interests are at stake, particularly concerning the disputes over islands and undersea resources in the East China Sea and the South China Sea, the possible collapse of North Korea, and, above all, the possibility of a declaration of independence by Taiwan.

But China is no longer a revolutionary power. It does not have fundamental complaints about the international economic and political systems from which it has benefited so much over the past 25 years. Moreover, its economic interdependence with the rest of the world will deter Beijing from military adventures unless such core interests are threatened. The rise of Chinese power, in turn, will deter China's neighbors from threatening its core interests. Beijing has drawn its red line in the Taiwan Strait so narrowly—a de jure declaration of independence by Taiwan—that it is unlikely ever to be crossed.

The real challenges from China are, therefore, far more subtle than alarmists would suggest. First, though China is willing to join the existing international order, it wants to play a larger role—as a rule-maker, not just a rule-taker. Fortunately, Washington's current policy of encouraging China to become a "responsible stakeholder" in the international system is largely compatible with Beijing's desire for greater influence.

A second challenge stems from the desire that Chinese firms gain the greatest market share domestically and join the ranks of large, profitable multinationals. China is a poster child for globalization, but Beijing's objective is to see that Chinese firms, and not foreign firms, are the winners of that global competition. Indeed, economic nationalism may pose a greater challenge for the world than any other form of Chinese power.

China is simultaneously rising on several dimensions—military, economic, diplomatic, ideological, and cultural. In that regard, it more closely resembles the United States of the 1950s than, say, 1930s Japan or Stalinist Russia. The greatest risk is not that Beijing will use its military power to attack other countries, but rather that it will use its growing resources to shift the overall balance of power in China's favor, especially in Asia. It is a strategic shift that has already begun.

References

For a comprehensive overview of current trends in China and their implications for the rest of the world, read *China: The Balance Sheet: What the World Needs to Know Now About the Emerging Superpower* (New York: PublicAffairs, 2006) by C. Fred Bergsten, Bates Gill, Nicholas R. Lardy, and Derek Mitchell.

The best account of China's daunting environmental problems can be found in Elizabeth Economy's *The River Runs Black: The Environmental Challenge to China's Future* (Ithaca: Cornell University Press, 2004). For insights on the future economic prospects for the Middle Kingdom, read *The Chinese Economy: Transitions and Growth* (Cambridge: MIT Press, 2007) by Barry Naughton.

For Foreign Policy's recent coverage of China, see "The Dark Side of China's Rise" (March/April 2006) by Minxin Pei, who argues that China's economic success has blinded the world to its weaknesses. In "The Virus Hunters" (March/April 2006), Karl Taro Greenfeld reveals how Beijing tried to cover up the 2003 SARS outbreak and asks whether China will be prepared to handle the next pandemic when it strikes.

Critical Thinking

1. According to Harry Harding, what are China's biggest risks? Why?
2. Is China politically stable? Will there be another Tiananmen Square?
3. Why is Chinese nationalism on the rise? What are the implications of this?
4. Will China's rise lead to military conflict with its neighbors and the United States?

Think Again China's Military

It's not time to panic. Yet.

Drew Thompson

"China's Military Is a Growing Threat"

Not yet. After two decades of massive military spending to modernize its armed forces, amounting to hundreds of billions of dollars, China increasingly has the ability to challenge the United States in its region, if not yet outside it. But the ability to project force tells us very little about China's willingness to use it.

Certainly, China has made moves over the last few years that have stoked the China-is-a-dangerous-threat crowd in Washington. In 2007, for instance, Beijing launched a missile that obliterated a communications satellite—a dramatic and unexpected display of capability—and then kept mum for 12 days before a Foreign Ministry spokesperson finally admitted it took place, stating: "This test was not directed at any country and does not constitute a threat to any country." In May 2008, satellite imagery revealed that China had constructed a massive subterranean naval base on the southern island of Hainan, presumably a staging point to launch naval operations into the Pacific. This January, China conducted another anti-missile test, shortly after the United States announced arms sales to Taiwan.

Similar developments have reliably shown up in annual Pentagon reports on China's military expansion, not to mention in articles such as Robert Kaplan's alarmist 2005 essay: "How We Would Fight China." Even Robert Gates, the mild-mannered U.S. defense secretary, warned last year that China's military modernization "could threaten America's primary means of projecting power and helping allies in the Pacific: our bases, air and sea assets, and the networks that support them." Last fall, Adm. Robert Willard, the new head of the U.S. Pacific Command, noted that "in the past decade or so, China has exceeded most of our intelligence estimates of their military capability," implying that maybe the alarmists are onto something.

At the same time, China's leaders vehemently denounce any suggestion that they are embarked on anything other than what they have referred to as a "peaceful rise" and haven't engaged in major external hostilities since the 1979 war with Vietnam. But they also don't explain why they are investing so heavily in this new arms race. Beijing's official line is that it wants to be able to defend itself against foreign aggression and catch up with the West, as it was famously unable to do in the 19th century.

When the late Chinese leader Deng Xiaoping began the process of reform and opening in 1979, he decided that bolstering the civilian economy would take precedence over military investments. But a dozen years later, the first Gulf War served as a wake-up call in Beijing, raising concerns about how quickly an inferior army could be demolished by better-equipped Western forces. In 1991, the Pentagon unleashed some of its most advanced weapons—including stealth technology and precision-guided munitions—against the Iraqi Army, the world's fourth largest at the time. U.S. and allied forces made short work of Iraq's Warsaw Pact military hardware, and the Chinese were duly shocked and awed.

It became immediately clear that Mao Zedong's doctrine of "human wave attacks"—having more soldiers than your enemy has bullets—would not meet China's defense needs in the 21st century. From the early 1990s, China's defense planners began intensively studying doctrine and sought to acquire superior foreign technologies for their People's Liberation Army (PLA). They also made a major strategic shift by cutting the size of their force to emphasize new technologies that would enable them to catch up with the United States and other possible foes.

Should the rest of the world be worried? Taiwan, long claimed as Chinese territory and well within range of Chinese ballistic missiles and conventional forces, certainly has cause to feel threatened. Even as cross-strait relations have warmed in recent years, Beijing has positioned more medium-range missiles facing Taiwan than ever. When asked why, Beijing demurs. India, Asia's other would-be superpower, also seems increasingly on edge. Last September, Indian analysts and media loudly worried over the publication of an article by Chinese analyst Li Qiulin in a prominent Communist Party organ that urged the PLA to bolster its ability to project force in South Asia.

But it's probably too soon for Americans to panic. Many experts who've looked closely at the matter agree that China today simply does not have the military capability to challenge the United States in the Pacific, though its modernization program has increased its ability to engage the United States close to Chinese shores. And the U.S. military is still, for all its troubles in Iraq and Afghanistan, the most capable fighting force on the planet.

"China's Armed Forces Are the Biggest in the World"

Yes, but it depends on how you count. The PLA has the most people on its payroll—2.2 million active personnel (though between 1985 and 2005, it shrank by 1.7 million soldiers and is still shrinking today). That's still far more than the 1.4 million active service members in the U.S. military.

Then again, the United States also has more than 700,000 civilian Defense Department employees and significant uncounted numbers of contractors. (In Iraq and Afghanistan, there are roughly equal numbers of contractors and uniformed personnel—about 250,000 contractors to 180,000 soldiers.) But in China, uniformed PLA soldiers carry out many of the same duties that contractors perform for the U.S. military.

Arguably, the more significant figure for comparison is defense spending. Here the PLA lags far behind the Pentagon. In 2009, the U.S. military spent $738 billion on defense and homeland security. Estimates for China's annual military budget vary considerably, ranging from $69.5 billion to $150 billion, but it's clear that U.S. military spending is still several times higher than China's, the world's second highest.

And the PLA's global range is much more limited. As of last June, the United States had 285,773 active-duty personnel deployed around the world. But China operates no overseas bases and has only a handful of PLA personnel stationed abroad in embassies, on fellowships, and in U.N. peacekeeping operations.

"China's One-Child Generation Will Weaken Its Military"

Probably. The PLA's hardware is improving, but what about its recruits? China's one-child policy is widely perceived as creating a generation of spoiled, overweight boys, dubbed "little emperors," who are doted on by four grandparents while their parents toil to support them in fields, factories, and offices. Although accounts are sometimes exaggerated (in practice, many families, particularly in rural areas, have managed to have more than one child), the dramatic demographic shifts brought about by this policy, started in 1979, certainly impact the PLA. By 2006, "only-child soldiers" made up more than half of the force, up from just 20 percent a decade earlier, giving China the largest-ever military with a majority of only-children.

"The PLA Is Slow, Conservative, and Backward"

Not anymore. Although it is still no match for the mighty U.S. military, the PLA has come a long way since Mao's ragtag army defeated the Nationalists in 1949. Over the last two decades in particular, China has improved the quality, technical capabilities, and effectiveness of its enlistees and officers, even as it has shrunk the total number of military personnel.

No longer are PLA soldiers rudimentarily equipped with second-rate Soviet technology, such as the outmoded Soviet T-55 tanks from the 1950s and 1960s that they used to have and that Iraqis fielded during the Gulf War. Although not every PLA unit has cutting-edge equipment, Chinese forces are continually integrating new weapons, doctrine, training, and command-and-control systems.

China's military today is, if not a near rival to that of the United States, at least a "fast-learning organization," in the view of many close foreign observers. It is deploying weapons that neutralize key U.S. advantages, such as ballistic missiles and supersonic sea-skimming missiles that can target U.S. aircraft carriers in the region; an enlarged submarine fleet; homegrown satellite reconnaissance and communications capabilities; and recently, the demonstrated capability to eliminate satellites and intercept ballistic missiles.

That said, not all technologies have proven easy to integrate. For instance, when China first bought Russian-made Kilo-class submarines, some were reportedly out of service for extended periods and in 2002 were even shipped back to Russia for repairs. China also reportedly experienced problems servicing imported fighter planes and engines.

Since those initial stumbles, China watchers think the PLA has made significant improvements in mastering the maintenance and operation of these imported platforms. Late in 2008, China dispatched a three-ship task force to the African coast to conduct anti-piracy missions and escort vessels through pirate-infested waters off the coast of Somalia. It was the first time in modern history that Chinese military vessels—in this case, two destroyers and a supply ship—had deployed outside local waters, armed and ready for combat. Using helicopters and special operations units, the PLA Navy successfully engaged pirates and escorted hundreds of civilian ships. It has now deployed its fourth task force, demonstrating its ability to maintain a presence for an extended time.

In a nod to the fact that enlistees are often the sole support for aging parents and grandparents, the PLA has shortened service commitments. In 1998, China reduced the time conscripts must serve to two years, lessening the economic and social burdens on rural families dependent on an only son. With a significantly shortened time to train conscripts and participate in exercises, many units will likely maintain low levels of readiness. Only-child officers are also more likely to leave the PLA to enter the private sector, where they are better able to support their parents and families.

Of course, it's difficult to really assess whether an army made up of only-child soldiers will be an effective fighting force, as the PLA has not been tested in combat since the late 1970s. The PLA has found that such soldiers have better communication and computer skills than their peers with siblings. However, they haven't performed as well in other ways. Only-child recruits are not as tough; they don't like to go through the pain of intense training; they call in sick more frequently; and they struggle to perform some simple chores like doing their own laundry. If too much hand-holding is required for these recruits, the PLA could well find itself all suited up for modern warfare—but without the soldiers ready to fight it.

"China Needs Its Army to Stamp out Domestic Unrest"

No. That's the job of the People's Armed Police. While the world watched in horror as armored personnel carriers and camouflaged soldiers suppressed riots in the Tibetan capital of Lhasa in 2008 and Uighur-dominated Urumqi in 2009, many assumed it was the Chinese army marching in the streets behind Plexiglas shields. But they were mistaken. A careful look at their insignia revealed that the units were part of the People's Armed Police, not the PLA.

The People's Armed Police is a paramilitary force with a wide range of responsibilities for public security. After June 1989, when the PLA was called upon to mobilize its tanks and clear protesters from Beijing's Tiananmen Square, the military sought guarantees from the Chinese leadership that it would no longer be tasked with suppressing domestic "incidents" that it was neither trained nor equipped to handle. The People's Armed Police was then given this specific job as well as significant increases in resources, personnel, and specialized training.

It is subject to many of the same military laws and regulations issued by the central government as its PLA counterpart. However, much of the armed police is under the command of the Public Security Ministry—China's civilian police force—and its bureaus, the largest unit of which is responsible for ensuring "internal security," including crowd control and riot response. When domestic disturbances arise, the armed police is called out to control crowds and put down riots, not the PLA. While China's 2008 defense white paper claims that 260,000 armed police are on daily guard duty, other official sources claim a total force of 660,000 officers. And though the armed police is tasked primarily with domestic security, it is also expected to support the PLA in a time of war.

"China's War Plans Are All about Invading Taiwan"

That was then. Chinese military leaders in the recent past did place intense focus on preparing their armed forces to fight a "limited war" over Taiwan, fully expecting that the United States would enter the conflict. Many weapons systems the PLA acquired or developed, as well as the exercises it trained for, were largely aimed at fighting a technologically superior enemy—with particular emphasis on developing tactics to keep the United States from bringing naval assets to China's shores, a strategy known as "access denial." In the past, massive annual amphibious-assault exercises, known derisively as the "million-man swim," defined the military experiences of hundreds of thousands of conscripts.

Although simulating a Chinese D-Day on Taiwan might be a tidy demonstration of the PLA's core mission, the armed forces today are developing capabilities and doctrine that will eventually enable them to protect China's expanding global interests. The PLA's Second Artillery Corps and science-and-technology units are increasingly capable in space and cyberspace operations, and they have honed the ability to launch and operate satellites to improve communications and intelligence collection. New air and naval platforms and capabilities, such as aerial refueling and new classes of ships, also increase the PLA's ability to deploy abroad.

Official Chinese military writings now pay increasing attention to a greater range of military missions, focusing not only on China's territorial integrity, but on its global interests. From oil rigs in Nigeria to a crude-oil pipeline under construction that will connect Yunnan's capital city to Burma's port of Sittwe on the Bay of Bengal, Beijing thinks it must be able to defend its people, infrastructure, and investments in some of the world's most volatile places—much as the British did in the 1800s.

"China's Military Has Global Aspirations"

Perhaps someday. At the height of the Cold War, Soviet military vessels prowled the world's oceans, and its aircraft patrolled international airspace. By contrast, China's navy rarely leaves its home waters; when it does patrol farther afield, it still does not cross the Pacific.

But there is little doubt that China is steadily building its ability to project power beyond its shores. Milestones such as the PLA Navy's around-the-world cruise in 2002 and its anti-piracy mission off the African coast indicate that China is looking to operate more globally.

Although Beijing has not yet sought to deploy combat-capable military units to the sites of international natural disasters, in the not-too-distant future Chinese military aircraft might be delivering Chinese-made disaster-relief supplies. Having recently commissioned a hospital ship, Chinese naval strategists have already identified disaster relief as a key mission for a future Chinese aircraft carrier, while military writers discuss how to conduct regional missions to protect China's interests outside its territorial waters.

Undoubtedly, Chinese war planners see a future in which China will be able to defend itself offshore and its navy will operate beyond what is sometimes referred to as the "first island chain" (an imaginary line stretching from Japan, through Okinawa and Taiwan, and south to the Philippines and the South China Sea), eventually encompassing much of the Western Pacific up to the "second island chain" that runs from Japan southward past Guam to Australia. But whether Beijing envisions one day establishing overseas bases, or simply having the capability to project power globally when needed, is unclear.

Some wonder whether China and the United States are on a collision course. Kaplan raised the ominous possibility in the *Atlantic* that when the Chinese navy does push out into the Pacific, "it will very quickly encounter a U.S. Navy and Air Force unwilling to budge from the coastal shelf of the Asian mainland," resulting in a "replay of the decades-long Cold War, with a center of gravity not in the heart of Europe but, rather, among Pacific atolls." Unquestionably, there is deep strategic mistrust between the two countries. China's rapid economic growth, steady military modernization, and relentless nationalistic propaganda at home are shaping Chinese public expectations and limiting possibilities for compromise with other powers.

This does not make conflict inevitable, but it is cause for long-term concern and will shape U.S. efforts to avoid hostilities with China. Military-to-military contacts lag far behind the rest of the U.S.-China relationship. Taiwan is an obvious point of disagreement and the one place where the two powers could conceivably come into direct conflict. U.S. maritime surveillance activities inside China's exclusive economic zone are another contentious point. There is, however, a growing recognition that the United States and China should engage one another and seek to avoid a conflict that would almost certainly be destructive to both sides.

Despite the goose-stepping soldiers at Chinese military parades, the PLA is far from a carbon copy of the Soviet threat. For all the jargon-laden, prideful articles about China's inevitable rise in the world, Chinese strategists are cautious not to openly verbalize aspirations to conquer the globe or establish distant bases, outposts, or supply stations.

Perhaps a generation from now, Chinese military planners might be strategizing more openly about how to acquire overseas basing rights and agreements with allies where they might station their forces abroad, just as the French and British have done since the Napoleonic wars and the Americans have done more recently. But with China, that process has not begun in earnest. At least, not for now.

Critical Thinking

1. Why has China increased its military budget since the early 1990s?
2. What are the different roles for the People's Liberation Army and the People's Armed Police?
3. Does China's military have global ambitions that challenge U.S. interests?

Drew Thompson is director of China studies and Starr senior fellow at the Nixon Center.

The China Model

Economic freedom plus political repression. That's the sinister, sizzling-hot policy formulation that's displacing the 'Washington Consensus' and winning fans from regimes across Asia, Africa, the Middle East, and Latin America. But, Rowan Callick asks, for how long?

ROWAN CALLICK

From Vietnam to Syria, Burma to Venezuela, and all across Africa, leaders of developing countries are admiring and emulating what might be called the China Model. It has two components. The first is to copy successful elements of liberal economic policy by opening up much of the economy to foreign and domestic investment, allowing labor flexibility, keeping the tax and regulatory burden low, and creating a first-class infrastructure through a combination of private sector and state spending. The second part is to permit the ruling party to retain a firm grip on government, the courts, the army, the internal security apparatus, and the free flow of information. A shorthand way to describe the model is: economic freedom plus political repression.

The system's advantage over the standard authoritarian or totalitarian approach is obvious: it produces economic growth, which keeps people happy. Under communism and its variations on the right and left, highly centralized state-run economies have performed poorly. The China Model introduces, at least in significant part, the proven success of free-market economics. As citizens get richer, the expectation is that a nondemocratic regime can retain and even enhance its power and authority. There is no doubt that the model has worked in China and may work as well elsewhere, but can it be sustained over the long run?

The Communist Party of China, or CPC, rose to power in the mid-20th century after decades of civil war, starvation, and eventually the invasion of the Japanese. But under Mao, communism fell far short of its economic promise. Then, after the bitter chaos of the Cultural Revolution, which began in 1966 and culminated with the death of Mao in 1976, Deng Xiaoping carefully devised and implemented the formula through which the CPC today retains its legitimacy: the party ensures steadily improving living standards for all, and, in return, the Chinese people let the CPC rule as an authoritarian regime. This economic basis for the party's power gives it a credibility that is being projected well beyond its own borders, with all the more success because of the recent decline in the international standing of the United States, focused as it is on its tough and increasingly lonely task in the Middle East.

The economic portion of the model works like this: open up the doors—*kai fang*—and let in foreign capital, technology, and management skills, guiding the foreigners to use China initially as an export base. Engage with global markets. Let your manufacturing and distribution sectors compete with the best. Give farmers control over their own land, and support the prices of staples.

Do everything you can to lift living standards. Give your middle class an ownership stake in the newly emerging economy by privatizing most of the government housing stock for well below the market price. Corporatize as much of the state sector as you can, and then list minority stakes on the stock market to provide a new outlet for savings. But don't let the central bank off the leash; use it to maintain a hold over the currency exchange rate and other key policy levers. Keep ultimate control over the strategic sectors of the economy; in China's case, these include utilities, transportation, telecommunications, finance, and the media.

The leaders of the Deng and post-Deng years have mostly been engineers, people of a practical bent with a particular enthusiasm for pouring cement and building infrastructure. The salient features of China's economic system, which is still evolving, include increases in inputs, improvement in productivity, relatively low inflation (with the state maintaining a grip on some prices while others are gradually exposed to the market), and rising supply, especially of labor. The country has a large pool of surplus rural workers, as well as many millions more who were laid off as state-owned enterprises underwent rapid reform, emerging from welfare-focused loss centers to become market-focused profit centers.

As it became easy to import sophisticated components from elsewhere in Asia and assemble them in China, most of the Asian neighborhood has been earning bilateral trade surpluses, becoming intimately enmeshed in the Chinese economy. The services sector remains undeveloped, a massive field awaiting investment and exploitation. Huang Yiping, chief Asia economist of Citigroup, says, "The mutually enhancing effects of reform and growth were probably the secret of China's success." The regime has succeeded in one of its prime goals, to generate sufficient surplus value to finance the modernization of the economy. China holds $1.3 trillion worth of foreign reserves.

Even after 30 years of the *kai fang* strategy, however, the Chinese economy remains only selectively open. For instance, although the currency, the yuan, can be converted on the current account, chiefly for trade, its conversion on the capital account, for investment, remains strictly controlled. China is still substantially a cash economy, with little use made of Internet banking or even of credit cards or mortgages.

The People's Bank of China remains a tool of government rather than an autonomous institution, as most Western central banks now are. A large range of core industries are, by policy, fully or majority-owned by the government, and although the four "pillar banks" have attracted massive investments from Western corporations and from international shareholders, their boards and management are regularly shifted according to the needs of the party-state. Foreigners are free to establish fully owned firms in a fast-growing range of activities from manufacturing, processing, and assembly to banking and leasing, rather than being required, as before, to enter joint ventures with local partners. But the regulatory hurdles to register such companies usually take many months to negotiate. Indeed, much of China's business environment is negotiable. There appear to be few absolutes.

Nontariff barriers to trade are declining, but they remain legion, especially in the services sector. Still, more and more foreigners are successfully doing business in China "below the radar" with small operations, such as restaurants, art galleries, and marketing firms, while the latest American Chamber of Commerce survey says that 73 percent of American companies operating in China claim they are operating profitably, with 37 percent adding that their profits in China are higher than their average global profits.

This steady but cautious opening of the economy to foreigners and to domestic entrepreneurs to a defined degree has ensured that as global liquidity has soared, much of it has found its way to China. The country's very scale, with a population of 1.3 billion, is a lure in itself, but it is the nation's convulsive arrival as not merely a receptacle but a driver of globalization that best explains its attraction to international business.

Many of China's global partners require transparent governance, independent courts, enforceable property rights, and free information. None of these is present in China today, or will be unless the party surrenders a degree of political authority it has so far regarded as inconceivable. Won't pressure for these four requirements in itself apply sufficient pressure to force liberalization? Not necessarily—because China meets all four, plus a freely convertible currency and a free port, in its own city of Hong Kong, governed under the "one country, two systems" format devised by Deng. Hong Kong is a valve to release pressures that might cause a rigid centralized economy to explode.

Nor does China seem especially vulnerable to outside shocks. Daniel Rosen, principal of China Strategic Advisory, says that at the time of the Asian financial crisis a decade ago, which China largely sailed through, "the country had not opened its capital account, relied on foreign debt, floated its currency, freed monetary policy from political control, or even relinquished the role of the state as a predominant force in financial flows. A decade later, with a new sort of financial crisis unfolding in the United States subprime mortgage market, many believe the factors that insulated China in the past still buffer it today." However, he concludes, this isn't the final word: "The macroeconomic outlook is strong not because China is immune from adjustment pressures amplified by global credit conditions, but because it has a demonstrated willingness to accept adjustment where necessary."

The China Model is demonstrating this cautious adaptability by shifting its focus from inward investment to outward investment—making its foreign reserves start to build the returns it will need as it faces the demographic jolt caused by the shift in policies from Mao's "populate or perish" to the one-child urban family. "In addition," says Italian journalist Francesco Sisci, "the party has shown itself adept at co-opting potentially troublesome private sector businesspeople by recruiting them into the National People's Congress and the Chinese People's Political Consultative Conference."

When previous leader Jiang Zemin opened the Communist Party to such people, many commentators saw capitalists taking over the party. The reverse has happened, with the party extending its controls into the thriving private sector, where growing numbers of party branches are being established. But, Sisci concludes, "it is very hard to believe that in 15 to 20 years, when the middle class could be asked to pay 30 percent or more of its income in taxes, and both Chinese society and the world at large have become more open, that this middle class will remain content to stay out of politics."

No one, however, is anticipating such a shift anytime soon. In the 1990s, a presumption grew that the crowds of well-connected young Chinese returning with their Ivy League MBAs would not acquiesce to the continued unaccountable rule of the cadres. But many of them instead joined the party with alacrity. A striking example is that of Li Qun, who studied in the U.S. and then served as assistant to the mayor of New Haven, writing a book in Chinese on his experiences. After his return to China, he became a mayor himself, of Linyi in Shandong Province in the Northeast. There, he swiftly became the nemesis of one of Chinas most famous human rights lawyers, the blind Chen Guangcheng. First, Chen was placed under house arrest and his lawyers and friends were beaten because of his campaign against forced sterilizations of village women. Then, Chen was charged, bizarrely, with conspiring to disrupt traffic when a trail of further arrests led to public protests. He was jailed for four years.

Thus, best of all, in the view of many of the international admirers of the China Model, is that the leaders, while opening the economy to foster consumption, retain full political control to silence "troublemakers" like Chen. Indeed, the big attractions of China to capital from overseas has been that the political setting is stable, that there will be no populist campaign to nationalize foreign assets, that the labor force is both flexible and disciplined, and that policy changes are rational and are signaled well ahead. Economic management is pragmatic, in line with Deng Xiaoping's encomium to "cross the river by feeling for stones," while political management is stern but increasingly collegiate, the personality cult having been jettisoned after Mao and factions having faded together with ideology.

The CPC is replacing old-style communist values with nationalism and a form of Confucianism, in a manner that echoes the "Asian values" espoused by the leaders who brought Southeast Asian countries through their rapid modernization process in Singapore, Malaysia, Thailand, and elsewhere. But at the same time, in its public rhetoric, the party is stressing continuity and is assiduously ensuring that its own version of history remains correct. Historian Xia Chun-tao, 43, vice director of the Deng Xiaoping Thought Research Center, one of China's core ideological think tanks, says, "It's very natural for historians to have different views on events. But there is only one correct and accurate interpretation, and only one explanation that is closest to the truth." The key issues, he says, are "quite clearly defined" and not susceptible to debate. "There is a pool of clear water and there's no need to stir up this water. Doing so can only cause disturbance in people's minds. . . . However much time passes, the party's general judgment" on such key events won't change.

The party, for example, required its 70 million members to view, late in 2006, a series of eight DVDs made by the country's top documentary producers about the fall of the Soviet Union. The videos denigrated the Khrushchev era because it ignored the crucial role of Joseph Stalin and thus "denied the history of the Soviet Union, which in turn triggered severe problems." Young Soviet party members who grew up in this atmosphere lacked familiarity with the party's traditions, and "it was they who went on to bury the party." According to the CPC version of history, Stalin was wrongly viewed in the Soviet Union as the source of all sins, "in spite of the glories of socialism."

But the documentary series ended on an upbeat note: the Russian people are rethinking what happened, and two-thirds of those surveyed now regret the fall of the Soviet Union. "When Vladimir Putin stepped in, he reestablished pride in the country," according to the videos. For Stalin, of course, read Mao. The CPC has no intention of taking Mao's vast portrait off the Tiananmen Gate, nor burying his waxed corpse, sporadically on view at the Soviet-style mausoleum whose construction destroyed the *feng shui,* the harmony, of Tiananmen Square that Mao himself created.

In the May/June edition of the American, Kevin Hassett, director of economic policy studies at the American Enterprise Institute, explained that evidence is emerging that developing "countries that are economically and politically free are underperforming the countries that are economically but not politically free." China, of course, is in the lead of the economically free but politically unfree nations. Hassett wrote, "The unfree governments now understand that they have to provide a good economy to keep citizens happy, and they understand that free-market economies work best. . . . Being unfree may be an economic advantage. Dictatorships are not hamstrung by the preference of voters for, say, a pervasive welfare state. So the future may look something like the 20th century in reverse. The unfree nations will grow so quickly that they will overwhelm free nations with their economic might."

The kleptocrats who have ruled many developing countries in past decades have tended to come unstuck when Western aid dwindles, their own economies falter and then fall backward, and all too often rivals emerge within their armies. The China Model presents the possibility that such rulers can gain access to immense wealth through creaming off rents while at the same time their broader populations become content, and probably supportive, because their living standards also are leaping ahead. This formula also entails hard work, application, policy consistency, and administrative capacity spread through the country—hurdles where followers of China are likely to fall. But for now, they're lining up in hope and expectation.

Even some Westerners are impressed by the new China. American swimming superstar Michael Phelps said on a visit to Beijing, the host city of the Olympic Games next August, "Going to the hotel, we see Subway, 7-Eleven, Starbucks, Sizzler, McDonald's. It's like a big American city. They have everything we have in the States." In fact, they don't. They lack basic freedoms.

It is true that the Chinese people are free to consume whatever they can afford. That's novel. They have also gained in the last three decades the freedom to travel where they want, at home and abroad. They can now work for whom they want, where they want. They can buy their own home, and live where they choose (the *houkou* system of registration is breaking down). A Chinese woman can marry the man she loves, though in the cities the couple can still only have one child unless they can afford the fine for having more. The Chinese can study at any institution that will have them. They can meet anyone they like, but not in a suspiciously large group.

The emerging middle class also benefits from a "gray economy" that, a recent survey by the National Economic Research Institute led by Wang Xiaolu has discovered, is worth a breathtaking $500 billion a year, equivalent to 24 percent of China's GDP. That's why new graduates clamor for government jobs ahead of those with glamorous international corporations—because the opportunities to get rich quick during this transitional period of asset transference from the state, and to benefit hugely from rent-seeking, are so great.

But Chinese citizens can't form a political party, or any other organized group, without official permission. They can't choose their leaders. Even the ordinary CPC members have no say in their hierarchy. Sisci, Beijing correspondent of the Italian newspaper La Stampa, writes in the China Economic Quarterly that "there is something like a 75 percent avoidance rate on personal income tax" because few people anywhere choose to concede taxation without any representation. "There is a political pact," he writes. "The government allows tax evasion in return for political obedience. So far, the middle class has acquiesced: it prefers to pay less taxes and not vote, rather than buy its right to elect the government by paying more taxes."

Phone calls, text messages, and emails are likely to be screened, and many Internet sites—such as Wikipedia and BBC News—are blocked or filtered by the 30,000 "net police." Bloggers must give their real names and identity card numbers to their Internet service providers, which must in turn make them available to the authorities when asked. Tim Hancock, Amnesty International's campaign director in the UK, says, "The Chinese model of an Internet that allows economic growth but not free speech or privacy is growing in popularity, from a handful of countries five years ago to dozens of governments today who block sites and arrest bloggers."

All books published in China must bear a license code from a state-owned publishing house. Books under question are sent out to groups of retired cadres to censor. Before art exhibitions, says painter Yao Junzhong, who now sells most of his work over-seas, the local cultural bureau usually sends a list of taboos to the organizers, and through them to the artists: "You will be told not to attack communism, not to attack the party. Sex is sensitive. So is violence. But not as much as politics." He recently had to send three photos of a painting, ready for an exhibition, to a gallery owner, the cultural bureau, and the exhibition organizer. The picture showed his young son holding a gun, set in a renminbi coin. "I named it 'Qianjin,' which means advance, and also means money." A cadre couldn't put his finger on it, but felt "there must be a political element," so the organizer was told it was not appropriate.

All films must be vetted by the State Administration for Radio, Film, and Television. All print media are government- or party-owned. The party's propaganda department has recently introduced a penalty scheme for media outlets that deducts points for defying government guidance. Twelve points means closure. Every inch of public territory remains tightly defended, although warnings are usually not explicit, leaving maximum space for artists to choose to censor themselves. Leading new-wave filmmaker Jia Zhangke, winner of the top award, the Golden Lion, at last year's Venice Film Festival, says, "If we don't touch the taboo areas, we will have a lot of freedom. But then those areas grow larger. If your tactic is to guess what the censors are thinking, and try to avoid their concerns, you are ruined as an artist."

Chinese people do not expect to obtain justice from the courts, which are run by the party, the judges answerable to the local top cadres. Ordinary people, the *laobaixing,* have to negotiate their way out of any troubles if they can. They have grown accustomed to, but not accepting of, widespread corruption. They are meant to report to the neighborhood police whenever someone new comes to stay with them. A file is kept on Chinese citizens, which follows their work and home moves, but they cannot see it. There are only two legal churches that the Chinese can join, the Catholic and the Three Self (Protestant) organizations; the leaders of both are ultimately responsible to the party. Evangelism is not permitted. There are no church schools.

Freedom House, in its annual survey, gives China a ranking of "7" for political rights—the organization's lowest rating and the same as that of North Korea, Burma, and Cuba (Japan ranks "1"). China ranks only slightly higher, at "6," for civil liberties, the same as Iran, Saudi Arabia, and Zimbabwe.

In the 1980s, wishful thinking on the part of some Western observers, combined with a form of historical determinism that was, in its way, a tribute to the thinking of Hegel and Marx, had China inevitably becoming more free and democratic as it became more of a market economy. The Tiananmen massacre caused some head-scratching for a while, but Western business, in particular, tended to take the public view, when pressed, that a semicapitalist country was bound to evolve in time into a democracy, because the emerging middle class would demand it.

Now, such views have faded. Premier Wen Jiabao said during the last annual session of China's version of a parliament, the National People's Congress, that the country would remain at the present "primary stage of socialism," during which it would require continued guidance by the party, for at least another 100 years. This model of the state has power going from the top down, and accountability from the bottom up. It also leans heavily on the seductive story of China's ancient uniqueness, its serial defiance of foreign prescriptions.

This story of cultural heroism, even though it is bound up in the China Model, with its far warmer embrace of globalization than most other developing nations have conceded, has acquired a glossy appeal because of the sheer, palpable success of China's modernization drive. The nation's gross domestic product has grown at an average annual rate of more than 10 percent since 1990.

When 21 leaders controlling three-fifths of the world's economy met at the latest Asia Pacific Economic Cooperation summit in Sydney in September, The Nation newspaper in Thailand editorialized: "One could easily spot who the real mover and shaker among them was. It used to be that what the leader of the U.S. said was what would count the most. That is no longer." The new mover and shaker is China. The entire piece was reprinted by The Statesman, an influential English-language newspaper in India. Developing nations believe that, as an ideal, the China Model has replaced the American Model, especially as embodied in the "Washington Consensus," a set of 10 liberal democratic reforms the U.S. prescribed in 1989 for developing nations.

Last November, 41 African heads of state or government were flown by China to Beijing for a summit hosted by President Hu Jintao. The government ordered most cars to stay off the roads as the leaders sped to meetings and banquets. One million security forces were deployed for almost a week to ensure the summit went smoothly. Abundant affirming slogans such as "La Belle Afrique" and attention to ingratiating detail—hotel staff learning African greetings, rooms decorated with African motifs, magnificent gifts even for the thousands in the presidents' and prime ministers' retinues—marked a contrast with concerns about corruption, crime, and cruel civil war that comprise most Western encounters with Africa.

Premier Wen said that two-way trade between China and Africa would double to $110 billion by 2010, after soaring tenfold in the last decade, with fuels comprising more than half of China's imports from Africa. China is canceling its debts due from the least developed countries in Africa, setting up a $5.5 billion fund to subsidize Chinese companies' investments in the continent, and increasing from 190 to 440 the number of items that Africa's poorest countries can export to China tariff-free. Already, by a large margin, China is the biggest lender to Africa, providing $8.9 billion this year to Angola, Mozambique, and Nigeria alone. The World Bank, by contrast, is lending $2.6 billion to *all* of sub-Saharan Africa.

The Western requirement that good-governance medicine must be consumed in return for modest aid is now not only unwelcome but also, as far as many African leaders are concerned, outdated. They are no longer cornered without options. Now they've got China, which is offering trade and investment, big time, as well as aid. And more than that, they've got the China Model itself.

This is no longer the communist program that Mao Zedong tried to export with little success except in places like Peru and Nepal, where Maoists have survived long after they have vanished from China itself. It is, instead, the program that gives business room to grow and make profits, while ensuring it walks hand in hand with big, implacable government. And, of course, the China Model holds out the promise of providing the leaders of developing nations the lifestyles to which they would love to become accustomed.

This is the China Model: half liberal and international, half authoritarian and insular. Can it last?

A few writers have become mildly wealthier by forecasting doom. The best known is lawyer Gordon Chang, whose book, *The Coming Collapse of China,* published in 2001 by Random House, concluded, "Beijing has about five years to put things right." Chang made clear that he did not expect the party to pull it off. His litany of likely triggers of collapse included entry into the World Trade Organization (which happened in December 2001), within whose regulatory structure, he said, China could not remain competitive; the impossibility of reforming the 50,000 state-owned enterprises, which he said sucked up 70 percent of domestic loans while producing less than 30 percent of the economy's output; the failure of the reform of the banking system, with its huge burden of nonperforming loans and planned restructuring through inexperienced asset-management companies; the government's lack of revenues; corruption; and the rush of new global information and views made available on the Internet. Chang's critique was plausible at the time, but now, six years later, it merely underlines how dangerous it is to bet against China's pragmatic economic reform program.

Randall Peerenboom of UCLA describes in his new hook—*China Modernizes: Threat to the West or Model for the Rest?*—the country's "paradigm for developing states, a 21st-century, technologically leap-frogging variant of the East Asian developmental state that has resulted in such remarkable success for Japan, South Korea, Hong Kong, Singapore, and Taiwan." China resisted the advice of foreign experts to engage in shock therapy, he says, and has persisted in gradual reform. The state has played a key role in setting economic policy, establishing government institutions, regulating foreign investment, and mitigating the adverse effects of globalization in domestic constituencies. And the strategy has broadly worked: "Chinese citizens are generally better off" than in 1989. "Most live longer, more are able to read and write, most enjoy higher living standards. China outperforms the average country in its income class on most major indicators of human rights and well-being, with the notable exception of civil and political rights."

Where China fails to match up, however, is in creativity and innovation, without which it may have to resign itself to remaining a net importer of new technologies, and a manufacturer under license. It has failed to produce a single global brand to compare with its neighbors. Japan has its Sony, Toyota, Panasonic, Honda, and the rest. South Korea has its Samsung and Hyundai. Taiwan has its Acer, BenQ, and Giant bicycles. China's Haier white goods and Lenovo personal computers remain, for now, wannabes. The controls that China deploys on use of the Internet, the battles it wages with its artists in every field, the focus in its education system on rote learning, the continuing failure to implement its own intellectual property rules, and now the embracing of a new Confucianism—all of these inhibit lateral thinking and invention.

As Maoism and Marxism lose their grip, the dangers of nationalism as a defining value system become apparent, and religion remains under suspicion as a potentially powerful rival to the Communist Party and the authoritarian state, Chinas leaders are eagerly rediscovering the country's 2,500-year-old Confucian tradition.

Contemporary philosophers claim to be reengineering Confucianism to suit the needs of 21st-century China by, for instance, focusing on the ecological potential in its advocacy of "the unity of heaven and humanity" and on its requirement of self-discipline. But the reasons that early-20th-century modernizers and artists, including China's greatest writer, Lu Xun, rejected Confucianism as essentially authoritarian and inimical to modernization remain unaddressed.

Lee Yuan-tseh, the president of Taiwan's top research institute, Academia Sinica, describes how, after a traditional Confucian upbringing, he shifted for postgraduate opportunities to the United States, first at the University of California at Berkeley, then Harvard, then Chicago, and back to California. He says that the inquisitive academic climate there "made me think bad thoughts: that my teacher was wrong." In time, he took what he describes as the biggest step of his life—telling the teacher so. In 1986, he won the Nobel Prize for chemistry. In 2000, Gao Xingjian won the Nobel Prize for literature, but only after he had exiled himself and become a French citizen. No person has ever won a Nobel Prize for work in China; the U.S., by contrast, has won nearly 300 Nobel Prizes, winning or sharing four of the six 2006 awards.

Even in entrepreneurship and wealth creation, the CPC retains its grip. Carsten Holz, an economics professor at the Hong Kong University of Science and Technology, wrote in the Far Eastern Economic Review that "of the 3,220 Chinese citizens with a personal wealth of 100 million yuan ($13 million) or more, 2,932 are children of high-level cadres. Of the key positions in the five industrial sectors—finance, foreign trade, land development, large-scale engineering, and securities—85 percent to 90 percent are held by children of high-level cadres."

Attempts are being made to shift the economy higher up the value-added chain by creating and importing more capital-intensive companies. For instance, on a vast industrial estate southeast of Beijing, Richard Chang, a Taiwan-raised American citizen, has built a $1.5 billion microchip-making factory for his company. Semiconductor Manufacturing International Corporation. The factory employs 2,000 staff who work in "clean rooms" constantly tested for dust and humidity. The machines there cost up to $30 million each. About 55 percent of the staff have undergraduate degrees, and 10 percent are hired from overseas. All have three months of in-house training before they begin work.

In 2006, manufacturers in China bought $64 billion worth of chips, but much of that hardware was imported. Whether SMIC—which Chang founded only in 2000, after working for 20 years with Texas Instruments—can build the research capacity and the skills needed to compete will provide an important test of China's ability to move its model on to a higher plane. China, in typical fashion, threw a heap of incentives Chang's way to ensure he got up and running: free land, syndicated loans, R&D aid, zero tax.

Like most other East Asian states, China's route to development placed economic reforms before democratization. But the China Model differs markedly from most of the region in that it has resisted taking any serious steps down that road to democracy. There is talk of "intraparty democracy" within continued one-party rule, but unsurprisingly, no champions of it have any real influence. There were also some token village elections, but they have remained dominated by the CPC and its cadres. Commentator Shu Shengxiang wrote in July, on the influential website Baixing ("the common people"): "The democratically elected cadres are gouging the people [by corruption] too. They do not know that a democracy which only has elections but not supervision is at best a half-baked democracy, if not a fake democracy. Halt-baked democracy not only harms the villagers' personal interests, but even gives them the misconception that democracy is not good."

At the same time, however, the party is refining its contract with the Chinese people to reflect shifting popular concerns about living standards: the quality of growth as well as the quantum. Leading this new agenda is the environment, which surveys show tends to top, together with corruption and access to health and education services, the concerns of most Chinese. The World Bank says China contains 16 of the world's 20 most polluted cities in their air quality, and anxieties about both water and air have triggered a large proportion of China's "mass events"—demonstrations and protests—which even official figures estimated at 87,000 in 2005.

The central government has effectively opened the environment to media commentary and to the establishment of NGOs, and has pinned the blame for the worst environmental disasters on avaricious or neglectful local officials. Ambitious targets have been set, in the five-year plan that began in 2006, for reduction of carbon dioxide and sulfur emissions, and for the use of energy per unit of production. The goals are not being met, and the central government appears determined to impress the whole apparatus that its new green program is not just rhetoric, but that it means business.

This shift from quantity to quality of growth will form the core of the agenda in the second term of the current "fourth generation" of leaders around Hu Jintao and Wen Jiabao. At the party congress, which occurs every five years and started on October 15, their aim is to ensure that only people on board this quality-of-life program get promoted, both in the central institutions and in the provinces. The key principle for their decade in power is the Confucian concept of harmony—as in stability and avoidance of dissent.

This China offers a seductive model that is being eagerly taken up by the leaders of countries that have not yet settled into democratic structures: Vietnam; Burma; Laos; the Central Asian dictatorships that were part of the Soviet Union; a growing portion of the Middle East, starting with the United Arab Emirates, including its glossy new centers like Dubai; Cuba; most of Africa, including South Africa; and even to a degree the hereditary cult that is North Korea. Beijing sometimes gives more than it receives to cement its developing world leadership, according most-favored-nation status to Vietnam, Laos, and Cambodia even before they join the World Trade Organization. In an unsettling way, the China Model is attractive to the leaders of some countries that had already become democratic, such as Venezuela. The model is even inspiring democratic India to compete with its own adaptive version.

The China Model is, of course, admired in the West, too, with business leaders' words (at platforms such as Forbes magazine conferences and the World Economic Forum, which has just instituted an annual summer session in China) providing great reinforcement for Chinese leaders. The World Bank is just one of the international institutions that champion China (its greatest client and in some ways its boss) as a paradigm for the developing world. Also fascinating is the appeal of the China Model to Russia, which as Azar Gat, professor of national security at Tel Aviv University, writes in Foreign Affairs, "is retreating from its post-communist liberalism and assuming an increasingly authoritarian character as its economic clout grows." China is exporting scores of Confucian Institutes, most of them at first just language schools but in the future offering platforms for extending Chinese influence.

But back to our question: Is the China Model sustainable? Two recent books come up with opposite answers. The British center-left economist Will Hutton, author of *The Writing on the Wall,* says China must accede to Enlightenment values or start to fall back again. American journalist James Mann, author of *The China Fantasy,* argues, in contrast, that the Chinese middle class is thoroughly behind the China Model as its major beneficiary, ensuring that the usual source of confrontation with the power elite is not just docile, but eagerly applauding.

It may take a non-Sinologist like Hutton to see how very strange it is that a one-party state that pays lip service (at least) to the doctrines of Marx, Lenin, and Mao should not only survive into the 21st century after all its principal totalitarian and authoritarian peers have collapsed, fallen apart, or been beaten in world wars, but actually prosper, to the degree that its system is starting to gain such currency in the developing world.

"The party is facing a growing issue of legitimacy," he writes. "If it no longer rules as the democratic dictatorship of peasants and workers, because the class war is over, why does it not hold itself accountable to the people in competitive elections?" The answer is in the phrase Hutton himself frequently uses; "party-state." The party won power by force of arms. The People's Liberation Army answers to the party, not to the government or the nation, insofar as those concepts can be levered apart from the party any more. The party's legitimacy, as viewed by most Chinese, lies in its history and past leaders, in its contemporary success at bringing prosperity to a thankful nation, and in its unyielding grip on the trappings of nationalism.

The large gold star in China's flag, for instance, represents the CPC, the four smaller stars the workers, the peasants, the petty bourgeois, and capitalists sympathetic to the party. In the early 1980s, there was some vestigial discussion about separating party and state, but the idea was abandoned as both impractical and undesirable. The party rules today through four pillars—the army, the legal apparatus including the courts and police, the administration, and the state corporations that dominate the "strategic" sectors of the economy. Without cutting away these pillars, without separating the powers, any attempt at "competitive elections" would be hollow. But the pillars appear, to use an understatement, firmly entrenched.

Hutton reiterates the old contention that China's middle class, "more internationalist than its poor," will ultimately insist that the party loosen its political control. James Mann replies that the middle class is doing very nicely, thank you, within the structure as it is. Its members are substantially incorporated into the party and are the structure's biggest supporters, not its underminers.

Hutton's most convincing critique is that "China has no business tradition that understands the moral facet of capitalism, . . . whose 'soft' institutions (a common culture and shared purpose) are as integral to growth and sustainability as the 'hard' processes." The difficulty Chinese enterprise has in understanding, let alone absorbing and practicing, the morality and trust that are at the root of capitalism and of successful globalization is on display ever more luridly as food, drugs, toothpaste, toys, and a growing list of other products ring international alarm bells and cast a shadow over the "Made in China" brand credibility.

His core conclusion is this: "Welfare systems, freedom of association, representative government, and enforceable property rights are not simply pleasant options. They are central to the capacity of a capitalist economy to grow to maturity. . . . The party can relax its political control to allow the economic reform process to be completed. Or it can retain political control, watch the economic contradictions build, and so create the social tension that may force loss of political control."

Hutton is not the first commentator to draw up a balance sheet of economic and social pluses and minuses for China, and figure that something has to give politically. "The clock is ticking," he warns. But time keeps passing unremarkably, and one diligent and uncharismatic group of leaders quietly makes way for the next, and the party-state not only remains intact but also appears to flourish.

James Mann, however, says that if China does retain a repressive one-party political system for a long time, this "may indeed be just the China that the American or European business and government leaders who deal regularly" with the country want. The "fantasy" in his book's title is the notion that commerce will lead to democracy or liberalization. One of his scenarios is that the party is still in power 25 years from now, though perhaps called the "Reform Party." But if it is to change names, the CPC will probably find that the "China Party" is the easiest sell.

Leading Australian economist Ross Garnaut, a former ambassador to China, adds an important element of historical perspective—that first Britain, and then the United States, industrialized in a helter-skelter way, which was crucially moderated and channeled by institutional accountability that prevented the industrial-baron entrepreneurs from losing proportion, alienating the population, and misallocating capital disastrously. The institutions that developed to meet this challenge of a rampant new power elite included parliaments, legal structures, and independent regulatory agencies. Will China follow suit, or will its go-it-alone party steer its economy, after three decades of success, into difficult waters (or onto the reefs) because it lacks the true self-confidence to expose itself to other sources of power?

It is almost certain that China will push on with its present structure, but with the prospect of broadening democratic competitiveness for posts within the party, and institutionalizing consultations with more diverse groups in Chinese society, some outside the party. The "ample evidence" that Randall Peerenboom catalogs in his book, "that other countries are looking to China for inspiration," reinforces the CPC's determination to persevere. Laos is following China's lead in implementing market reforms and producing higher growth. Iran and other Middle East countries, including Syria, have invited experts on Chinese law, economics, and politics to lecture to senior officials and academics. They are all attracted by what they see as China's pragmatic approach to reform. The official newspaper China Daily recently hosted on its website a reader discussion on the theme: "China is a role model to all developing nations. After centuries of oppression and domination by Western nations, most developing nations are trying to pull themselves up from poverty. They look at China's rapid progress as an example. China also gives aid and technical help to these nations." The theme attracted a host of supportive responses, such as, "China has shown that you can be successful by expanding through commerce and diplomacy, not by the imperialism demonstrated by the U.S. and UK."

Vietnam, Cuba, Burma, and Venezuela provide good examples of the China Model's attraction. Vietnam, whose economic reform program, *doi moi,* began 20 years ago, has followed China closely, especially replicating its outward-looking foreign investment regime. As a result, strong links have been created between the two communist countries, which are also the fastest-growing economies in Asia. In 2006, China had 377 direct investments in Vietnam. Since China and Vietnam resumed official economic relations in 1991, after Vietnam had allied itself with the Soviet Union, bilateral Chinese-Vietnamese trade has grown at an annual average of 40 percent. Meanwhile, Vietnam maintains a political system as authoritarian as China's, with a ranking of "7" for political rights from Freedom House.

Vietnam's new prime minister, Nguyen Tan Dun, says he wants to ramp up economic cooperation with China, and that the countries "should increase their cooperation to accelerate trade promotion and investment, plus organize trade fairs and exhibitions, to help each other seek more trade and investment opportunities." In Vietnam's north, close to China's booming Guangdong Province, average wages and real estate are much cheaper than in coastal China. Vietnam is the most successful economically of the countries using the China Model, and its entrepreneurial talents suggest that in some areas it could in time even leapfrog it.

China's ability to honor Mao, even as it tears down the economy he set in place, could provide a model for Cuba, says William Ratliff, a research fellow at Stanford University's Hoover Institution who is an expert on both countries. "During the past 15 years, important members of the Cuban political, military, and business elite, including Fidel and Raúl Castro and two-thirds of the members of the Communist Party Politburo, have visited China and remarked with great interest on the Chinese reform experience," Ratliff says. After Raul's visit to China, Zhu Rongji, a leading architect of economic reforms who was then premier, sent one of his chief aides to Cuba, where he lectured hundreds of leaders, with substantial impact.

Ratliff cites a Cuban intelligence official as saying: "Once Fidel Castro is out of the game, other areas of the Chinese experience will most probably be implemented in Cuba rather quickly." Besides the economic model, the Chinese concept of an orderly succession of leaders within an authoritarian system also holds a deep attraction.

Former Chinese Foreign Minister Tang Jiaxuan, often sent as an emissary by President Hu to neighboring countries, said in September that "China wholeheartedly hopes that Burma will push forward a democracy process that is appropriate for the country." The statement underlines China's crucial support for the ruling military regime, which usurped the election won convincingly in 1990 by Aung San Suu Kyi's party, with 390 of 492 seats. But Tang adds a note of disquiet; China is no longer blasé about being viewed as the main backer of dictatorships, including Sudan and Zimbabwe, and would rather that Burma, like those other outcast countries, worked harder to establish better relations with the rest of the world. In his speech. Tang was supporting the Burmese rulers' plan to introduce a new constitution, and at the same time move toward a market economy, reinforcing the influence of China as their key model.

Since the start of 2007, there has been a surge in diplomatic and business visits by leaders between the countries, with the intention of strengthening economic and strategic ties. As in other countries committing themselves to the China Model, the exchanges entail Chinese businesspeople, technicians, and workers coming to live in Burma for lengthy periods. The South China Morning Post has described three Burmese cities—Lashio, Mandalay, and Muse—as "virtually Chinese cities now." China is building a tax-free export zone for its own industries next to the port of Rangoon.

Meanwhile, Venezuela's regime has become the leader of the hard left's opposition to Western-led globalization. In August 2006, President Hugo Chávez said on arriving in Beijing: "This will he my most important visit to China, with whom we will build a strategic alliance. Our plans are to create a multipolar world, and to challenge the hegemony of the United States." His attempts to enmesh China in his high-stakes campaign against the U.S. were deflected by courteous formalities, but China's economic support—chiefly through investment in energy projects and purchase of oil, despite Venezuela's heavy crude being costly to refine and expensive to transport across the globe—considerably aided Chávez's election campaign last December. Chávez praised China as an economic model for the world: "It's an example for Western leaders and governments who claim that capitalism is the only alternative. One of the greatest events of the 20th century was the Chinese revolution."

Joshua Kurlantzick, the author of *Charm Offensive: How China's Soft Power Is Transforming the World,* writes: "No one has experience with today's China as a global player. . . . In a short period of time, China appears to have created a systematic, coherent soft power strategy, and a set of soft power tools to implement it"—particularly public diplomacy, aid, and trade—"though it is still in a honeymoon period in which many nations have not recognized the downsides of Beijing's new power." Those downsides might include a cavalier approach to the environment—other people's as well as China's own—growing military clout, the migration of large numbers of Chinese workers and businesspeople accompanying its trade and investment, harsh labor standards on its projects, and in general a new variant on old colonialism.

Kurlantzick writes: "As China becomes more powerful, other nations will begin to see beyond its benign face to a more complicated reality. They will realize that despite Chinas promises of noninterference, when it comes to core interests, China—like any great power—will think of itself first.

"China could create blowback against itself in other ways, too. Still a developing country itself, China could overplay its hand, making the kind of promises on aid and investment that it cannot fulfill. And in the long run, if countries like Burma ever made the transition to freer governments, China could face a sizable backlash for its past support for their authoritarian rulers. 'We know who stands behind the [Burmese] government,' one Burmese businessman told me last year. 'We'll remember.' "

The U.S., Japan, and other countries have been urging China to become more transparent about the rapid development of its military capacity, underlined last January by its missile shot that successfully destroyed an aging satellite, and by the sudden surfacing of a submarine, earlier this year, within five miles of the American aircraft carrier *Kitty Hawk*. At least some of the increased military budget is intended to compensate the People's Liberation Army for its lost revenues when it was required, by forceful former Premier Zhu Rongji, to sell off most of its considerable business portfolio.

The principal foreign policy goals appear clear: preventing Taiwan from converting its de facto independence to de jure independence, maintaining a constant capacity to attempt an invasion, extending China's capacity to open up new forms of access to reliable sources of energy and other commodities, and helping safeguard such routes.

China's capacity to project adventurist military power far beyond its borders, or to offer significant help for other countries to do the same—for instance in Venezuela, unsettling the U.S.'s immediate neighborhood—is limited today both by its resources and by its reluctance to leave its heartland short of the muscle the party may need to quell domestic disturbances like those that swept the country in 1989.

To prevent a growing fear of China's economic power, Beijing wants to demonstrate, Kurlantzick points out, that as it grows, it will become a much larger consumer of other nations' goods, creating—in a favorite phrase of the current leadership—"win-win" economics. Chinese leaders constantly talk up the value of likely investments and trade, with total outward investment rising, according to official statistics, 1,000 percent in 2005, though this figure includes mere commitments and the total is only $7 billion, compared with $60 billion in foreign direct investment (excluding the finance sector) flowing to China.

China's soft power offensive and the lure of the China Model remain, however, entirely official government programs. Where soft power has worked durably and has permeated connections among nations and nationalities, it has also involved civil society and the media, the arts, cultural attraction—the broad range of informal human contacts. Beijing will not let such areas of life off the leash at home, let alone license them for export. Thus, its charm in the developing world remains that of the official with his jacket still on, the limousine with darkened windows waiting outside fully paid for—and the critics regularly, clinically rounded up and removed beyond earshot.

Critical Thinking

1. What is the most distinctive feature of the so-called China model?
2. Does the Chinese government advocate such a model?
3. How has the international community responded to China's development path?
4. How sustainable is the China model?

Rowan Callick is the Beijing-based China correspondent of *The Australian* newspaper.

China: A Threat to or Threatened by Democracy?

EDWARD FRIEDMAN

How can one know whether China will or will not democratize? In general, as Karl Popper showed in *The Poverty of Historicism,* political futures in even the middle distance are unknowable because of the inherently uncertain and contingent dynamics of politics. Therefore, an analyst should focus on the multiple factors that make different futures more or less likely.

In *The Black Swan,* Nassim Haleb shows that in the post–Bretton Woods age of unregulated financial globalization, an extraordinary volatility is ever more likely. Thus, practical wisdom suggests a need to hedge against the unknowable and gargantuan risks of sudden booms and busts. Not even the hedge funds know how much to hedge. Unless one can create new international institutions to regulate the new monies created since the dollar floated in 1971 and since new instruments (non-bank banks) were invented in the middle 1980s, the global forces at work will produce unimaginable futures.

The almost impossible problem is how to imagine China's democratization potential in relation to the out-of-control and unpredictable workings of the new global economy. Are there ways to conceive the issue that might be more fruitful than others?

Despite the conventional wisdom, China is not a market-Leninist system in which the economic imperatives of wealth expansion are in contradiction with the political imperatives of control-oriented, anti-market Leninist institutions. China has already evolved politically into a non-Stalinist authoritarianism. Somewhat similar transitions occurred in nineteenth-century Imperial Germany and Imperial Japan, producing regimes that were readily compatible with sustained rapid growth. There are no hidden forces of history guaranteed to undermine China's resilient authoritarianism. China is a successfully risen superpower out to shape the world in a direction consonant with the priorities and imperatives of its authoritarian ruling groups—and, more especially, to preserve the Chinese Communist Party's (CCP) monopoly of power without accountability.

In order to deal with a superpower—antidemocratic China—democracies feel compelled to become less democratic. The democratic tide, therefore, is ebbing. India constrains demonstrators. Japan fears to speak with the Dalai Lama. The European Union pulls back from democratic Taiwan and considers selling arms to China. Chinese security forces are allowed to police the Olympics torch run even in Western democracies. Chinese leaders visiting the West are protected from seeing or hearing people protesting rights abuses by the Beijing regime. Publishers hesitate to bring out works critical of the CCP.

As with the Japanese graphic novel (*manga*) *China Has One Less Bone* (a reference to the symbol for bone, which in Chinese has one less stroke), China's success is leading to a diminished appreciation and defense of freedom. Author Oda Sora misleadingly tells readers that Japan, too, restricts freedom just as China does. The distinction between authoritarianism and democracy blurs. Even the Chinese come to think that their authoritarianism is in fact just as democratic as liberal constitutionalism. In such a Chinese-defined context, the global struggle for democracy can lose its impetus and inspiration.

The People's Republic of China is not a fragile polity. It is an authoritarian success story. And authoritarian China is winning. African countries lean toward China. Westerners compete to do business there, on Chinese terms. And yet, the human desire not to live a life of fawning and scraping to arrogant and unaccountable ruling groups inevitably ignites a desire for political freedom.

The forces of potential democratization are defined by the particulars of an era and the peculiarities of a region. Barrington Moore, Jr.'s classic study *The Social Origins of Dictatorship and Democracy* offers a model for general analysis, even though his book is about the age in which agrarian empires came to an end and industrialization created new groups with very different interests. China's potential for democracy, in this type of sociohistorical logic, will be largely shaped by the dynamics of a region (Asia)—and by the groups and interests created by this socioeconomic moment: rapid industrialization and urbanization combined with post-Fordist globalization, the increasing importance of services, tourism (Macao attracts more gamblers than Las Vegas), advanced information technology, biotechnology, and a cohort that will live for another generation beyond the industrial-era retirement age of sixty-five.

To comprehend the likelihood of democratization in China, therefore, an analyst should look first at regional factors, then at the groups and interests shaped by rapid industrialization and a looming postindustrial future, and then relate both to the nature of the Chinese state.

Let's start with Asia, which is not, as many believe, a collectivist world with an overarching culture reflecting authoritarian "Asian values." Three of the six most populous democracies in the world are in Asia: India, Indonesia, and Japan, societies that are, in turn, majority Hindu, Muslim, and Buddhist. Confucianism, likewise, is not an insuperable obstacle to democratization. The most Confucian country in the world, South Korea, has democratized (and Confucianism in China is far more liberal than it is in Korea: think of the Chinese as New England Unitarians and the South Koreans as Southern Baptist Evangelicals). Similarly, Confucianism exists in Japan and Taiwan, which are both democracies. Confucian Hong Kong would have democratized if not for its retrocession to authoritarian China. In sum, Chinese culture is not a permanent barrier to the victory of democracy in China, although the CCP strives to make it so.

People in China have been struggling and sacrificing for democracy since the 1898 reform movement for a constitutional monarchy. The 1911 anti-imperialist Chinese republican revolution that toppled the Aisin Gioro ruling family's Manchu monarchy led to nationwide parliamentary elections in 1912. The May 4th movement of 1919 relegitimated democracy as China's better alternative to the chaotic warlordism that engulfed the nation after the 1913 assassination of the prime minister-to-be.

Although Chinese military mobilization against Japanese emperor Hirohito's imperial military deflected the democratic cause until an American-led coalition compelled the surrender of General Hideki Tôjô and the withdrawal of his armies from China, the post–Second World War competitors for national power appealed to the people in terms of contrasting democratic agendas. Mao Zedong's Red Armies pointed to village elections as proof that the CCP would deliver a New Democracy. Chiang Kai-shek's Nationalists drew up a new Constitution and chose a new Parliament.

When Mao's CCP, in power, actually imposed Stalinist political institutions, instead of the promised democracy, hundreds of thousands of Chinese protested Mao's betrayal of his democratic platform in the so-called Hundred Flowers liberalization of 1956–1957. During the next year, the Stalinist regime sent a million-plus democratic oppositionists to slave labor camps, where many died.

Nonetheless, a new generation joined Mao's so-called Cultural Revolution in 1966–1967, agreeing that the arbitrary rule of an unaccountable CCP was bad for China. Rebels for democracy were popular, but they were quickly suppressed.

After Mao died, a Democracy Wall movement exploded in 1978–1979. Its titular leader, Wei Jinsheng, insisted that Supreme Leader Deng Xiaoping's program of Four Modernizations needed to be supplemented by a Fifth Modernization, democracy. Wei was then imprisoned by Deng.

Nonetheless, a nationwide democracy movement grew in 1989. It was larger than many of the democracy movements in Stalinist polities in East Europe that democratized from 1989 to 1991. The attraction to democracy, the hope of joining a prosperous Western Europe, was uniquely powerful in Eastern Europe. In East Asia, by contrast, democratic Japan, the wartime invader of China, was not an attractive democratic magnet for patriotic Chinese. Dictator Deng, a survivor of the first revolutionary generation, had sufficient support within the CCP and the military to crush the democracy movement headquartered at Tiananmen Square in the Beijing Massacre of June 4, 1989. Despite constant repression of democrats after that and a ceaseless antidemocratic propaganda campaign, a Chinese Democratic Party was formed in 1998. Its leaders were imprisoned. Nonetheless, a Rights Defense movement, introduced in the writings of Merle Goldman and attracting courageous and principled lawyers, journalists, and activists to aid victims of the regime, has grown. In sum, for over a century, Chinese people have been struggling for constitutional liberties to end the humiliations, degradations, and inhumanities of selfish and arbitrary power. Chinese are quite capable of implementing a democratic project.

Why, then, cannot the democracies of the Asian region join together so that authoritarian China becomes the odd nation out, and why cannot China democratize in order to have legitimate "soft power" in the Asian region? The answer is mainly that the region will not organize and present itself as democratic. India would not welcome such a regional policy. As with democratic South Africa's policy toward Robert Mugabe's corrupt and disastrous government in Zimbabwe, Indian anti-imperialist passions preclude human rights activism against an Asian government that is seen as having struggled against colonialism. In addition, the CCP works ceaselessly to discredit the credentials of democratic Japan based on Japan's Second World War-era militarism. Consequently, Japan cannot lead other Asian democracies. Democratic Australia's huge economic gains from China's rapid economic rise preclude Canberra from joining an effort on behalf of a democratic China.

The best hope for a rights-oriented politics in Asia might come from Indonesia. Jakarta does not wish to be subordinated to an authoritarian and hegemonic China. To preempt a spread of democracy in Southeast Asia, Beijing supports the military dictatorship in Burma and the authoritarian regime in Cambodia. As with its embrace of the Uzbekistan tyrant who crushed a burgeoning democratic movement in Central Asia and as with its opposition to a united, democratic Korea in Northeast Asia, the CCP expends great energies in the southeast to preclude the spread of democracy. China's military has even created the capability for anti-democratic regional interventions. One should expect China to be militantly and militarily opposed to the spread of democracy in its geopolitical neighborhood.

These CCP antidemocratic policies are significant. Democratization tends to occur regionally—for example, after 1974–1975 in Southern Europe, subsequently in Latin America, in the late 1980s in East Asia (the Philippines, South Korea, and Taiwan), and after November 1989 in Eastern and Central Europe. The CCP regime, in contrast, aims to create an Asian region where its authoritarian ruling groups are unchallenged, in which regional institutions are inoculated against democratization. China's successes in that direction make it hard to imagine Asia, in any foreseeable future, becoming defined by a democratic ethos that makes authoritarian China seem the odd nation out.

An exception is democratic Taiwan. Starting in the 1990s, Beijing has portrayed Taiwan as a trouble-making polity and a chaotic society. But the basic interests of China's economic modernizers are to move as quickly as possible into advanced technology and Information Technology (IT). This requires improving economic relations with Taiwan, a world leader in IT. Good relations between Beijing and Taipei would increase exchanges of students, tourists, families, and entrepreneurs across the Taiwan Strait. Democratic Taiwan, over time, could come to seem to Chinese victims of a repressive, greedy, corrupt, and arbitrary political system to be China's better future.

If Singapore, in a post–Lee Kuan Yew era, would then democratize, that, too, could help make democracy seem a natural regional alternative to politically conscious Chinese. For the CCP is trying to solve its governance problems, in part, by evolving into a Singapore-type authoritarianism, a technocratic, professional, minimally corrupt, minimally cruel, one-party, administrative state. In sum, although the CCP's foreign policy works against the spread of democracy, there are some ways in which regional forces could yet initiate a regional democratization. The future is contingent on unknowable factors.

One key is Indonesia. There are political forces in Jakarta that oppose Beijing's efforts in Southeast Asia to roll back the advance of democracy. If Indonesia were to succeed, and if nations in South Asia, Pakistan, and Bangladesh, were also to democratize, it is possible to imagine politically conscious Chinese seeking to ride a wave of regional democratization, especially if Taiwan and Singapore were both admirable democratic alternatives. Although regional factors make all this unlikely, enough wild cards are in play that China's democratization is not impossible.

Having examined regional forces, we must then ask about the political possibilities inherent in the way economic forces create new social groups that interact with the different interests of state institutions. First, China's growth patterns have polarized the division of wealth such that China may soon surpass Brazil as the most unequal (but stable) major country in the world. All students of democratic transitions agree that great economic inequality makes ruling groups resistant to a democratization that they believe would put their ill-gotten gains at risk. This consensus hypothesis, that democratic transitions are more likely where economic polarization is limited, is formalized in a rational-choice model in Daron Acemoglu and James A. Robinson's *Economic Origins of Dictatorship and Democracy.*

Too much economic inequality is a huge obstacle blocking a democratic transition. The rising urban middle classes prefer to be defended by the authoritarian state rather than risk their status and fortunes in a democratic vote, where the majority is imagined as poor, rural, and vengeful against economic winners, imagined as an undeserving and traitorous upper stratum.

To be sure, there are democratic tendencies that result from the move from collective farming to household agriculture and from the rise of property rights, a new middle class, literacy, wealth, and so on—as Seymour Martin Lipset long ago argued. But an adaptable and resilient CCP regime that continues to deliver rapid economic growth is not going to be abandoned by rising classes worried about vengeance by the losers in a polarized society.

Still, China is combining rapid industrialization with a climb into postmodern service and high-technology-based growth in which industrial workers can seem a dying breed, an albatross to further growth. Core areas of industrialization are beginning to hollow out. It is possible to imagine the losers from China's continuing rapid growth—for example, sixty million laid-off former State Owned Enterprise (SOE) workers—turning against the regime. Should a global financial shock cause China to lose its export markets, instability might threaten the regime. As Haleb's *Black Swan* suggests, a full exploration of democratic possibilities should look into all the wild-card factors. The regime's economic reformers, however, could be portrayed as having sold the nation's better future to Western imperialism if Chinese lost their jobs because of an economic virus spreading from New York and London to Shanghai. And then, opponents of the government would not back a move to democracy.

The West would be seen as a fount of evil, and then both the people and the ruling groups might choose a transition to a more chauvinistic and militarist order that would renounce China's global openness as a betrayal of the nation's essence. History suggests that left nationalists within the regime, who largely control the security and propaganda apparatuses, would be militantly against any opening to democracy.

Such a neofascist ruling coalition might turn to military adventures or close China's doors in order to appeal to nativists—in ways, however, that would lose China the sources of continuing high growth. That is, neofascist hardliners might implement policies that would alienate many people in China and in Asia, and thereby create a counterforce that might find democracy attractive. But such imaginings rest too much on long-term speculations about concatenating factors leading to distant futures. Such meanderings of the mind should not be confused with confident predictions about a democratic outcome.

Still, it is clear that much depends on how the post-Mao right-authoritarian populist system relates to social contradictions. The CCP is moving toward presidential succession rules similar to what Mexico institutionalized in its earlier era of a one-party dominant presidential populism. Mexico had a one-term president for six years who chose his successor; China has a president who serves two five-year terms and chooses his successor at the close of the first. Chinese analysts fear that as economic stagnation, corruption, and debt delegitimated Mexico's presidential populism, so the same could happen with China. The danger is dubbed Latin Americanization.

Anxious analysts worry about the entrenchment of greedy local interests that resist the many adaptations required for the continuing rapid growth that wins legitimacy and stability for the regime. Ever less charismatic and weaker presidents in

China will lack the clout to defeat the vested interests who will act much as landed elites acted in the days of the *ancien régime* to block the changes required for economic growth. Resultant stagnation would create a regime crisis, as occurred in Latin America in the 1960s and 1970s, leading there to a wave of military coups, but also, in the 1980s, to a democratic opening in Mexico—because, among other things, Mexico uniquely abutted the United States and wished to benefit from greater access to the U.S. market. China has no similarly large and attractive democratic neighbor, unless globalization so reduces distance that the two sides of the Pacific seem no further apart than the English Channel did in the eighteenth century. This is a real possibility in our age of transportation and communication revolutions.

The internal Chinese analysis of a future crisis brought on by Latin Americanization should be treated seriously. But East Asian economic growth seems to me to be of a different order than Latin America's. Region is decisive. In addition, household agriculture and physical mobility in China make it likely that Kuznets curve factors, in which the economic gap narrows after an initial widening as a country develops, will operate in China in the future. That is, the forces of polarization will be reversed. Chinese household agriculture is very different from the world of the landed elites that emerged out of slave-plantation Latin America. Perhaps there will turn out to be truth to the analogy of a feudal-like CCP-type system rooted in Russian czarist feudal institutions with the repressed labor relations of plantation slavery and its aftermath. My own hunch, however, is that anxiety about Latin Americanization in China is an indicator that the regime remains preemptive, flexible, and responsive to threats and will, therefore, head off dangers to the regime, nipping them in the bud. It is a resilient regime, not a fragile one.

Although we may be seeing through a glass darkly to try to locate forces of regime instability or democratization in China, what is clear is how to analyze the forces at work that will decide whether it is more or less likely that China will democratize. An analyst should try to understand how the forces of region, of groups and interests fostered by the economic moment globally and at home, and of the state, comprehended in terms of the strength and weakness of its diverse and conflicting elements, interact. My own reading of this interaction is that democracy is not impossible, but that a far more likely outcome is either continuity, that is, evolutionary change toward a dominant-party populist presidentialism imagining itself as becoming more like authoritarian Singapore, or a transition in a more chauvinistic and militaristic direction. China is not likely to democratize in any immediate future, but it is not inconceivable.

China is a superpower probing, pushing, and pulling the world in its authoritarian direction. Japan is out of touch in imagining a superior Japan leading China into an East Asian Community, with Japan showing China the way in everything from environmentalism to shared high standards of living. For Confucian China, China is the core, apex, and leader of an Asian community. The CCP intends for authoritarian China to establish itself as a global pole.

China will similarly experience it as a threatening American arrogance for the U.S. government to assume that an incredibly successful China, imagining itself as a moral global pole leading humanity in a better direction, needs to be saved by American missionaries of democracy. The democracies might be able to promote an end to systemic abuses of human rights in China, but Americans will not be heard in Chinese ruling circles unless they abandon a democratization agenda in which change for the better in China presupposes ending the leadership role of the CCP. Appeasement is the price of long-term good relations. The alternatives seem too costly.

There is no other long-lasting basis for trustful cooperation with the government in Beijing than to accept the regime's legitimacy. CCP ruling groups imagine foreign democracy-promotion as a threat to China's—and the world's—better future, identified, of course, as at one with the interests of CCP ruling groups. Can the world afford not to treat China as the superpower it is? The CCP imagines a chaotic and war-prone world disorder of American-led democracy-promotion being replaced by a beneficent Chinese world order of authoritarian growth with stability. There may be far less of a challenge to China from democracy than there is a challenge to democracy from China.

Democracy-promoter Larry Diamond concludes in his recent book *The Spirit of Democracy* that democracy is in trouble across the world because of the rise of China, an authoritarian superpower that has the economic clout to back and bail out authoritarian regimes around the globe. "Singapore . . . could foreshadow a resilient form of capitalist-authoritarianism by China, Vietnam, and elsewhere in Asia," which delivers "booming development, political stability, low levels of corruption, affordable housing, and a secure pension system." Joined by ever richer and more influential petro powers leveraging the enormous wealth of Sovereign Investment Funds, "Asia will determine the fate of democracy," at least in the foreseeable future. Authoritarian China, joined by its authoritarian friends, is well on the way to defeating the global forces of democracy.

Critical Thinking

1. Why is today's China often compared to Imperial Germany and Imperial Japan in the 19th century?
2. Is Confucianism an obstacle to democratization in China?
3. Will an authoritarian China defeat the global forces of democracy?

Edward Friedman is a professor in the Department of Political Science at the University of Wisconsin, Madison, where he specializes in Chinese politics. A recent book is *Revolution, Resistance and Reform in Village China* (Yale University Press, 2005).

Article 5

Is China Afraid of Its Own People?

The diplomatic tussle over the East China Sea has calmed down, but a bigger foreign-policy problem awaits: China's newly empowered masses won't take 'no' for an answer, and Beijing is right to be scared.

WILLY LAM

China and Japan's recent showdown over the Diaoyu (or Senkaku, to the Japanese) archipelago seems to have cooled down with the release of the captain of a Chinese fishing vessel who was detained by the Japanese coast guard earlier this month [late summer 2010]. Quite a few official Chinese media outlets ran big headlines proclaiming that the Japanese had capitulated. Yet it's by no means clear that China was the victor.

Indeed, the extraordinary lengths to which Beijing has gone to rein in public protests over the alleged Japanese occupation of the Diaoyu, as the islands are called in China, has exposed a critical shortcoming of the so-called China model: the Chinese Communist Party leadership's inability to make effective use of public opinion to advance domestic as well as diplomatic goals. Instead of leading public opinion, these days Chinese leaders are sometimes pushed into uncomfortable stances that reduce their options.

The row with Japan is a case in point. At the height of the dispute, Chinese authorities pulled out all the stops to prevent patriotic Chinese from airing their views. Protest organizers of protests, such as the editors of www.cfdd.org.cn, a website well-known for its advocacy of Diaoyu-related issues, were given warnings by the police "not to break the law" by holding demonstrations and other radical actions.

The few hundred activists who joined rallies on Sept. 18, 2010—which marked the 79th anniversary of the Japanese invasion of China's northeastern provinces—in cities including Beijing, Guangzhou, Shanghai, and Shenzhen, were subjected to tight surveillance by police, who outnumbered the demonstrators by at least four to one. The protesters were dispersed by law-enforcement agents within an hour or so.

On Sept. 12, Chinese police prevented a group of nationalist activists from renting a boat to sail from Fujian province to the Diaoyu islets to proclaim Chinese sovereignty. A similar action 10 days later by a patriotic NGO in Hong Kong was foiled by the local administration, which stopped the fishing vessel on the grounds that it was not licensed to carry passengers.

One reason Beijing is so nervous about demonstrations is that based on past experience, "troublemakers" often take advantage of such rare occasions to air grievances regarding nondiplomatic issues, especially corruption within party and government departments. That explains why at least nine activists, according to the watchdog Chinese Human Rights Defenders, were detained or warned not to participate in the rallies in Beijing and Guangzhou. Among them were Xu Zhiyong, a lecturer at Beijing University of Posts and Telecommunications, and Teng Biao, a lawyer. Xu and Teng are well-known NGO activists who have stood up for victims of official corruption.

Yet the most important reason why party authorities are paranoid about public protests is that aside from casting aspersions Tokyo's way, demonstrators might also zero in on Beijing's failure to do anything substantial to recover the lost territory. Sino-Japanese wrangling over the Diaoyu/Senkaku islands dates back to the early 1970s, when Washington returned the archipelago to Japan, but Beijing's actions have never gone beyond rhetorical assertions of its "sovereignty since time immemorial."

Nor are they likely to. Despite the leaps-and-bounds development of the Chinese Navy, a military solution seems out of the question. The islets fall within the Japanese-American mutual defense treaty, a fact that was reiterated by U.S. Secretary of State Hillary Clinton when she met visiting Japanese Foreign Minister Seiji Maehara in New York last week [2010].

A more realistic solution is the one advocated by late patriarch Deng Xiaoping when he visited Japan in 1978: seeking joint development of the islands, which are rich in natural resources, while shelving sovereignty concerns. Deng said on the occasion that it might be better to let "future generations, which may be wiser" to tackle the sovereignty imbroglio. Deng's statement, which could be interpreted as legitimizing the status quo of the Diaoyu being run by Japan on a de facto basis, has never been given much publicity in China. It is also not mentioned in high-school history textbooks.

Why? Why does China fear its own people so much?

Apart from the party leadership's well-known tradition of undemocratic governance, the main reason behind "black-box diplomacy" is to avoid taking responsibility for failing to stand up to foreign powers such as the United States or Japan. Despite the relative efficacy of the Great Firewall of China, fast-growing numbers of nationalists have frequently been able to use the Internet to express their views, including negative ones about Beijing's foreign and security policies. These increasingly vocal nationalists generally believe that rising China has become a mature power and deserves a place in world affairs to match its burgeoning economic clout.

It is out of fear of a nationalist backlash that China's negotiations with the United States and other countries regarding its accession to the World Trade Organization for instance, were wrapped in secrecy. Beijing apparently worried that should ordinary Chinese learn about the considerable concessions that it had made in areas including tariff reductions, senior cadres including former Premier Zhu Rongji would be labeled "traitors" by WTO opponents.

The same fears shrouded negotiations with Russia regarding a treaty that ended decades of disputes over the two countries' shared 2,700-mile border. The pact, which was officially signed in 2008, was mainly negotiated between former Presidents Jiang Zemin and Boris Yeltsin. It legitimized Russian ownership of huge chunks of Chinese territories—estimated to be 40 times the size of Taiwan—that had been taken away from China in the days of the czars.

Yet in other cases, party leaders' refusal to engage the public—including China's increasingly well-educated and sophisticated middle class—in the formulation of foreign policy has considerably reduced the room to maneuver of officials and diplomats.

For example, President Hu Jintao and then Japanese Prime Minister Yasuo Fukuda reached a theoretical accord in mid-2008 to settle sovereignty disputes over the East China Sea. The agreement was largely based on the principle of "seeking joint development while shelving sovereignty."

Again, Beijing made no efforts to explain to its citizens the rationale behind the potentially win-win solution. When the East China Sea accord was announced a couple of weeks after Hu left Tokyo, Chinese netizens expressed massive disapproval, even on official websites. Since then, Chinese diplomats have dragged their feet in negotiations on transforming the Hu-Fukuda theoretical agreement into a formal treaty.

But never has Beijing tried to persuade the Chinese public of the wisdom of compromise. And in the past year, hard-liners including military hawks have openly expressed disapproval of the formula of "seeking joint development while shelving sovereignty disputes."

Yet another drawback of Beijing's nontransparent foreign policy making is that China tends to resort to dubious if not irrational measures to appease nationalists. During this recent dispute, Beijing brandished economic cards that included discouraging Chinese from visiting Japan as tourists and, reportedly, threatening to cut off the export of rare-earth metals to Japanese companies.

These tactics were in essence no different from the familiar rallying cry to "boycott Japanese products," which has been frequently raised by Chinese nationalists. Yet as Ambassador Wu Jianmin, former president of China Foreign Affairs University, pointed out last week, "In this day of globalization, 95 percent of Sony's products are made in China. Isn't it stupid to call for 'boycotting Japanese products'?"

More broadly, the Communist Party leadership's recent assertiveness has stoked the flames of the "China threat" theory—and prompted countries including Japan, South Korea, India—and several Southeast Asian countries to join the "anti-China containment policy" supposedly spearheaded by Washington. Beijing's apparent effort to pre-empt criticism from nationalist Internet users has resulted in a radicalization of Chinese diplomacy that might undercut China's global clout.

What now? Before Beijing can effectively navigate a host of sensitive sovereignty issues with its neighbors, President Hu and his Politburo colleagues must first seek an understanding with the Chinese public on the parameters of China's national interests—and how to achieve them through well-recognized international norms. In the long run, continuing to treat the Chinese people like yet another threat to be neutralized will only create a self-fulfilling prophecy.

Critical Thinking

1. Why is the Chinese government nervous about public protests, even if they are not aimed at the communist party?
2. What is the main purpose of China's "black-box diplomacy"?

Five Reasons Why China Will Rule Tech

Recent development points to growing concern in Washington about China's tech moves, but here's why it may be unstoppable.

PATRICK THIBODEAU

Washington—China's focus on science and technology is relentless, and it's occurring at all levels of its society. Its labor pool is becoming increasingly sophisticated, its leadership is focused on innovation, and the country is adopting policies designed to pressure U.S. firms to transfer their technology.

The trend is causing increasing worry in Washington, but there are five reasons why China may yet succeed in its goal to achieve world dominance in technology.

1. China's Leadership Understands Engineering

In China, eight of the nine members of the Standing Committee of the Political Bureau, including the Chinese president, Hu Jintao, have engineering degrees; one has a degree in geology.

Of the 15 U.S. cabinet members, six have law degrees. Only one cabinet member has a hard-science degree—Secretary of Energy Steven Chu, who won the Nobel Prize in physics in 1997, has a doctorate in physics. President Barack Obama and Vice President Joe Biden have law degrees.

2. China's Leadership Wants to Out-innovate the U.S.

China's political leadership has made technological innovation a leading goal in everything from *supercomputers* to nanotech. One highlight of this is China's investment in clean energy technologies.

In March, the Pew Charitable Trusts reported that China led the U.S. in clean energy investments. Last year, the country invested $34.6 billion in clean energy, nearly double the U.S. total of $16.8 billion, Pew said.

"It's very sad that Americans spend more on potato chips than we do on investment in clean energy R&D," said John Doerr, a partner at venture capital firm Kleiner Perkins Caufield & Byer, at a forum in June with *Microsoft* Chairman Bill Gates. He *warned of a threat to U.S. future* if the country doesn't increase its contribution to clean energy research.

3. China's Science and Technical Talent Pool Is Vast

The technical labor pool in China is so large that Shanghai-based offshore *outsourcing* company Bleum Inc. can use an IQ test to screen applicants, with a *cutoff score for new computer science graduates in China of 140.* Less than 1% of the population has a score that high.

Bleum has started hiring a U.S. workforce but sets an IQ score of 125 as a screening threshold because of the smaller labor pool. The company employs 1,000 people in China.

One *data point* to note: In 2005, the U.S. awarded 137,500 engineering degrees, while China awarded 351,500, according to a workforce study last year.

4. The U.S. Is Failing at Science and Math Education

A stark assessment of the U.S. failure in science and math education was made *by U.S. Sen. Kay Bailey Hutchinson (R-Texas)* at a Senate hearing in May, when she compared the performance of students in Texas to those in China.

"In my home state of Texas, only 41% of the high school graduates are ready for college-level math (algebra), and only 24% are ready for college-level science (biology)," said Hutchinson.

"Furthermore, only 2% of all U.S. 9th-grade boys and 1% of girls will go on to attain an undergraduate science or engineering degree. In contrast to these troubling numbers, Mr. Chairman, 42% of all college undergraduates in China earn science or engineering degrees," she said.

5. China Is Getting U.S. Technology, All of It

In 2008, Sony Corp. closed what was identified as the last television manufacturing plant in the U.S., in Westmoreland, Pa. It shifted work to an assembly plant in Mexico, but the vast majority of TVs' electronics components are made in Asia. (*Dell sources $25 billion annually alone* in components from China, for example).

One year prior to the television plant's shuttering, Alan Blinder, a professor of economics at Princeton University and former adviser to the Clinton administration, told lawmakers at a congressional hearing that TV sets had become a commodity and that the loss of the manufacturing jobs *was an indication of economic success,* since it demonstrated that the U.S. had moved on to the production of higher-value goods.

"If we are to remain big exporters as the rest of the world advances, we must specialize in the sunrise industries, not the sunset ones," he said.

But Andy Grove, co-founder of *Intel,* wrote in *an article this month for Bloomberg* that he believes Blinder got it wrong.

The loss of the TV manufacturing wasn't a success, Grove contended. "Not only did we lose an untold number of jobs, we broke the chain of experience that is so important in technological evolution," he wrote.

China's goal is not to just build TV sets and computer components. It has established what it calls an indigenous innovation policy, meaning it wants Chinese-origin technology that is owned by Chinese companies.

This policy, "designed to encourage technology transfer and force U.S. companies to transfer R&D operations to China, will force U.S. companies to transfer technology in exchange for access to its markets," U.S. Commerce Secretary Gary Locke testified at a Senate hearing in June.

China's indigenous innovation policy may be showing results.

One Chinese-owned company, Dawning Information Industry Co., which makes servers for China's market and some foreign markets, *just built the world's second-fastest supercomputer.* The company includes a photo on its website of President Hu Jintao during a visit, illustrating the attention China's government is giving to supercomputing.

China built this system, which it called Nebulae, using Intel chips, but China has its own developing chip technology, and if it follows through on its innovation policy, it's only a matter of time before a Chinese-origin chip is used in future supercomputers.

Critical Thinking

1. What are the five reasons that China may succeed in its goal to achieve world dominance in technology?
2. What is the purpose of China's "indigenous innovation policy"?

Patrick Thibodeau covers SaaS and enterprise applications, outsourcing, government IT policies, data centers and IT workforce issues for *Computerworld.* Follow Patrick on Twitter at *@DCgov* or subscribe to *Patrick's RSS feed.* His e-mail address is pthibodeau@computerworld.com.

- In your view, what measures should the U.S government adopt to maintain technological leadership of the U.S?

Article 7

How Being Big Helps and Hinders China

DAVID PILLING

After a great deal of time spent travelling in China, reading about China and thinking deep thoughts about China, I have come to the conclusion that the most profound thing one can say about it is this: China is exceedingly big. This may seem like an awfully trite observation and, frankly, a terrible waste of the Financial Times' money. But China's sheer size helps to explain much about the country, from its impact on global commodity markets to the fact that one of the world's poorest countries is now routinely mentioned in the same breath as the (still) mighty US. El Salvador, a nation with roughly the same standard of living as China, barely gets a look-in.

China's 1.3bn multiplier effect makes almost everything it does seem extravagantly important. In some cases, its scale changes the very nature of the obstacles that confront it and the opportunities it can create.

China's size is sometimes a distinct drawback. Take the controversy over the renminbi and China's trade surplus with the US. In fact, China's path to economic take-off has been fairly standard. Like Japan, South Korea and Taiwan before it, it has relied on external demand to kick—start industrialisation, bending the rules in its favour when that has suited its development needs. But unlike those countries, it has been "found out" much earlier. By the time Japan was causing serious trade friction with the US in the 1980s, it had all but caught up with western living standards. China is provoking anger on Capitol Hill at a time when its per capita income, even on a purchasing power parity basis, is just one-seventh of US levels. If only China were a 10th the size, no one would even have noticed the level of its (non-convertible) currency.

In other areas, too, size counts against Beijing. Its outsized need for oil, iron ore, bauxite, lumber and so on affects the very commodity markets on which it relies. To get an idea of the scale of Chinese demand, look at coal, where China imports just 3 per cent of its needs. Even so, it accounts for roughly one-fifth of global seaborne trade in that commodity. Similarly, it consumes roughly half the world's cement, a third of its steel and a quarter of its aluminium. Each year, it adds 105GW of power to its electricity grid, greater than the entire generating capacity of India. All of this has a material—not to say decisive—effect on the prices of the commodities on which it depends. Chinese demand, for example, helped to propel oil to an uncomfortable $140 a barrel in 2008 and has kept a floor under it ever since.

China's hunger for commodities also has business, as well as diplomatic, consequences. Beijing's chagrin at paying what it considers monopoly prices for iron ore led it to pounce on Rio Tinto, $19.5bn in hand, causing friction with Australia. It has backed Sinochem's attempt to trump BHP Billiton's $39bn hostile offer for Canada's PotashCorp because of its concerns about food security. Likewise, it has scoured Africa, central Asia, Latin America and sometimes-hostile corners of the globe for oil and other resources, thrusting it into tricky areas of foreign policy before it might have wanted.

There are also genuine questions about whether the world has enough resources—at whatever price—to satisfy China's gargantuan appetites. If every Chinese person lived like an American, not only would there be a terrible shortage of Hawaiian shirts and loud trousers, but, according to the Earthwatch Institute, it would also mean raising global oil production by 20m barrels a day to 105m and increasing the output of grain and meat by between two-thirds and four-fifths.

The vast scale of China threatens to constrain its growth. But its size also confers great advantage. China's huge internal market gives it the economies of scale to develop globally competitive industries from cars to high-speed rail. The sheer size of its economy has made Hong Kong a globally important financial market and could, in just a few years, see Shanghai become a top-three exchange in terms of market capitalisation of its listed companies.

China's size has been crucial in getting it this far. Its seemingly limitless supply of cheap labour was a magnet for foreign investment and technology. Now, its potentially endless queues of shoppers are having the same effect. As more and more Chinese are sucked into the urban workforce, China has the opportunity to turn them into purchasers of its own products, weaning its economy off the export-dependency that still afflicts Japan. China's size means it can't do anything without making waves. But it also gives it a much better chance of riding out the rough seas ahead.

Critical Thinking

1. How does China's size affect its impact on global affairs?
2. Do Western concerns about China stem from its size only?

China's Team of Rivals

A financial meltdown in China promises to test the Communist Party's power in ways not seen since Tiananmen. But theirs is a house divided, as princelings take on populists and Pekinologists try to make sense of it all. Will this team built for economic success implode once the money dries up? An insider's guide to the leaders at China's controls.

CHENG LI

The two dozen senior politicians who walk the halls of Zhongnanhai, the compound of the Chinese Communist Party's leadership in Beijing, are worried. What was inconceivable a year ago [2008] now threatens their rule: an economy in freefall. Exports, critical to China's searing economic growth, have plunged. Thousands of factories and businesses, especially those in the prosperous coastal regions, have closed. In the last six months of 2008, 10 million workers, plus 1 million new college graduates, joined the already gigantic ranks of the country's unemployed. During the same period, the Chinese stock market lost 65 percent of its value, equivalent to $3 trillion. The crisis, President Hu Jintao said recently, "is a test of our ability to control a complex situation, and also a test of our party's governing ability."

With this rapid downturn, the Chinese Communist Party suddenly looks vulnerable. Since Deng Xiaoping initiated economic reforms three decades ago, the party's legitimacy has relied upon its ability to keep the economy running at breakneck pace. If China is no longer able to maintain a high growth rate or provide jobs for its ever growing labor force, massive public dissatisfaction and social unrest could erupt. No one realizes this possibility more than the handful of people who steer China's massive economy. Double-digit growth has sheltered them through a SARS epidemic, massive earthquakes, and contamination scandals. Now, the crucial question is whether they are equipped to handle an economic crisis of this magnitude—and survive the political challenges it will bring.

The year 2009 marked the 60th anniversary of the People's Republic, and the ruling party is no longer led by one strongman, like Mao Zedong or Deng Xiaoping. Instead, the Politburo and its Standing Committee, China's most powerful body, are run by two informal coalitions that compete against each other for power, influence, and control over policy. Competition in the Communist Party is, of course, nothing new. But the jockeying today is no longer a zero-sum game in which a winner takes all. It is worth remembering that when Jiang Zemin handed the reins to his successor, Hu Jintao, in 2002, it marked the first time in the republic's history that the transfer of power didn't involve bloodshed or purges. What's more, Hu was not a protégé of Jiang's; they belonged to competing factions. To borrow a phrase popular in Washington these days, post-Deng China has been run by a team of rivals.

This internal competition was enshrined as party practice a little more than a year ago. In October 2007, President Hu surprised many China watchers by abandoning the party's normally straightforward succession procedure and designating not one but two heirs apparent. The Central Committee named Xi Jinping and Li Keqiang—two very different leaders in their early 50s—to the nine-member Politburo Standing Committee, where the rulers of China are groomed. The future roles of these two men, who will essentially share power after the next party congress meets in 2012, have since been refined: Xi will be the candidate to succeed the president, and Li will succeed Premier Wen Jiabao. The two rising stars share little in terms of family background, political association, leadership skills, and policy orientation. But they are each heavily involved in shaping economic policy—and they are expected to lead the two competing coalitions that will be relied upon to craft China's political and economic trajectory in the next decade and beyond.

One thing is for sure: They have the profoundly difficult task of quickly and effectively transforming the country's long-standing model of export-led development. That task will require a delicate balance of innovative reforms, further market liberalization, and occasionally, strong government intervention to reshape China's economy into one driven largely by domestic demand. It is a daunting challenge, particularly when the men at the helm differ so profoundly. There are bound to be power struggles. But there is also a good chance that these everyday rivals, understanding that the party's survival hangs in the balance, will put aside infighting to guide China out of the crisis.

The team of rivals arrangement is not a choice, but a new necessity for the Chinese leadership. In elevating both Xi and Li in 2007, Hu signaled the importance of the different constituencies each represents and the belief that only consensus-building will successfully forestall serious political

The Next Generation

The Populists

Protégés and confidants of President Hu Jintao, populists are primarily concerned with addressing social problems and political tensions. They are also well versed in organizational skills and propaganda. The leaders are known as *tuanpai,* for the Communist Youth League through which most of them rose in the ranks.

Li Keqiang

Executive Vice Premier and Likely Successor to Premier Wen Jiabao

Born into a humble family in which his father was a low-ranking local official, Li began his career as a farmer and then rose through the ranks of the Chinese Communist Youth League. Li went on to attend the prestigious Beijing University and earn an undergraduate degree in law and a PhD in economics. The 53-year-old is known for his enthusiasm for a "harmonious society"; he frequently stresses the importance of helping the unemployed, providing affordable housing, and improving access to healthcare.

Li Yuanchao

Director of Party Organization

Li Yuanchao hails from a high-ranking family, and his father served as the vice mayor of Shanghai in the early 1960s. A graduate of Beijing University, Li studied briefly in the United States as a visiting scholar at Harvard University's Kennedy School of Government. Now head of the powerful Department of Organization, responsible for staffing positions within the party, the 58-year-old Li has been known to call for bolder democratic reforms, such as allowing the public to evaluate local officials.

Wang Yang

Guangdong Party Secretary

Party secretary of China's richest province, Wang grew up in Anhui Province, serving the local provincial leadership in the early 1980s. Some China watchers attribute his rapid rise through the ranks to Hu Jintao's mentorship. Last year, the 53-year-old Wang used the phrase "thought emancipation" four times in his inauguration speech, signaling an intention to make political development, not economic growth, Guangdong's principal objective.

The Elitists

Having cut their teeth on finance early in their careers, the core group of the elitists is called the "princelings" because its members come from some of China's most politically powerful families. They are tasked with representing the interests of entrepreneurs and the coastal business communities.

Xi Jinping

Vice President and Heir Apparent to President Hu Jintao

The son of a powerful former Politburo member, Xi was a farmer and party secretary in a Shaanxi Province village for six years during the Cultural Revolution. His family's station allowed him to later serve as a *mishu,* or personal secretary, to the then minister of defense. The 55-year-old Xi's résumé is filled with leadership posts in prosperous coastal cities and provinces, and he is known to favor pro-market reforms and the private sector.

Wang Qishan

Vice Premier

A former banking chief, Wang took over as mayor of Beijing during the 2003 SARS epidemic. He is credited with reassuring a jittery public that the government was in control, while acknowledging the severity of the crisis for the first time. The 60-year-old Wang, who labored during the Cultural Revolution in rural Shaanxi Province, is also the son-in-law of a conservative party elder and the protégé of former Premier Zhu Rongji.

Bo Xilai

Chongoing Party Secretary

Bo is secretary of the most politically important of the inland municipalities. The son of one of China's "Eight Immortals," elderly members of the Communist Party who held great power in the 1980s and 1990s, the 59-year-old Bo is media savvy and often politically controversial. As the former head of the Ministry of Commerce, Bo raised his profile by attracting record foreign investment.

upheaval in the so-called fifth generation of leaders, of which Xi and Li are members. The idea of turning rivals into allies "for the sake of the greater good," as Abraham Lincoln put it, has been widely cited in the Chinese media. A recent article published in *China Youth Daily,* one of the most popular newspapers in the country, called the "team of rivals" (*zhengdi tuandui*) a "brilliant idea to achieve political compromise in order to maximize common interest and political capital for survival."

The two groups can be identified as the "populists" and the "elitists." The populists are currently led by President Hu Jintao and Premier Wen Jiabao. Members of their core group, including Li Keqiang, Director of Party Organization Li Yuanchao, and Guangdong Party Secretary Wang Yang, are known as *tuanpai,* after the Chinese Communist Youth League through which they advanced their careers. Most tuanpai—they now make up 23 percent of the Central Committee and 32 percent of the Politburo—served as local and provincial leaders, often in poor inland provinces, and many have expertise in propaganda and legal affairs. President Hu is himself a tuanpai, and the leaders of this faction are widely regarded as his longtime confidants; most of them worked directly under Hu in the early 1980s, when he headed the youth league. Tuanpai are known for their organizational and propaganda skills, but they are lacking when it comes to handling the international economy. Their credentials weren't as highly valued in the Jiang Zemin era, when foreign investment and economic globalization were stressed above all else, but they are considered critical now as the risks of social unrest and political tensions increase.

The elitist coalition was born in that Jiang era, and though its two current leaders—Wu Bangguo, chairman of the national legislature, and Jia Qinglin, head of a national political advisory body—are little known outside China, they are among the country's highest-ranking political leaders. Members of the core group of the fifth generation elitists, including Xi Jinping, Vice Premier Wang Qishan, and Chongqing Party Secretary Bo Xilai, are known as princelings because they are the children of former high-ranking officials. The fathers of Xi, Wang, and Bo, for example, all once served as vice premiers. Princelings command 28 percent of the seats in the Politburo today. Most princelings grew up in the richer coastal regions and pursued careers in finance, trade, foreign affairs, and technology. Although patron-client ties are not always strong among the princelings themselves, the shared need to protect their interests, especially in a time of growing public resentment against nepotism, is what binds them together.

Of the six members of the fifth generation serving on the Politburo today, three are tuanpai and three are princelings. The policy differences between these factions are as significant as the contrasts in their backgrounds. To a great extent, their differences reflect the country's competing socioeconomic forces: Princelings aim to advance the interests of entrepreneurs and the emerging middle class, while the tuanpai often call for building a harmonious society, with more attention to vulnerable social groups such as farmers, migrant workers, and the urban poor.

The platforms of Xi Jinping and Li Keqiang, for example, are strikingly divergent. Xi's enthusiasm for market liberalization and the continued development of the private sector is well known to the international business community. Not surprisingly, his primary policy concerns include making the economy more efficient, keeping GDP growth high, and deepening China's integration into the world economy. Xi is particularly interested in keeping wealthy elites in China's eastern coastal region happy.

By contrast, Li Keqiang is more concerned about the plight of the country's unemployed. He has made affordable housing more widely available and understands the importance of developing a rudimentary social safety net, beginning with the provision of basic healthcare. The rejuvenation of the northeastern provinces, China's old industrial base and one of its most labor-intensive areas, appears to be Li's regional focus. For Li, reducing economic disparities is far more urgent than enhancing economic efficiency. These diverging policy priorities between Xi and Li will likely grow in importance as the men respond to pressing economic questions, such as how China should react to foreign pressure on the value of the yuan and how the government should proceed with its stimulus plan.

Despite their many differences, the fifth generation of tuanpai and princelings share a common trauma: They are part of China's "lost generation." Born after the founding of the People's Republic, they were teenagers when the Cultural Revolution broke out in 1966. They lost the opportunity for formal schooling as a result of the political turmoil, and many of them were the "sent-down youths," young men and women who were moved from cities to rural areas and who worked for many years as farmers.

Princelings Xi Jinping and Wang Qishan were sent from Beijing to Yanan, in Shaanxi Province, where they spent years on farms. Tuanpai Li Keqiang and Li Yuanchao labored in some of the poorest rural areas in Anhui and Jiangsu provinces. Such arduous and humbling experiences forced these future leaders to cultivate certain traits, such as endurance, adaptability, foresightedness, and humility. They not only had the unusual opportunity to come to know rural China, but they also had to adjust to a completely different socioeconomic environment. This adjustment forced them to learn at an early age how to handle challenges and how to compromise. Xi Jinping recently told the Chinese media that his time in Yanan was a "defining experience," a "turning point" in his life.

If there is another event that approaches the importance of the Cultural Revolution in the lives of these men, it is undoubtedly the Tiananmen Square incident in 1989. We don't have much information about how the incident affected them individually, but they are a generation older than many of the protesters, and at the time, several were municipal leaders or chiefs of the youth league. It is clear that they appreciate, as a group, that China's leadership during Tiananmen was deeply divided over how to respond to the unrest. They also realize that the internal struggle aggravated the crisis and ultimately culminated in a brutal response.

These events taught the fifth generation two lessons: First, they must maintain political stability at all costs, and second, they should not reveal their fissures to the public. Although these leaders wear their differences on their sleeves, there is solidarity at the highest level, inspired by past unrest, to avoid any sign of a split in the leadership, which would be dangerous for the party and for the country.

So, what do these profound differences and influential shared experiences tell China watchers about how the next generation will steer the Chinese economy? The economic prowess of the princelings will be essential to responding to the macroeconomic challenges the country will face this year [2009] and beyond. And the sensibilities of the tuanpai, versed as they are in organization and propaganda, will be invaluable as China responds to social problems born of—and exacerbated by—economic stagnation.

The rise of the team of rivals arrangement may result in fewer policies aimed at maximizing GDP growth rates at all costs. Instead, it might give way to policies that provide due consideration to both economic efficiency and social justice. Already, the ongoing global financial crisis has driven the leadership to change its emphasis from export-led growth to encouraging domestic demand, which means addressing rural needs. An ambitious land reform plan, which was adopted in the fall of 2008, promises to give farmers more rights and market incentives to encourage them to subcontract and transfer land. This strategy aims to increase the income of farmers, reduce economic disparity, promote sustainable urbanization, and ultimately end the century-long segregation between rural and urban China. Some analysts think that this land reform, along with a nearly $600 billion stimulus plan announced in November that favors railroad construction and rural infrastructure development, will greatly boost the country's domestic economy and hopefully propel China through the current economic crisis.

Although the land reforms largely reflect President Hu's agenda and the influence of the populists, leaders from the elitist camp have also been supporters of these policy initiatives. Political compromise and consensus-building, not zero-sum factional infighting, have shaped the rural development and stimulus plans.

But China's new game of elite politics may fail. What will happen, for instance, if economic conditions continue to worsen? Factionalism at the top might grow out of control, perhaps even leading to deadlock or outright feuding. Different outlooks over many issues—including how to redistribute resources, establish a public healthcare system, reform the financial sector, achieve energy security, maintain political order, and handle domestic ethnic tensions—are already so contentious that the leadership might find it increasingly difficult to build the kind of consensus necessary to govern effectively.

Barring something entirely unexpected, though, the populist policy platform will prevail over the next three to four years, and the ongoing global financial crisis will likely push Chinese leaders to increase government intervention in the economy. Yet there may be a swing in the opposite direction in 2012 as princeling Xi Jinping succeeds Hu Jintao, similar to the transition from Jiang to Hu. The establishment of such shifts during transitions at the top can create a healthy political dynamic that prevents one faction from wielding excessive power. Because of new leaders' differences in expertise, credentials, and experiences, contending coalitions will realize that they need to find ways to coexist in order to remain in power. They do, after all, have a common interest in social stability and the shared aspiration to further China's rise on the world stage. Given China's long history of arbitrary decision-making by one individual leader, this "one party, two coalitions" practice represents a major step forward—for the party and the people.

Critical Thinking

1. What is the source of legitimacy of the Chinese Communist Party today?
2. What are the two elite groups that compete for power? Who are their respective leaders?
3. What are the consequences of the two teams competing for power?

Cheng Li is research director and senior fellow at the Brookings Institution's John L. Thornton China Center and, most recently, editor of *China's Changing Political Landscape: Prospects for Democracy* (Washington: Brookings Institution Press, 2008).

China Offers Direct Line to Its Leaders

Citizens invited to e-mail party bosses Web strategy part of Beijing PR push.

KATHRIN HILLE

The Communist party of China has launched an online bulletin board on which citizens can leave messages to their senior political leaders. The innovation is the ruling party's latest attempt to meld propaganda with the public relations techniques of the internet age.

Direct Line to Zhongnanhai, referring to the compound housing the party leadership's offices and homes in central Beijing, features a box in which users can type messages before clicking on the name of Hu Jintao, the party chief, or any other member of the politburo inner circle, in order to send it to him.

The move is part of a broad and growing effort of the leadership to demonstrate responsiveness and transparency as it tries to deal with a myriad social and economic pressures without ceding its grip on power.

That effort has become more urgent as the internet has diluted the state media's traditional information monopoly. Corruption, abuse of power and the other ills of one-party rule are now being revealed online every day.

"The main task [of the new site] is publicity," said Dong Guanpeng, a media and PR expert who has undertaken extensive advisory work and media training for the government and the party.

That was clearly visible on the new page's first day in operation. The 37,746 messages that it said had been received by Mr Hu were a well-balanced mix of expressions of support and mild expressions of concern on economic and social issues.

"I suggest to strike out harder against arbitrary village officials and against corruption," said one.

Several messages complained that property prices were too high.

Others praised the leadership for its decision to launch the board and its work in general.

Attempts to leave messages accusing the Communist party of a lack of democracy in China could not be posted, and the computer from which the attempt had been made could no longer access Mr Hu's mailbox.

The message board, hosted on the website of People's Daily, the party's mouthpiece, is the culmination of a long series of attempts by Chinese officials to use the internet proactively rather than waiting to be surprised by the fallout from it.

Provincial and city governments have built an internet presence, some of which also have an interactive element, in recent years.

Both Mr Hu and Wen Jiabao, the premier, have held chats with internet users before. The senior leaders' message board takes these efforts to a new level.

"For years now, the Chinese government has come out and asked PR companies and journalists how it can be more responsive," said Scott Kronick, president of Ogilvy China. The PR firm undertakes media training work for Chinese government officials.

"This is a step forward. The jury is out, though, on how it will be utilised. Having one billion people e-mail the president is not that practical," he added.

Web Connection

- The website features a box in which users type messages before clicking on a member of the politburo inner circle to send it to
- Options for message recipients include Hu Jintao, the party chief, Wu Bangguo, head of the national people's congress. Wen Jiabao, Chinese premier, Jia Qinglin, head of the China people's political consultative conference, and Li Changchun, who is responsible for party propaganda
- On the website's first day in operation, Mr Hu alone received 37,746 messages
- Media experts see the site as one more example of what Mr Hu has called "public opinion channelling" or "public opinion guidance"

Critical Thinking

1. How does Direct Line to Zhongnanhai work?
2. Why do Chinese leaders want to establish such a direct line?

Seven Notches on the Chinese Doorpost

DAVID PILLING

The world has grown used to the restless pace of change in China. Another year, another 10 percent. But like a parent who monitors an adolescent's growth spurt by making notches on the doorpost, there is something to be said for keeping track. This is not merely a matter of recording China's growth. In fact, the milestones I have in mind are harder to measure. Certainly this was the year that, in dollar terms, China's economy surpassed Japan's, a moment that will ensure 2010 a place in the history books. But in other ways too, the past year will be seen as crucial in China's renaissance. Here, in no particular order, are seven notches on the doorpost.

Frown diplomacy: This was the year China's regional "smile diplomacy" turned into a frown. This can be overstated. Beijing's more assertive stance came partly in response to Washington's attempt to re-engage more actively in the region, exemplified by its offer to mediate in disputes centred on the South China Sea. Beijing sees this as interference. It has allowed Chinese commentators to raise the idea that the whole South China Sea is a "core interest", making it non-negotiable on a par with Taiwan and Tibet. In a separate dispute, over the Senkaku/Diaoyu islands in the East China Sea, Tokyo was taken aback by Beijing's shriller tone after its arrest of a Chinese fisherman for ramming a Japanese vessel. Some Asian diplomats—scared of being left alone with a robust China—are trying to draw the US more fully into the region's fledgling institutions.

Google: China was prepared to call Google's bluff when the US company threatened to withdraw from the country. Put another way, Google was prepared to call Beijing's bluff by pulling back from the world's biggest—though not most profitable—internet market. Either way, China has begun to take a different attitude to the multinational companies whose investments and technology it once craved. Having absorbed much foreign technology, it now seems to be tilting the playing field in favour of its own companies.

Trains: China did not only grow this year. It also shrank. Thanks to rapid investment, China's high-speed rail network is already more extensive than the rest of the world's combined. A 1,068km journey from Wuhan to Guangzhou takes just three hours, against 10 hours in 2009. As Lex points out, a similar length journey, between Chicago and New York, takes 18 hours. So 2008.

Renminbi: Beijing continued to resist pressure to revalue the renminbi. In 2010, the Chinese currency appreciated all of 2.5 percent against the dollar. Far more important were the measures China took to speed up internationalisation of its currency. Most significant was a change that allowed offshore banks and central banks to invest in China's interbank bond market, giving them a reason to hold renminbi in the first place. The amounts of offshore renminbi are small, but exploding. Malaysia's central bank is probably not the only one to hold part of its balance sheet in "redbacks".

Liu Xiaobo: If China is a teenager—in some ways an unhelpful metaphor for one of the world's oldest civilisations—then its reaction to this year's Nobel Peace prize was its door-slamming moment. It would have been better simply to have ignored the award. Better yet, Beijing could have explained (with much justification) how much freer China's population has become. Best of all would have been to release Mr Liu, something it could have done without risking its legitimacy in the eyes of most Chinese.

Rare Earths: This year, the world woke up to the fact that China controls 97 percent of rare earths, an important input in the electronics, automotive and arms industries. Chinese companies are now on an unprecedented buying spree of other mineral resources, picking up deposits of coal, copper, iron ore and oil. Analysts say its cleverest tactic is its quiet accumulation of prospecting rights, less controversial than working mines and oilfields. In 10 years' time we may discover China dominates more than just rare-earth production.

Foxconn: The deepest notch of all. A spate of suicides at a Foxconn site employing 300,000 in southern China not only triggered wage rises of 20–25 percent across parts of the Chinese labour force, it also raised questions about the cheap migrant-labour model that has turbocharged growth for 30 years. Other countries from Bangladesh to Indonesia are seeing a pick-up in manufacturing—and manufacturing wages—as a result. For China, there will be more industrial development in poorer inland provinces and less emphasis on export-led growth.

Some of these trends are causing anxiety—in defence ministries of Asian neighbours and boardrooms of US multinationals. But let us not overdo it. China's rise has actually caused remarkably little friction. Its success in drawing hundreds of millions out of poverty is to be celebrated. Recently, it has become fashionable to talk about a risen China. This is premature. China is only now recovering from the torpor of the past 200 years. When China has truly risen, there will be no need to make notches on the doorpost. You will know all about it.

Critical Thinking

1. Why was 2010 a significant year for China?
2. What are the "seven notches" mentioned in the article?
3. Are these developments causing concerns outside China? Why or why not?

---

http://www.ft.com/cms/s/0/1f09e1a2-0882-11e0-80d9-00144feabdc0.html#ixzz1EtVr4Vp4

China's Complicit Capitalists

KELLEE S. TSAI

Until the late 1970s, China did not even keep official statistics on private enterprises because they were illegal and negligible in number. Today there are over 29 million private businesses, which employ over 200 million people and generate two-thirds of China's industrial output. The private sector's spectacular growth has led many observers to speculate that China is developing a capitalist class that will overthrow the Chinese Communist Party and demand democracy based on the principle of "no taxation without representation."

Inspired by the experience of a handful of Western countries, this expectation is based on a two misguided assumptions: first, that private entrepreneurs comprise a single, consistent class; and second, that these entrepreneurs would support a regime change. Although China's capitalists are not poised to demand democracy, they have had a structural impact on Chinese politics. In order to run their businesses in a transitional and a politically charged regulatory environment, private entrepreneurs have created a host of adaptive strategies at the grass-roots level. The popularity and relative success of these strategies have, in turn, enabled reform-oriented elites to justify significant changes in the country's most important governing institutions.

Entrepreneurs Divided

China's private entrepreneurs should not be regarded as a coherent "class" that shares similar identities and interests. Business owners come from all walks of life, and as such, they bring different resources to bear when they have operational or policy grievances. The private sector now includes people as varied as laid-off state workers running street stalls, factory owners producing exports for the global marketplace and rags-to-riches capitalists on the Forbes annual list of China's wealthiest individuals. The sociopolitical composition of private entrepreneurs is further complicated by the emergence of "privatized" (technically "corporatized") state-owned enterprises, frequently operated by their former managers. The proprietors of newly privatized state entities are much more likely than regular private entrepreneurs—who have built up their businesses *de novo*—to be local elites with well-established social and political networks.

At the same time, China's capitalists face different operating realities at the local level. Observers who focus on aggregate statistics showing private-sector growth tend to overlook the significant variation in local political and economic conditions. Governments in areas that opened to foreign capital earlier on in the reform era, for example, have discriminated against local private businesses by offering foreign investors favorable tax rates and privileged access to bank loans and land use. It has similarly been more difficult for private businesses to thrive in localities that inherited a large state-owned industrial base as local officials have been too preoccupied with the challenges of subsidizing local factories and maintaining social stability to address entrepreneurship.

Then there are areas such as Wenzhou in Zhejiang province where the local government looked the other way as private entrepreneurs engaged in capitalist practices before it was officially sanctioned and even collaborated with local entrepreneurs to allow vibrant underground financial markets to flourish. Given the vast differences among local governments in their orientation towards the private sector, it is overly simplistic to assume China's private entrepreneurs face similar business conditions and concerns.

One might counter that the predictive logic of capitalists pushing for democracy is only meant to apply to the highest economic tier of business owners, i.e., that we would only expect the most successful entrepreneurs (not street vendors) to have both the ability and the means to agitate for political change. However, even if we set aside small retail vendors, the fact is that the wealthiest capitalists remain divided by region, sector and most significantly, by their previous backgrounds. A prerequisite of class formation is class identity, and a prerequisite of class identity is a sense of shared values and interests. Real-estate tycoons born out of party-state patronage have little in common with the owners of manufacturing conglomerates who still remember what it was like to grow up hungry in mountainous areas with little arable land. Social and political identity in China is defined by more than ownership of private assets and net worth.

Further evidence of private entrepreneurs' limited desire for democracy can be seen in their nonconfrontational modes of dispute resolution. When private entrepreneurs are disgruntled with policy issues, they are much more likely to use informal channels for solving their problems than the legal system or political participation. Based on a national survey of private entrepreneurs and extensive fieldwork in 10 provinces, I found that only 5% of business owners regularly rely on more assertive modes of dispute resolution—such as "appealing to the local government or higher authorities" or "appealing through judicial courts." Moreover, among the entrepreneurs who believe that there is a need to strengthen rule of law in China, few associate legal reform with democratization. Instead of aspiring for a more liberal political system, most entrepreneurs fear that democratic reforms would lead to instability, which would jeopardize the prospects for continued economic growth.

Indirect Political Influence

Although China's capitalists have not politically organized themselves, the business environment for private firms and their owners has improved dramatically since the late 1970s. After the political crisis of 1989, capitalists were banned from joining the Communist Party and there were a few years of uncertainty about whether economic reforms would continue. But the fact is, once unthinkable changes have occurred in both party rhetoric and official governmental regulations. Capitalists are now *encouraged* to join the Communist Party and the constitution of the People's Republic of China now protects private property rights (at least in principle). In fact, according to official surveys, 33.9% of private entrepreneurs are now members of the CCP, and conversely, 2.86 million or 4% of party members work in the private sector. Yet the really remarkable part about these changes is that private entrepreneurs themselves never lobbied the state or party directly for these macro level changes. Instead, Beijing has been surprisingly responsive to the adaptive, informal strategies created by entrepreneurs to get things done in the context of a transitional socialist economy.

For example, before 1988 it was illegal for "individual businesses" to hire more than eight employees because Marx's *Das Kapital* indicated that businesses with more than eight employees were "exploitative capitalist producers." Private entrepreneurs found a way of getting around this restriction by simply registering their businesses as "collective enterprises." This adaptive strategy became commonly known as "wearing a red hat." By the time private enterprises with more than eight employees were permitted to operate, there were already over 500,000 red-hat enterprises. In effect, the center sanctioned, post hoc, what was already going on.

A similar dynamic occurred with the Communist Party's incorporation of private entrepreneurs. Wearing a red hat enabled party members to become red capitalists, which changed the occupational composition of the party from within. As employees of the state began running their own businesses, albeit disguised as collective ventures, the party's ban on private entrepreneurs became increasingly unrealistic, if not anachronistic. By the early 2000s, the spread of red capitalists presented the party with the critical dilemma of whether to condemn their economic activities or embrace them: 19.8% of entrepreneurs surveyed by official entities in 2000 indicated that they were already CCP members.

After consulting with provincial and subprovincial officials throughout the country, the party's core leadership decided that it was in the interest of economic growth, as well as party rejuvenation and survival, to legitimize the existing red capitalists and co-opt other private entrepreneurs. Within a relatively short period of time, the party line shifted from banning capitalists to welcoming them. Such a policy reversal would have been difficult to justify in the absence of pre-existing grass-roots deviations from the party line.

Private sector development has clearly had a structural effect on Chinese politics, but not in the manner expected. Economic growth during the reform era has been associated with urbanization, higher rates of literacy and the emergence of economic elites. Moreover, China's political system has become more inclusive and institutionalized. But the people driving the country's growth, private entrepreneurs, never mobilized as a class to pressure the regime for these changes, and it is unlikely that they would do so in the future. Instead, in the interest of staying in power, China's leaders have proven to be remarkably responsive, if not overtly attentive, to the unarticulated needs and interests of private capital. Neither capitalists nor communists are interested in disrupting the implicit pact that has emerged in the last two decades: continued growth for continued communism.

Critical Thinking

1. How fast has China's private sector grown?
2. Why do private entrepreneurs have limited desire for democracy in China?
3. What can capitalists do to promote democracy in China?

Article 12

Bye Bye Cheap Labor

Guangdong Exodus

Higher taxes, a new labor law and the growing demands of China's increasingly sophisticated workers are forcing manufacturers either up the value chain or toward the exits.

ALEXANDRA HARNEY

Chinese factories, whose ultra-low prices have been blamed for millions of job losses and countless plant closures around the world, are falling on hard times. A confluence of unfavorable factors—rising energy, material and payroll costs, an appreciating currency, higher tax rates and tougher environmental and labor regulations—are driving thousands of factories in southern China's Guangdong province out of business. Some plants are reopening in cheaper areas in inland China; others are packing up and moving to countries like Vietnam and Cambodia. Still others are closing their doors for good.

The Federation of Hong Kong Industries estimates that 10% of the 60,000–70,000 factories Hong Kong-owned factories in Guangdong will close this year [2008]. In 2007, nearly 1,000 shoe factories left the region. In any other country, an exodus on this scale would be a national political issue. There would be angry pickets by laid-off employees and complaints from labor unions about how the government's trade policies were crippling manufacturing.

But in China, the popular response seems to have been relatively muted. "It won't be too big a problem for the workers," says Liu Kaiming, executive director of the Institute for Contemporary Observation, a labor advocacy and consultancy group in the southern city of Shenzhen.

How could this be? The short answer is that China's economy is growing at such a staggering pace that it can absorb the loss of even thousands of factories. While it is difficult to determine precisely how many factories have left Guangdong, those that have closed appear to be small by Chinese standards, employing hundreds, rather than thousands of workers each. It is likely they were not the region's most efficient or profitable plants.

The longer and more surprising answer is that there are plenty of people who actually wanted these factories to leave anyway. How Guangdong came to be weary of the same factories that Western workers still fear says much about China today. The country's export manufacturing sector is in the midst of a historic transition as the government reins in preferential policies and costs spiral higher. While this shift is likely to cause some disruption, it is mostly good for China, if not the rest of the world.

In the late 1970s, as China began to reform its economy after decades of turmoil and relative isolation, Guangdong was among the first to see the opportunity. Beijing gave the province more freedom to manage its economy and to attract foreign investment. Chinese leaders also put three out of four of the first "special economic zones" in Guangdong. They hoped these zones, which offered preferential tax rates and exemptions on import duties, would serve as a kind of Venus fly trap for foreign technology and investment.

Their plans worked. Hong Kong investors, facing rising labor costs in the then-British colony, poured millions of dollars into the region, setting up factories and workshops near the border. Tens of millions of workers flooded out of the countryside and into Guangdong. By the mid-1990s, Guangdong was a booming light industrial center, producing a growing share of the world's consumer goods. Its success also persuaded Taiwanese businessmen like Terry Gou to invest. Today, Mr. Gou's Shenzhen factory, owned by Hon Hai Precision Industry, employs some 270,000 people and counts Apple, Hewlett-Packard and Nintendo among its customers.

The export-processing industry made Guangdong one of China's wealthiest regions. But it also brought serious social and environmental problems. With so many cities vying for foreign investment, local officials often looked the other way when factories violated labor and environmental laws to keep investors happy. The tens of millions of migrant workers, living for years at a time in factory-owned dormitories, tested the public infrastructure and the management skills of their employers.

Labor protests and strikes are now common in Guangdong. A yawning income gap and growing pool of disgruntled migrant workers have lifted crime rates. Factories in Guangdong have been struggling to find staff for five years, driving up wages at double-digit rates. Turnover is so high that some factories have to replace their entire workforce every year. The province's air and water are now filled with the noxious side-effects of its industrial success. And a generation of factory owners from Hong Kong and Taiwan is reaching an age and a standard of wealth that allows for weekday golf games. Many of their children see their future in finance, not factories.

It's hardly surprising, then, that Guangdong's leaders, like many senior leaders in Beijing, want to propel the economy up the value chain, away from polluting, resource-draining, labor-intensive light industry and towards innovative, high-technology and service businesses. Over the past two years, Beijing has rolled out a series of policies that effectively end the last three decades of preferential policies toward many export manufacturers. It has slashed export tax rebates, the lifeline of many otherwise unprofitable factories. And it has allowed the yuan to float higher. The Chinese currency rose almost 7% against the dollar in 2007.

And while China still has a long way to go to improve law enforcement, local governments have started monitoring factories' environmental impact more closely and creating new regulations to better protect workers' rights. On Jan. 1, Beijing introduced a new contract-labor law which tightened requirements for employers and gave more power to the state-backed union. Foreign investors in southern China say some local governments are now refusing to license highly polluting industries such as leather tanning. Soaring raw material prices have added to the pressure on factories. Even in labor-intensive industries, raw materials can account for 70% of production costs. For the first time in years, manufacturers of many consumer goods are raising their prices to foreign buyers, who are in turn raising retail prices.

None of this means that China will cease to be the workshop of the world. Its advantages—modern infrastructure, a large pool of relatively cheap labor compared to developed countries, and an ecosystem of raw material and parts suppliers—cannot be quickly replicated elsewhere. And the lure of producing for China's 1.3 billion customers in their own market remains.

The rising costs in Guangdong do mean that export manufacturing will be dispersed more evenly around the country. Kenneth Chan, whose company, Gates 2 China, manages design, supply chains and logistics for multinational companies, says he relies increasingly on factories in the northern city of Tianjin as well as the eastern cities of Ningbo, Wenzhou and Nanjing. Goods from factories in those areas are cheaper than Guangdong, Mr. Chan says, but only by 5% to 10%. As these areas develop, wages have started to rise. In Wuhan, in Hubei province, the urban minimum wage has nearly tripled since 1995.

Nor are workers in inland China pushovers. One of the pillars of Guangdong's success in export manufacturing has been its reliance on migrants. Because their *hukou* or household registration was in their rural hometowns, these farmer-workers had no access to state-subsidized health care, education or housing. Living far from their families, migrants have been willing to log long hours on the assembly line for low pay. Their 18-hour days have been one of China's key advantages in producing goods so cheaply. But factories that employ local laborers in inland China are less likely to work those hours. Their employees live at home, rather than in dormitories. They have children and parents to care for. Working conditions tend to be better in inland areas, says Mr. Chan, in part because both factory managers and employees are local. It's harder to crack the whip on somebody you grew up with.

At the same time, China's younger generation of workers is increasingly willing to stand up for itself. Born after Beijing introduced its one-child policy in 1979, China's Generation Y comes from smaller families and has grown up in a more prosperous economy. Factory managers and labor advocates say that workers born after 1980, in particular, tend to be more selective about where they work, more assertive and more interested in developing a career instead of just earning money as their parents did.

These workers, while undeniably harder to manage, augur well for working conditions in China's manufacturing sector because they are more willing to voice their opinions. Employees who care more about their workplace might be tomorrow's whistleblowers, raising the alarm about product-safety problems or labor and environmental violations.

The current transition should be good for China's factories in other ways. Many sectors still struggle with excess capacity, which holds prices—and margins—down for everyone. "In the household appliances industry, where I have 30 years' experience, I still can't count all of the brands in this sector," says Yu Yaochang, deputy vice president of Galanz, the world's largest microwave manufacturer. "But there are certainly hundreds, if not thousands." Knocking out the least profitable tier of manufacturers should help those left standing to survive. It might even help improve the quality of Chinese exports.

In short, the transition in China's manufacturing sector will make it seem less exceptional. The challenges facing China's manufacturing sector and industrialized areas will begin to more closely resemble those of more developed countries: How can we attract the best talent? How do we motivate these people to perform? How do we move from being a producer of commoditized consumer products to design and development, technology and services? What is our competitive advantage?

Guangdong province is already asking these questions. "We have been trying to put quality over quantity in economic development," the China Daily, the government's official mouthpiece, quoted Guangdong governor Huang Huahua as saying in February 2007. But innovation is hard to achieve by diktat. Guangdong could, however, do more to protect intellectual-property rights to persuade more high-tech firms to invest there. And to keep the factories it has, it will need to improve its image with migrant workers, who have been moving to other provinces in pursuit of better working conditions.

As Guangdong and other parts of China invest more in higher value-added industries, they will need more engineers, skilled technicians and managers. But China's labor shortage in this area is more severe than among semi-skilled factory hands in Guangdong. For China's economy is developing more quickly than its universities. In a 2005 report, McKinsey & Co. argued that though China had 1.6 million young engineers, their education's emphasis on theory rather than practice left only 160,000 who were suitable to work at a multinational company.

So far, China's size has been an asset to its progress. But it must balance the need to move into more sophisticated industries with the political and social imperative of keeping the masses gainfully employed.

Critical Thinking

1. Why is labor no longer as cheap as it had been in China?
2. How does the rising labor cost affect foreign investment and China's continued development?

Liu Xiaobo and Illusions about China

FANG LIZHI

I heartily applaud the Nobel Committee for awarding its Peace Prize to the imprisoned Liu Xiaobo for his long and nonviolent struggle for human rights in China. In doing so, the committee has challenged the West to re-examine a dangerous notion that has become prevalent since the 1989 Tiananmen massacre: that economic development will inevitably lead to democracy in China.

Increasingly, throughout the late 1990s and into the new century, this argument gained sway. Some no doubt believed it; others perhaps found it convenient for their business interests. Many trusted the top Chinese policy-makers who sought to persuade foreign investors that if they continued their investments without an embarrassing "linkage" to human rights principles, all would get better at China's own pace.

More than 20 years have passed since Tiananmen. China has officially become the world's second largest economy. Yet the hardly radical Liu Xiaobo and thousands of other dissidents rot in jail for merely demanding basic rights enshrined by the United Nations and taken for granted by Western investors in their own countries. Human rights have not improved despite a soaring economy.

Liu Xiaobo's own experience over the last 20 years ought to be enough evidence on its own to demolish any idea that democracy will automatically emerge as a result of growing prosperity.

I knew Mr. Liu in the 1980s when he was an outspoken young man. He took part in 1989 in the peaceful protests at Tiananmen Square and was sentenced to two years in prison for his efforts. From then until 1999 he was in and out of labor camps, prisons, detention centers and house arrest. In 2008, he initiated the "Charter 08" petition calling for China to comply with the United Nations Universal Declaration of Human Rights. Consequently, he was again arrested, this time sentenced to a particularly harsh 11 years in prison for "inciting subversion of state power"—even though China is a signatory of the U.N. declaration.

According to human rights organizations, there are about 1,400 political, religious and "conscience" prisoners in prison or labor camps across China. Their "crimes" have included membership in underground political or religious groups, independent trade unions and nongovernmental organizations, or they have been arrested for participating in strikes or demonstrations and have publicly expressed dissenting political opinions.

This undeniable reality ought to be a wake up call to anyone who still believes the autocratic rulers of China will alter their disregard of human rights just because the country is richer. Regardless of how widely China's leaders have opened its markets to the outside world, they have not retreated even half a step from their repressive political creed.

On the contrary, China's dictators have become even more contemptuous of the value of universal human rights. In the decade after Tiananmen, the Communist government released 100 political prisoners in order to improve its image. Since 2000, as the Chinese economy grew stronger and stronger and the pressure from the international community diminished, the government has returned to hard-line repression.

The international community should be especially concerned over China's breach of international agreements. Besides the U.N. Declaration on Human Rights, China also signed the U.N. Convention Against Torture in 1988. Yet, torture, maltreatment and psychiatric manipulation are extensively used in detention and prison camps in China. This includes beatings, extended solitary confinement, severely inadequate food, extreme exposure to cold and heat and denial of medical treatment.

As the regime's power grows with prosperity, the Communist Party feels confident in its immunity as it violates its own Constitution. Article 35, for example, says that "citizens of the People's Republic of China enjoy freedom of speech, of assembly, of association, of procession and of demonstration." Yet who can doubt that the government regularly violates these rights.

As the unfortunate history of Japan during the first half of the 20th century illustrates, a rising economic power that violates human rights is a threat to peace.

Thankfully, the courageous Nobel Committee has exposed this link once again in the case of a prospering China. The committee is absolutely right to make a connection between respect for human rights and world peace. As Alfred Nobel so well understood, human rights are the prerequisite for the "fraternity between nations."

Critical Thinking

1. Who won the Nobel Peace Prize in 2010? Why did he win?
2. According to Fang Lizhi, human rights have not improved (in China) despite a soaring economy. Do you agree or disagree? Why?

Fang Lizhi, a professor of physics at the University of Arizona, was a leader of the pro-democracy movement in China before fleeing the country in 1989.

China Winning Renewable Energy Race

STEVE HARGREAVES

Five miles off the coast of Shanghai, the Chinese recently completed the country's first offshore wind farm.

The project was completed before construction on the first American offshore wind farm has even begun.

The Shanghai project is not just another wind farm. It's the next generation in wind power technology and the latest example of how China is jumping ahead of the United States.

Earlier this month, the accounting firm Ernst & Young named China the most attractive place to invest in renewables, knocking the United States out of the top position.

The study ranked countries on such things as regulatory risk, access to finance, grid connection and tax climate. It cited the lack of a clear policy promoting demand for renewables in the United States—a product of Congress' failure to pass an energy bill—as one of the main factors for the dethroning.

China has already surpassed the United States in the amount of wind turbines and solar panels that it makes. China is also gaining on the United States when it comes to how much of their energy comes from renewable energy sources.

The country that leads in the renewable energy industry, is opening the door to more home-grown jobs.

Cash Is Pouring In

From an investment point of view, the trend is clear.

In 2009, nearly $35 billion in private money flowed into Chinese renewable energy projects, including factories that make wind turbines and solar panels, according to the research firm Bloomberg New Energy Finance. The United States attracted under $19 billion.

"Within the past 18 months, China has become the undisputed global leader in attracting new investment dollars," Ethan Zindler, head of policy analysis at New Energy Finance, recently told a congressional committee.

Zindler said the money came from not only the Chinese government and banks, but also Western private equity funds and individual investors buying publicly-traded Chinese stocks.

Jobs Growth, for China

The result of all this investment money is jobs.

In wind power, China-based companies are on track to make 39% of the turbines sold worldwide in 2010, according to New Energy Finance. U.S.-based companies will make just 12%.

In solar, China-based firms will make 43% of the panels. U.S. firms will make 9%.

"Countries that make the most investments will create the most jobs," said Chris Lafakis, an economist at Moody's Analytics, an economic consultancy.

Lafakis, citing a Pew Charitable Trusts study, noted that the overall "green" economy is still pretty small—in the U.S. it employees 770,000 people. But it's growing rapidly—three times as fast as the overall economy.

"It's important to invest in this sector because the jobs of tomorrow will be created here," he said.

Why China

Most analysts feel the investment money is flowing to China because that country has stable policies that encourage the construction of renewable energy power projects.

"China is really taking clean energy to be part of their national strategy," said Gil Forer, one of the analysts at Ernst & Young that ranked China the best place to invest. "They are trying to create a competitive advantage in what many believe will be the industry of tomorrow."

China has instituted a policy that requires a modest 3% of its electric power to come from renewable resources by 2020. This stimulates demand for renewable energy.

In the United Sates, about half the states have such a goal, often with more ambitious targets. But a federal standard has been held up, largely by lawmakers in the Southeast, where wind energy isn't as available.

On Tuesday [September 2010] a new bill with just such a standard was introduced in the Senate with bipartisan support, but analysts say the Senate's busy schedule means the bill has a slim chance of passing this year (2010).

China also requires utilities to buy renewable power at a higher rate than conventional power, a system known as feed-in tariffs.

The United States, by contrast, has a patchwork system of state and federal tax incentives for renewable energy production, which need to be renewed every few years. This is often confusing for businesses in the sector, and does nothing to stimulate demand.

"What we have today is uncertainty, and that is not good," said Forer. "We need to look at this as a strategic decision."

Not Buying It

While China may be increasing wind turbine and solar panel manufacturing to grow its economy, some people feel that it isn't as interested in applying the more expensive technology at home.

"It is impossible to tell why they are investing anything at all in [wind farms and solar power plants], which are not competitive with coal," said Myron Ebell, an energy analyst at the Competitive Enterprise Institute, a conservative research and advocacy group. "My guess is that it's window dressing for the West."

Ebell said China is excelling at making renewable energy parts for the same reason they are excelling at making all sorts of other things: low cost, high skilled labor; weak environmental laws; and cheap energy.

The Chinese, Ebell believes, are simply using the West's fascination with wind and solar as a chance to sell its products.

Indeed, while China has surpassed the United States when it comes to making the equipment, it has not yet caught it in terms of renewable energy production.

In 2009 China could produce 25 gigawatts of wind power and under 1 gigawatt of solar, according to Ernst & Young. The United States could produce 35 gigawatts of wind and 2 gigawatts of solar.

One gigawatt can power 780,000 U.S. homes. The world used 4,428 gigawatts of electricity in 2007, according to the U.S. Energy Information Administration.

China isn't expected to produce more renewable power than the United Sates until sometime between 2020 and 2025, according to EIA.

We're No. 2

In talking about Chinese renewable energy, almost all analysts wanted to make two things clear:

One, it's not really a game of the United Sates versus China. Each solar panel or wind turbine contains parts from all over the world, and an expanded market and lower costs benefit everyone.

And two, despite losing the number one spot, the United States is still a pretty good place to invest.

"You can't forget that the United States is still No. 2," said Forer.

Critical Thinking

1. Why is the investment money flowing into China's renewable energy industry?
2. Is China winning the renewable energy race with the United States? Why **or why not**?

- Support your view w/ information in the article and from the internet sources

Mania on the Mainland

Think the U.S. real estate bubble was bad? China's could be worse.

Dexter Roberts

Li Nan has real estate fever. A 27-year-old steel trader at China Minmetals, a state-owned commodities company, Li lives with his parents in a cramped 700-sq.-ft. apartment in west Beijing. Li originally planned to buy his own place when he got married, but after watching Beijing real estate prices soar, he has been spending all his free time searching for an apartment. If he finds the right place—preferably a two-bedroom in the historic Dongcheng quarter, near the city center—he hopes to buy immediately. Act now, he figures, or live with Mom and Dad forever. In the last 12 months such apartments have doubled or tripled in price, to about $400 per square foot. "This year they'll be even higher," says Li.

Millions of Chinese are pursuing property with a zeal once typical of house-happy Americans. Some Chinese are plunking down wads of cash for homes: Others are taking out mortgages at record levels. Developers are snapping up land for luxury high-rises and villas, and the banks are eagerly funding them. Some local officials are even building towns from scratch in the desert, certain that demand won't flag. And if families can swing it, they buy two apartments—one to live in, one to flip when prices jump further.

And jump they have. In Shanghai, prices for high-end real estate were up 54% through September [2009], to $500 per square foot. In December alone, housing prices in 70 major cities rose 7.8%, while housing starts nationwide rose 34%.

The real estate rush is fueling fears of a bubble that could burst later in 2010, devastating homeowners, banks, developers, stock markets, and local governments. "Once the bubble pops, our economic growth will stop," warns Yi Xianrong, a researcher at the Chinese Academy of Social Sciences' Finance Research Center. On Dec. 27, China Premier Wen Jiabao told news agency Xinhua that "property prices have risen too quickly." He pledged a crackdown on speculators.

Unaffordable Prices

Despite parallels with other bubble markets, the China bubble is not quite so easy to understand. In some places, demand for upper middle class housing is so hot it can't be satisfied. In others, speculators keep driving up prices for land, luxury apartments, and villas even though local rents are actually dropping because tenants are scarce. What's clear is that the bubble is inflating at the rich end, while little low-cost housing gets built for middle and low-income Chinese. In Beijing's Chaoyang district—which represents a third of all residential property deals in the capital—homes now sell for an average of nearly $300 per square foot. That means a typical 1,000-sq.-ft. apartment costs about 80 times the average annual income of the city's residents. Koyo Ozeki, an analyst at U.S. investment manager Pimco, estimates that only 10% of residential sales in China are for the mass market. Developers find the margins in high-end housing much fatter than returns from building ordinary homes.

How did this bubble get going? Low interest rates, official encouragement of bank lending, and then Beijing's half-trillion-dollar stimulus plan all made funds readily available. City and provincial governments have been gladly cooperating with developers: Economists estimate that half of all local government revenue comes from selling state-owned land. Chinese consumers, fearing inflation will return and outstrip the tiny interest they earn on their savings, have pursued property ever more aggressively.

Companies in the chemical, steel, textile, and shoe industries have started up property divisions too: The chance of a quick return is much higher than in their primary business. "When you sit down with a table of businessmen, the story is usually how they got lucky from a piece of land," says Andy Xie, an independent economist who once worked in Hong Kong as Morgan Stanley's (MS) top Asia analyst. "No one talks about their factories making money these days."

Homes Built on Sand

Newly wealthy towns are playing the game with a vengeance. Ordos is a city of 1.3 million in China's Inner Mongolia region. It has gotten rich from the discovery of a big coal seam nearby. An emerging generation of tycoons, developers, and local officials will go to any length to invent a modern Ordos. So 16 miles from the old town, a new civic center is emerging from the desert that could easily pass for the capital of a mid-size country. An enormous complex houses City Hall and the local Communist Party headquarters, each 11 stories tall with sweeping circular driveways. Nearby loom a fortress-like opera house and a slate-gray, modernist public library. Thousands of villas and apartment towers stretch into the distance, all built by local developers in the hope that Ordos' recently prosperous will buy the places to be near the new center of power. Workers get bused daily to the new city hall, but the housing is still largely unoccupied. "Why would anyone go there?" asks Zhao Hailin, a street artist in the old town. "It's a city of empty buildings." (Ordos officials would not comment for this story.)

The central government now faces two dangers. One is the anger of ordinary Chinese. In a recent survey by the People's Bank of China, two-thirds of respondents said real estate prices were too high. A serial drama with the ironic name *The Romance of Housing,* featuring the travails of families unable to afford apartments, was one of the most popular shows on Beijing Television until broadcasting authorities pulled it off the airwaves in November. The official reason was that the show was too racy (one woman got an apartment by becoming the mistress of a corrupt local official), but online chat rooms speculated that the show was cut because it was upsetting to people unable to afford apartments.

The debate has become even more charged following injuries and deaths related to real estate. A woman from Chengdu committed suicide by torching herself when her former husband's three-story factory and attached living space were demolished to make way for a new road. A man in Beijing suffered severe burns in a similar protest over his home. In early December [2009] five professors at Peking University wrote to the National People's Congress calling for changes to a land seizure and demolition law and accusing developers of usurping the government's role when taking land for construction. The law is leading to "mass incidents" and "extreme events," the professors warned.

The second danger is that Beijing will try, and fail, to let the air out of the bubble. Pulling off a soft landing means slowly calming the markets, stabilizing prices, and building more affordable housing. To discourage speculation, the State Council, China's cabinet, is extending, from two years to five, the period during which a tax is levied on the resale of apartments. Tighter rules on mortgages may follow. Beijing also plans to build apartments for 15 million poor families.

Key to Growth

The government is reluctant to crack down too hard because construction, steel, cement, furniture, and other sectors are directly tied to growth in real estate; in November, for example, retail sales of furniture and construction materials jumped more than 40%. At the December [2009] Central Economic Work Conference, an annual policy-setting confab, officials said real estate would continue to be a key driver of growth.

The worst scenario is that the central authorities let the party go on too long, then suddenly ramp up interest rates to stop the inflationary spiral. Without cheap credit, developers won't be able to refinance their loans, consumers will no longer take out mortgages, local banks' property portfolios will sour, and industrial companies that relied on real estate for a chunk of profits will suffer. It's not encouraging that the Chinese have been ham-handed about stopping previous real estate frenzies. In the 1990s the government brutally ended a bubble in Shanghai and Beijing by cutting off credit to developers and hiking rates sharply. The measures worked, but property prices plunged and economic growth slowed.

Analysts are divided over the probabilities of such a crash, but even real estate executives are getting nervous. Wang Shi, chairman of top developer Vanke, has warned repeatedly in recent weeks about the risk of a bubble. In his most recent comments he expressed fear that the bubble might spread far beyond Beijing, Shanghai, and Shenzhen.

Profit vs. Soul

One difficulty in handicapping the likelihood of a nasty pullback is the opacity of the data. As long as property prices stay high, the balance sheets of the developers look strong. And no one knows for sure how much of the more than $1.3 trillion in last year's bank loans funded real estate ventures. Analysts figure a substantial portion of that sum went into property, much of it indirectly. Banks often lend to state-owned companies for industrial purposes. But the state companies can then divert the funds to their own real estate businesses—or relend the money to an outside developer. Meanwhile, the big banks may be cutting back on their real estate risk by selling loans to smaller local banks and credit co-ops.

For now, the party continues. On Dec. 12, [2009] Beijing developer Soho China celebrated a record-breaking year with a gala at the China Central Place JW Marriott (MAR). Guests dined on crab and avocado timbale, white bean soup, and beef tenderloin with wild mushrooms (Soho would not comment for this story). After a dance performance, a panel debated "The Balance Between Profit and Soul." When a writer joked he could not afford an apartment—and was still waiting for Soho Chairman Pan Shiyi to give him one—the crowd of 600 well-heeled developers, entrepreneurs, and consultants laughed appreciatively. If the bubble bursts, few will be laughing.

Critical Thinking

1. Where is Ordos? What can we learn about China's real estate problem from Ordos?
2. Despite the potential real estate bubble, why is the Chinese government reluctant to crack down too hard?

- In spite of the danger of the bursting of the real estate bubble, why is the Chinese government reluctant to crack down too hard?

China's Reform Era Legal Odyssey

JEROME A. COHEN

Thirty years ago, China was a legal shambles. The "antirightist" campaign of 1957–58 and the Cultural Revolution of 1966–76 had demolished the system that the People's Republic of China had initially imported from the Soviet Union. The country had virtually no contemporary legislation. The procuracy or public-prosecution offices and the courts had been decimated, and the Ministry of Justice abolished. The modest legal profession launched in the mid-1950s had not functioned for two decades. Legal education was only beginning to resume after a long hiatus. Experts were few and out of date. Bookstores had no law section.

But things began to look up during the first week of December 1978. Every day the People's Daily reported the momentous decisions being made in Beijing at the Chinese Communist Party's work conference and Third Plenum. Deng Xiaoping was confirmed as China's leader. The Party was going to abandon "class struggle" and modernize the country. This would require a "socialist rule of law." It would also mean opening the country to the world.

On Dec. 15, [2008] the P.R.C. and the United States announced the long-awaited normalization of their diplomatic relations. Within weeks, American officials were in China negotiating bilateral banking, trade, dispute resolution and consular arrangements, and American lawyers and law professors were lecturing attentive Chinese officials and scholars about contracts, technology transfer, joint ventures and tax credits. Even more astonishing, P.R.C. delegations went to Washington seeking guidance from the U.S. Treasury Department, the Internal Revenue Service and other agencies about the legislation China would need to attract major oil companies and other foreign investors. European governments, United Nations organizations and many individual experts joined the effort to fill China's legal vacuum.

There were many reasons for the P.R.C.'s belated acknowledgement of the need to construct a credible legal system. Market transactions, including those with foreigners, required it, as did coherent and efficient operation of government. It was essential to the recovery of social stability after many years of chaos and crime. Reliable institutions for settling all kinds of disputes had to be fashioned, and criminal laws and procedures had to be developed to cope with a growing crime problem. People, including leaders like Deng himself, were demanding protection of basic rights against both mob rule and oppressive officials. Without a credible legal system, the Party's legitimacy would continue to be in doubt.

Thirty years later, China plainly has a formal legal system incorporating many norms and concepts taken from continental European legal systems. It bears the earmarks of its Soviet origins, of more recent Anglo-American influence and, perhaps most important, of the country's imperial past. The National People's Congress, the State Council's executive departments, and their provincial and local counterparts have promulgated a huge volume of legislation, regulations and other norms covering most spheres of human activity. The P.R.C. has also gradually formulated most of the domestic legal rules required by the many multilateral and bilateral international agreements that it has concluded, including its World Trade Organization commitments. Its failure to enact the comprehensive prohibitions required by the U.N. Convention Against Torture constitutes a glaring exception.

The P.R.C. has revived and strengthened its courts, its procuracy, its Ministry of Justice, its legal profession, its commercial arbitration organizations and other relevant institutions. Legal education and scholarship have proliferated. China has well over 600 law schools of various kinds, several legal newspapers and a flourishing law-publishing industry. It has roughly 200,000 judges, 160,000 procurators, 150,000 lawyers and hundreds of thousands of legal specialists working as law teachers, legislative aides, government agency experts, "in house" corporate counsel, publishing editors and even journalists. The country now has a series of burgeoning legal elites that did not exist in 1978.

Does all this impressive accomplishment add up to a credible legal system? What does a "socialist rule of law with Chinese characteristics" amount to? Certainly laws and lesser norms have played an important role in guiding the conduct of officials, economic actors and ordinary people, and providing standards for the settlement of disputes. Especially in the area of economic activity, they have also given people many new freedoms. Yet the P.R.C. has had difficulty establishing the legal institutions necessary for the formal application of its norms—principally its courts and officially sponsored organizations for arbitrating domestic, international and foreign-related legal disputes.

To be sure, so far as outsiders can observe a largely non-transparent system, the courts deal with huge numbers of ordinary civil and criminal cases in an apparently impartial manner, and they have sometimes been responsive to administrative suits of individuals who have been harmed by the arbitrary acts of government officials. But in broad swaths of cases they fail to meet the expectations of impartiality usually associated with the word "courts."

Politically sensitive criminal cases provide the most obvious disappointments. For example, prosecutions against those who attempt to organize a democratic political party, openly challenge the views of the CCP, freely practice their religion or mount public protests almost uniformly result in convictions. No Falun Gong adherent can expect due process. Indeed, most criminal trials are one-sided affairs, even though procedures have moved part of the way, at least in principle, from the inquisitorial to the adversarial system. Witnesses seldom appear in court and therefore cannot be crossexamined, and when defense lawyers take part, as they do in most sensitive cases, their participation is still restricted in both law and reality.

As the Nov. 21 [2008] report of the U.N. Committee Against Torture reconfirmed, in China torture frequently produces coerced confessions. These are rarely overturned by courts that are dominated by police, procurators and Party officials. In sensitive criminal cases, the media can also play an influential role, and generally the only doubt concerns the sentence to be imposed, which is decided not by the judges who tried the case but by their court superiors or even higher judicial or Party officials.

The recent conviction and execution of Yang Jia was a bizarre illustration of how the lust for punishment can overcome considerations of law and judicial fairness. In the fall of 2007, Yang, a Beijing native, was reportedly abused for six hours by Shanghai police who had erroneously detained him for questioning about an alleged bicycle theft. On July 1, 2008, the 87th anniversary of the Party's founding, Yang, having failed in his efforts to win compensation from the police for the mental anguish they inflicted, launched a revenge attack on the local police station, killing six officers.

Yang, who had no previous criminal record, was an obvious candidate for the full psychiatric examination prescribed by Chinese law for determining the degree of his responsibility for his heinous acts. Yet, although the details remain shrouded in the secrecy that enveloped the proceedings, the trial court contented itself with a cursory certification of the defendant's mental state despite the fact this was a capital case, and the death sentence was sustained on appeal. Moreover, Yang's mother, who could otherwise have testified to the mental condition of her son, was herself secretly detained by the police in a mental hospital for four months until shortly before his execution.

Cop-killers usually enlist little sympathy in any country, and the Chinese people vigorously support the death penalty. But in Yang's case, despite strong official and media pressure for execution, widespread resentment of police misconduct and doubts about several aspects of his trial produced extraordinary public unease about his impending execution.

Chinese courts frequently display their shortcomings in other types of sensitive cases. Some observers underestimate the distorting influences that can plague decision-making in civil and commercial cases, especially if local business or government interests are involved. Even a mundane child-custody battle can generate a contest over which of the divorcing parents can muster superior *guanxi* (connections) with the court. Although some scholarship suggests that judicial corruption is not especially serious in China, compared with other countries at similar levels of development, at this stage of our knowledge such claims should probably be treated with skepticism. The recent detention by the Party Discipline and Inspection Commission of the highest-ranking Chinese judge ever removed from office on corruption charges, former Supreme People's Court Deputy President Huang Songyou, has only hardened widespread popular suspicions about judicial integrity.

China's judicial system is politically weak. Throughout his 10 years at the helm, recently retired SPC President Xiao Yang strived to strengthen the professionalism and autonomy of the courts. Yet he was never a member of the Party Politburo and had limited political power. The Minister of Public Security Zhou Yongkang, by contrast, did serve in the Politburo. Mr. Zhou has now been promoted to the charmed circle of the Politburo Standing Committee and to leadership of the national Party Political-Legal Committee that sets policy for and coordinates the activities of all government legal institutions. Yet neither the new procurator general nor the new SPC president has been elevated to Politburo status.

Mr. Xiao's successor as SPC chief, Wang Shengjun, is not a legal professional but a career police and Party administrator who has been assigned the task of further tightening Party control over the judiciary. His highly ideological public statements have faithfully followed the new Party line on law proclaimed by Party General Secretary Hu Jintao just a year ago [2007]. Mr. Hu told the National Conference on Political-Legal Work that judges and procurators "shall always regard as supreme the Party's cause, the people's interest and the Constitution and laws." This instruction soon became a doctrine known as "the Three Supremes." SPC President Wang then followed up by ordering the courts, when deciding capital cases, to heed "the feelings of the masses," social conditions and the Constitution and other laws. This has already set back recent judicial efforts to reduce the country's appalling number of executions.

The Party, of course, is the only authoritative interpreter of the feeling of the masses, and its cause is understood to be "more supreme" than the Constitution and the laws. In recent months, judges have been subjected to intense "study" sessions of this newly minted doctrine, which actually echoes the "mass line" first developed in the simple rural conditions of the pre-1949 Communist-controlled "liberated areas" of China. In accordance with this "democratic" line, courts are being instructed, not for the first time, to emphasize mediating disputes and to adopt social welfare measures to assuage people's grievances instead of concentrating on less "harmonious" aspects of their work—i.e., adjudication.

P.R.C. courts have never had the power to interpret and apply the Constitution, which explicitly allocates that function to the NPC Standing Committee. Because the NPC Standing Committee has been reluctant to exercise that power despite numerous formal requests to do so, the courts have been bombarded by plaintiffs seeking to tempt them into the constitutional thicket. The few judges who have openly flirted with constitutional decision-making have probably regretted their adventure. Huang Songyou, currently detained on bribery charges, was the major figure associated with the SPC's ill-fated 2001 effort to encourage the lower courts to apply constitutional provisions. Some jaded observers question whether Mr. Huang's present embarrassment, which occurred after the retirement of his mentor, Mr. Xiao, is a belated retaliation by conservative Party leaders to deter similar initiatives.

From time to time the courts have been denied the power to deal with certain types of civil cases, such as claims that securities legislation has been violated. The Party currently has been dithering over whether to allow the families of Sichuan earthquake victims or poisoned milk powder victims to seek relief through litigation. Courts frequently deny plaintiffs access to justice on an ad hoc basis in controversial matters, such as damage suits against the police, and without offering reasons, as they should.

One can hardly blame Chinese judges for welcoming opportunities to avoid decision-making. Those who hear disputes rather than engage in administration or research operate under multiple pressures, including time limits for disposing of each case. Despite on-the-job training, older judges—who were often recruited, without legal education, from the military, the police and other nonprofessional departments—feel technically inadequate to handle increasingly diverse and complex matters. Such cases challenge even the recent law-school graduates from whom judges are now recruited. Many judges are puzzled by the P.R.C.'s recent efforts to replace the inquisitorial trial system that seemed congenial to Chinese traditions with the unfamiliar and more demanding Anglo-American adversarial system. They usually get little help from lawyers, who themselves are still getting accustomed to the new procedure.

Moreover, judicial caseloads are often too great to allow careful compliance with procedures, determination of the facts, analysis of the law and writing of judgments. For example, in 2006, Beijing's Chaoyang District Court reportedly received some 30,000 cases, requiring each civil judge to conclude at least several hundred cases before year's end. Because of the strain on judicial resources, many cases that should be tried by a three-judge panel are instead handled by a single judge under summary procedure. This adds to the judge's sense of personal responsibility, which is intensified by his awareness that he serves at the will of court leaders and may be punished for mistaken judgments. Indeed, the entire elaborate system for evaluating judges, who are regarded as ordinary civil servants and treated as such, enhances their personal insecurity and their incentives to minimize risks.

Judges have many risk-reducing techniques for disputes that get beyond the court's gatekeeping acceptance division. They encourage parties to settle disputes out of court or accept a court-mediated compromise. If successful, this makes a decision unnecessary. If required to make a decision, they embrace the practice of taking the advice of their division chief or the court's vice president or president. They may also ask the opinion of a senior judge, technical experts and law professors. They may rely on how similar cases were handled or even informally consult the higher court to which their decision might be appealed in the hope of avoiding reversal!

Of course, the system does not permit them to decide important or complex cases that come before them. Those must be reported to and decided by the court's adjudication committee composed of the president and other court leaders. None of these behind-the-scene consultations for obtaining the opinions of personnel who did not hear the case is made known to the parties or their counsel.

Trial judges or their court superiors are also sometimes the target of extrajudicial pressures from local government officials, legislators and Party cadres. Judges are, after all, locally appointed, promoted, paid and fired. Proposals by Mr. Xiao to expand local judicial autonomy by moving the power to manage court personnel and finances to the central or provincial levels did not get off the ground.

Yet the judicial situation is not completely depressing. Demand for court positions, at least in urban areas, remains high among talented law school graduates. Experience suggests that the current conservative Party line emphasizing "red" over "expert" may not long withstand the demands of modernization and globalization. A better-informed and more dynamic public seems to be increasingly asserting its rights and dissatisfactions. Younger, better-educated, more professional judges just below the leadership level are waiting for their opportunity to create a credible rule of law. And the Politburo has just issued an announcement promising further unspecified "judicial reforms" to alleviate rising social tensions and support economic development. Thirty more years may witness significant improvements!

Critical Thinking

1. When and how did the PRC begin to build its legal system?
2. What does the Yang Jia case tell us about China's legal system and Chinese society in general?
3. How can China improve its legal system?

Mr. Jerome A. Cohen is a New York University law professor and an adjunct senior fellow at the Council on Foreign Relations.

Article 17

China's Final Frontier

The Chinese are latecomers to space, and desperate to catch up. Two years after shooting down a satellite, they stand accused of stealing US secrets. A new arms race has begun.

Sophie Elmhirst

Dongfan Chung had lived in Orange County, California, for 45 years. The 72-year-old, known as Greg to his friends, led a quiet life with his artist wife and son. Quiet, that is, until dawn on 11 February 2008, when the FBI came to his home to arrest him on eight counts of espionage.

Chung, who had worked for Rockwell International and then Boeing—both companies involved in operating the Space Shuttle and the International Space Station for Nasa—is accused of sending confidential information on the US space programme to China over a 30-year period. His trial begins on 6 May. If convicted, he could face spending the rest of his life in jail.

What could have made him do it? The indictment issued by the District Court of California includes extracts from a letter Chung wrote in 1979 to a professor at the Harbin Institute of Technology in China: "I don't know what I can do for the country. Having been a Chinese compatriot for over 30 years and being proud of the achievements by the people's efforts for the motherland, I am regretful for not contributing anything . . . I would like to make an effort to contribute to the Four Modernisations of China."

A list found in Chung's possession showed the extent of the knowledge to which he had access; it included manuals on aircraft and space shuttle design as well as military specifications. It seems he would simply take documents out of the office, hide them at his home, and then travel to China to present the information, sometimes using his wife as a foil; he pretended on one occasion that they were going there at the invitation of a Chinese art institute. His hosts were grateful. Gu Weihao, an official of the ministry of aviation in Beijing, signed off a letter to Chung saying: "It is your honour and China's fortune that you are able to realise your wish of dedicating yourself to the service of your country." Chung was playing his patriotic part in the construction of the new China, ensuring the motherland gained that defining accessory of a great power: a space programme.

The country's space story begins, as the China National Space Administration white paper puts it, "50 splendid years" ago under Chairman Mao with the development of a ballistic missile programme. Over the next generation, space and nuclear research continued and expanded. By 2003, China became the third country, after the United States and Russia, to launch a manned mission into space; the first spacewalk by a Chinese astronaut took place last September. Footage of the event shows Zhai Zhigang waving a Chinese flag as he drifts against the black sky, attached by an umbilical cord to the Shenzhou VII spacecraft. The red flag catches the sunlight reflecting off the earth. Zhai's voice crackles: "My country, please have faith in me. I and my team will finish this mission."

Zhai became a national hero. He had shown the world how quickly China was progressing. In a speech shortly afterwards, the then Nasa administrator, Michael Griffin, acknowledged the achievement. "I personally believe China will be back on the moon before we are," he said. "I think that when that happens Americans will not like it. But they will just have to not like it."

Forty years on from Neil Armstrong's famed first steps, moon landings still capture the imagination. They give countries geopolitical status, prized membership of an elite club. But China's lunar aspirations tell only half the story. All space research develops technology that can have civil or military uses—satellites, for example, can monitor weather patterns or troop movements. The lack of distinction between the two in China causes the US "quite a bit of concern," according to Jing-dong Yuan, director of the East Asia Non-proliferation Programme at the Monterey Institute of International Studies. There is, he says, "no organisational separation between the civilian and military" parts of the Chinese space programme. Other space-faring nations, such as the US or India, make the division institutionally clear, but in China the whole show is run by the People's Liberation Army. As Yuan says: "The Chinese military understands that modern warfare depends on how you use space."

No wonder the case of Greg Chung prompted a strong reaction. Ken Wainstein, then assistant attorney general for US national security, warned of "the threat posed by the relentless efforts of foreign intelligence services to penetrate our security systems and steal our most sensitive military technology and information." It was, he said, "a threat to our national security and to our economic position in the world." Says Alan Paller, a cyber security expert who advises the US government: "We're talking about the equivalent of the following thing happening at every major defence organisation: a guy is walking into the building, copying files and taking them away. They're not taking 25 files, or 50 files, they're taking millions of files."

In a report to Congress last November, the United States-China Economic and Security Review Commission claimed that there are about 250 organised hacker groups tolerated and possibly encouraged by the Chinese government. In one year, China was said to have downloaded between 10 and 20 terabytes of data from US government and contractor websites (roughly the equivalent of all the text in the British Library). So what is Beijing doing with it all? Catching up, for one thing. As Yuan says, the Chinese were "latecomers" to space and they want to avoid reinventing the wheel. But the apparent scale of the espionage campaign is making the Americans anxious. Chris Shank, until recently director of strategic communications at NASA, drops his voice when asked about its effect: "It's deeply disconcerting . . . They know that we know what is going on and they know it's hurt relations."

Beijing dismisses the commission's report as "unworthy of rebuttal." At the time of its release, the foreign ministry spokesman, Qin Gang, was defiant. "The commission always sees China through distorted colour spectacles, and intentionally creates obstacles for China-US co-operation through smearing China deliberately and misleading the general public," he said. The official line is that China "is unflinching in taking the road of peaceful development, and always maintains that outer space is the common wealth of mankind."

Alan Paller, for one, doesn't buy it. He has no qualms about saying that China and the US are in an "arms race," pure and simple. Observers in Washington point to the first Gulf War as the moment when the Chinese realised that developing sophisticated technology in space was synonymous with being a major military power. They watched how the Americans used satellite systems for all aspects of warfare—navigation, communications, imagery and early missile attack warnings—and realised that if they were to have any hope of matching US military weight they would need to shape up in space. The Chinese were also being realistic. They knew that closing the gap with the US in conventional military force was impossible. But US dependence on space systems was what Yuan calls their "soft rib." If the Chinese could develop the capability to threaten the US 500 miles above the earth, it wouldn't matter how many tanks they had.

That capability was made dramatically apparent when, in January 2007, a "kinetic kill vehicle" was propelled into space from a base in the remote Sichuan Province. Travelling at 18,000 miles an hour, it successfully hit its target, a Chinese weather satellite. It took almost two weeks for the Chinese to confirm they had done it, despite the international outcry over the "weaponising" of space. "There's no need to feel threatened about this," said their foreign ministry spokesman at the time. To anyone outside the space business, China blowing up one of its own weather

Timeline: The Long March into Space

1958
Tiuquan, China's first satellite launch centre, is founded

1966
The country tests its first guided nuclear missile

1970
Launch of the first Chinese satellite, the Dong Fang Hong I

1987
The Chinese become involved in the international space industry, providing services for the European aerospace manufacturer Aérospatiale-Matra

1990
China launches its first communications satellite

1992
The Chinese officially begin the country's manned space flight programme

1999
The first unmanned space flight completes its 21-hour voyage

2001
The US unilaterally withdraws from the Anti-Ballistic Missile Treaty. China's muted response ignites fears of a new space race

2003
Launch of China's first manned mission, making it the third country to send a man into space

2007
China shoots down an old satellite using anti-ballistic missiles, prompting warnings in the US of a future "star wars"

2008
Dongfan Chung is indicted for passing US space secrets to the Chinese government

2008
The former fighter pilot Zhai Zhigang carries out China's first ever spacewalk

—Kate Ferguson

satellites doesn't seem like such a big deal. But a China expert and analyst for US defence organisations, Dean Cheng, says that "it made pretty much everyone think differently."

It was, if you like, another Sputnik moment. When the Soviets launched the first artificial satellite into orbit in 1957, they demonstrated an ability that worried the US. Sputnik showed that the Soviets could use ballistic missiles to carry nuclear weapons from Europe to the US. The Chinese anti-satellite test was, similarly, a muscle-flex. For a start, it hadn't been done since the last US exercise in 1985. More importantly, it proved that if the Chinese wanted, they could take out satellites at will. It was, according to Scott Pace, former associate administrator at NASA and now director of the Space Policy Institute at George Washington University, a "surprisingly dirty" move. Dirty in all senses: a cloud of debris from the collision now stretches hazardously for hundreds of miles in space.

Pace, Shank and many other American Establishment observers studiously avoid talking of a new space race. So do the Chinese. After the anti-satellite test, they insisted that their intentions were innocent: "Neither has China participated, nor will it participate in an arms race in outer space in any form." In any case, according to Pace, China is years behind—"roughly in the mid- to late-Sixties period"—in the technology it is creating. He refers to the Chinese space programme, and that of the other emerging giant, India, with a kind of avuncular benevolence. The United States "would be happy to see them on the moon," he says, yet still he is wary. "There are aspects that look benign, and there are aspects that look worrisome . . . There's not a lot of insight into who everyone is and how decisions are made."

This seems to be an understatement. Cheng says no one "had a clue" what was going on when the 2007 anti-satellite test happened. The Chinese are not forthcoming about their space programme; no one was willing to be interviewed for this article. When Michael Griffin went on the first official NASA trip to China's space facilities in 2006 he didn't get anywhere near a launch site. In his version of events, it was like "a first date, if you will," each side coyly sizing up the other. Others saw it as a clear message from the Chinese that there are aspects of their space programme which are not for sharing.

The stakes could be very high. Yuan believes it is not inconceivable that there could be war at some point between China and the United States, possibly provoked by US support for Taiwan's democratic system, a policy that has long riled Beijing, which insists the island is part of China. The new Taiwanese government has improved relations but, says Yuan, "the problem has not been solved . . . [The Chinese] still have to prepare for a potential conflict."

What this conflict might actually look like is a question that intrigues John Sheldon, an ebullient professor at the US air force's graduate school for air and space power strategy in Alabama. "When I'm teaching US officers I tell them, 'Whatever you imagine space war is going to look like, you're wrong. Darth Vader, *Star Trek*—get it out of your head . . .'" Instead, he says, it will be "very real, and at the same time rather subtle and mundane," because nobody actually knows how a space war might start, or if we would even know that it had. It could be the jamming of a signal to a satellite, or a software virus that disrupts enemy communications. Or it might simply be "six guys who hide in the bushes and eat snakes for two weeks and kick down the door of your ground station." Either way, "We'll be looking over our shoulders and wondering, 'What the hell happened there?'"

A deterioration in Sino-US relations is in nobody's interests. On a visit to Beijing at the start of last month to commemorate 30 years of diplomatic ties between the two countries, the former US president Jimmy Carter described their bond as being the most important relationship in the world today. Many observers hope that the new president will handle that relationship differently from his immediate predecessor, whose administration's anti-Chinese sentiment one insider characterised as "visceral."

For their part, the Chinese clearly want President Barack Obama to sit up and listen: they released their latest defence white paper describing (though somewhat opaquely) their nuclear capability on the day of his election. Even before he entered the Oval Office, however, Obama's transition team was talking to the Pentagon and NASA about speeding up production of new military rockets. Recent reports speculate that Obama might merge the two organisations' space programmes—a move that, paradoxically, would mimic the Chinese arrangement. Like it or not, this space race is on.

Critical Thinking

1. Who is Zhai Zhigang? What's the significance of Zhai's move?
2. Why do some people think the space race has begun between the United States and China?

China's Land Reform: Speeding the Plough

Tom Orlik and Scott Rozelle

For China's 700 million strong rural population, the light of property rights may finally be appearing at the end the Communist-era tunnel. Thirty years of economic reform have left the countryside significantly better off than before, but neither farmers nor migrants are yet able to participate fully in the benefits of China's ongoing modernization. That may start to change if positive signals from China's leaders translate into concrete policies for turning farmers' contractual land-use rights into legal title to land.

China's peasant farmers, each tilling a fraction of a hectare, have a tough life. An average annual disposable income of 4,140 yuan ($572) compares unfavorably to urban net incomes of 13,786 yuan ($1,907), and the gap is widening. Under the current land-contracting system, there is still a barrier to achieving either a rapid improvement in standards of living on the farm, or an enterprising move to the big city. Without formal legal title to their land, China's peasant farmers cannot raise the money for, and have little incentive to invest in, soil testing, a tractor, or other productivity-enhancing technology. Neither can the rural masses sell up and raise the funds necessary to set up shop on the urban East coast. For some, requisition of farmland by the government for industrial projects, for which the occupants are never adequately compensated, is a constant threat. Yan Junchang, a 67-year-old peasant farmer from a village in Anhui province, puts the problem succinctly: "If land rights are returned to us, we will have peace of mind, we can plan for the future. Without land rights, we will always feel troubled, always fear our land may be taken away."

China's reform-era agricultural policies have been a striking success. With the right incentives in place, farmers have consistently increased supply to meet rising demand from a still growing population. But for the agricultural sector, the consequence of the current land-contracting system is productivity lower than where it could be. Millions of small farmers, laboring under uncertain conditions, are unable to raise the level of their on-farm incomes and produce less-than-optimal harvests. For society as a whole, the consequence is rural poverty, food prices that are higher than they need to be, and an increased risk of food-safety disasters. While no one is blaming the farming population, poisoned dumplings, tainted pet food, and—most recently—contaminated baby milk, are the natural consequences of a food chain with millions of small players, minimal traceability and zero accountability.

Raising agricultural productivity and contributing to food safety are compelling arguments for land reform. Even more compelling, however, is the role that strengthened land rights could play in China's transition to a modern, urban society. This transition is a work in progress: in unprecedented numbers, China's rural laborers are streaming out of agriculture into the off-farm sector. Even the State Statistical Bureau's figures that 60% of the rural labor force (all of those aged between 18 and 65 with a rural *hukou*) have already made the transition, disguise the magnitude of the labor transformation. A recent study by Stanford University, MIT and the Chinese Academy of Sciences found that when looking at younger cohorts of workers (those between 16 and 35), there is virtually full employment, with most workers already working in the city. Those left in the countryside are fully engaged as small businessmen or are raising families.

The labor transformation from agriculture to industry and services may be nearly complete, but China's transition to a modern, urban society is still far from over. There are virtually unlimited employment opportunities for young, able-bodied rural workers in China's cities. But because they are unable to purchase a flat or afford a long-term lease on an apartment, and because high entry costs and risks prohibit investment in a small business, many rural workers subsist in the netherworld of urban villages, suburban slums and cramped dormitories. For these unfortunates, work in the city has yet to transform into a life in the city. They are unable to launch into their new lives—lives that will necessarily be tough for them no matter what the circumstances—but which will give hope for their children, who will be able to benefit from an urban education and entitlements. This transition, cementing the integration between rural and urban populations, is the missing piece of China's development puzzle and bringing it about is the most compelling argument for land reform.

What role does rural land reform play? Securely under control of China's rural families, land becomes an asset to produce a stream of income (by leasing it out) or a significant sum of cash (by selling it), which can help households finance the difficult and expensive move to the city. Under the current system, poor land rights, which are a barrier to the straightforward and secure sale or leasing of rural land, is one of the major factors hindering the transition to a modern, urban society—a transition that all successful developing countries go through.

The argument for land reform appears compelling. In early October [2008], speaking to a gathering in Yan Junchang's village, President Hu Jintao seemed to recognize as much, promising long-term stability in land contracts, and legally enforceable land rights for farmers. Reports following the recent meeting of Communist Party leaders appeared to confirm that the government would press ahead with reforms, aiming to "construct a healthy market for the transfer of land-contract rights . . . based on the principles of legality, free will and adequate compensation for the peasants." The first land use rights exchange has been established in Chengdu, Sichuan province, to allow farmers to sell or rent out the rights to use their land. Thirty years after Deng Xiaoping's agricultural reforms opened the door to increased harvests and higher standards of living for China's peasants, China's fourth generation of leaders appears to have bitten the land-reform bullet. But there is more to be done to make a rural-reality of the government's high-level commitment to land reform—a process that is under intense debate in Beijing today.

First, in the absence of a land registry, a thorough survey is required. Establishing a public record of the extent and ownership of land will be no easy task. With 200 million farming families, each tilling on average five plots of land, there are around one billion plots to be demarcated. With multiple local boundary changes over the past 30 years, informal agreements on land sharing and leasing, and rapacious local officials conducting the survey, the scope for disputes and rent-seeking behavior is immense.

Second, on the basis of the completed survey, the government will have to issue binding contracts (or land titles) to the households. The new policy directives are clear on how to do this: new titled contracts are supposed to be allocated according to the current distribution of use right (which for most villages took place in the late 1990s in the form of 30 year contracts). However, if as it appears, this is the last and final redistribution of land among households, there will be intense pressure on village leaders to find an equitable solution and not leave any current households out (or give some households more than others). Many villages are considering the possibility of again redividing the land according to some egalitarian rule—probably equal division amongst households.

Third, and the most important priority at this stage, is a change in the law. Under the current system, farmers' collective land rights are inferior to state land rights and subservient to them. In effect, this allows any level of authorized government to acquire land by taking the collective land and converting it by executive fiat into state land. Throw greedy developers and rent-seeking local officials into the mix, and the result is a helter-skelter development process: ill-conceived urban sprawl built on the backs of dispossessed peasant farmers. The key to resolving this cluster of problems is to make title to land the primary proof of ownership. If this were the case, no freeway, or residential development, or industrial park could be built on land till all the land title certificates had been brought up and transferred to the new owner's (or government's) name. Where, in the past, the peasant farmer was tossed aside by the development juggernaut, this new system would give them a substantial stake in the process, and the capacity to negotiate for a share in the profits that flow from it.

Finally, rules of ownership must be agreed and announced. In particular, Mr. Hu's ambiguous assurance of "long-term" stability in land-use rights must morph into a concrete commitment. Fifty-year land use rights, 100-year rights, or indefinite rights (as is the case in Vietnam) are all possibilities. In addition, to work effectively the new system will have to introduce clarity regarding the procedures for sales of land, leasing of land, the use of land as collateral, and the bequeathing of land to children, the village, or anyone else.

All of these pieces need to be in place to make land reform work. In Vietnam, the survey and division of land alone took 15 years to complete. In China, rapid agreement on a change in the law and rules of ownership would line all parties up behind the need for more rapid completion of the package of reforms. Even so, it is clear that the road ahead is long. The devil of policies announced with a fanfare in Beijing is always in the detail of implementation at the provincial and, in this case, village level.

The final outcome is not yet completely certain; this is one of the reasons why the initial policy pronouncement was not clearer. There are powerful interests ranged against reform. Local government, industrial ministries, and the construction and property development sectors all have a stake in the status quo. Development and career-minded local officials know that land reform may slow the process of acquisition of land for industrial projects. Construction companies and property developers, many of whom have strong links to the government in Beijing, know that reform will drive up the cost of acquiring land. At the same time, there is also the fear that a small number of ill educated and credulous peasants will be duped into selling their land for a song, forming a new landless and impoverished underclass.

Perhaps more realistic, there is a risk that a much larger group of peasants would sell their land at reasonable prices but would be thrown into landless poverty by the first serious recession. In the past, this combination of political economy constraints and genuine concern for farmers' future well-being won the argument against land reform. Reading the signs in the Chinese press, and the statements from government insiders, it is still not completely clear who has won this round of the reform battle. But for reformists in the government, the hope is that this time the stars are aligned in their favor: in terms of economic development and agricultural productivity, the time is right for land reform.

Land reform would give farmers a valuable asset they can collateralize or sell to finance a move to the city. A low-cost urban housing program introduced alongside land reform would allow Mr. Yan, or perhaps his children, and others like them, to sell their plot of land in the village in Anhui province and move to a housing development in Hefei—the provincial capital. Permanent migration, with capital to fall back on, and perhaps even some entitlement to the benefits of urban citizenship, would strengthen migrants' position in the labor market. More important, it would mean that migrants' children will grow up with an urban education—and the opportunities and aspirations that brings with it.

In the long run, one of the most fundamental consequences of the newly proposed "third land reform" will be the reshaping of China's agriculture into a modern sector with farmers that are able to produce high enough incomes to support their families comfortably. A successful land-reform policy will also support an agricultural sector able to produce sufficient supplies of safe food for the hundreds of millions of their compatriots that live in the city.

This will not happen quickly or seamlessly. Land reform will not bring the same step-change in harvests that followed the green revolution of the 1970s or the introduction of the household responsibility system in the early 1980s. Farmers will be reluctant to sell. Farm households are small, poor and risk adverse; owning land is a big part of their risk management strategy and no one will sell their land without a great deal of thought. Increasing the size of farms is, therefore, going to be a slow process.

Since the beginning of the reform era, innovation in land use and agricultural technology has enabled China to confound expectations and consistently increase food supply to meet growing demand. Land reform will speed the plough of consolidation and modernization in the agricultural sector, and at the same time do more than any other policy to improve the standard of living for China's 700 million rural population.

Critical Thinking

1. Why does China need to carry out land reform now? Which specific aspects need to be reformed?
2. What are potential obstacles to land reform?
3. How many of China's population are rural?
4. How can the Chinese government address the *san nong wen ti?*

Mr. Tom Orlik is a free-lance writer based in Shanghai. **Mr. Scott Rozelle** is a professor of economics at Stanford University.

Chinese Acquire Taste for French Wine

Pigs' feet with Saint-Emilion and duck tongues with Margaux are set to be popular.

PATTI WALDMEIR

China has overtaken the UK and Germany to become the top export market for Bordeaux wines in value for the first time this year [2010]—and the Chinese are increasingly learning how to savour good wine, not just how to buy it, according to the Conseil Interprofessionnel du Vin de Bordeaux.

Sales of Bordeaux wines in China have doubled every year for the past five years but, in the first half of this year [2010], China and Hong Kong surpassed the UK to take top position for exports of Bordeaux wine by volume—and by value, now totalling €90m ($118m), according to Thomas Jullien, Asia marketing director of the CIVB.

Germany was relegated to second place, in terms of volume, and the UK to second place in terms of value. China and Hong Kong—where many sales are also destined for the mainland—were fourth in terms of volume last year.

But wine consumption habits in China are becoming more sophisticated, says Mr Jullien.

"Five or six years ago, importers did not even want to taste the wines, they just wanted to look at the price list," he says.

"Now when you go to a wine fair, people are tasting the wines, and they realise wine is not just about brand and price."

Bordeaux winemakers recently published a recipe book pairing various Bordeaux wines with traditional Chinese dishes, including perhaps unlikely combinations such as pigs' feet with Saint-Emilion and duck tongues with Margaux.

Shaun Rein, head of China Market Research in Shanghai, says many Chinese people are still unfamiliar with the traditional conventions of red wine consumption. "They either put ice cubes in it, or they drink it in shots," he says.

"I've seen people drink $1,000-plus bottles as shots."

But Rupert Hoogewerf, publisher of the China Rich List, says such habits may not last long.

"The speed of refinement, whether it is understanding watches, or private jets, or private cars, has taken the market by surprise each time and I don't expect it to be any different for red wine," Mr Hoogewerf says. He expects demand will continue to grow—and not just for consumption.

"Over the past five or six years, the Chinese entrepreneur has moved his residence from an apartment into a villa," he adds—and with villas come wine cellars. "People are just in the process of building themselves a cellar and that is another driver for red wine."

Even hosts at Chinese official banquets—including at Zhongnanhai, the compound where China's top leaders work and dine—are increasingly serving red wine instead of baijiu, the traditional Chinese liquor so famous for rendering banqueting businessmen senseless in China.

More than $15m-worth of first-growth Bordeaux and Burgundies will go under the hammer as the wine auction season kicks off in Hong Kong, **Reuters writes.**

New York-based Zachys began the auction season in Hong Kong last week and sold a 1990 La Tache Domaine de la Romanée Conti for $50,262, and cases of 1989, 1990 and 1995 Chateau Petrus, which fetched $40,837, $40,837 and $23,560 respectively.

On Friday and Saturday Acker Merrall & Condit will auction a six-pack of magnums of 1971 Romanée Conti for about $185,000 and three cases of 1982 Chateau Petrus, which each could sell for $72,000.

Christie's offers 300 lots in Hong Kong on Saturday, including 80 cases of Lafite-Rothschild, with estimates of between $11,000 and $24,500 a case.

Critical Thinking

1. How fast have the sales of Bordeaux wines grown in China in recent years?
2. Why is there such a growth?
3. How are red wines changing China's consumer culture and political culture, respectively?

China Won't Revalue the Yuan

No amount of hectoring by Barack Obama is going to change the calculus of Chinese leaders. An undervalued currency may be critical to their very survival.

JOHN LEE

President Barack Obama had an **intensive discussion** about the yuan with Wen Jiabao Sept. 23, 2010, on the sidelines of the U.N. General Assembly meeting in New York, if White House officials are to be believed. In response, the Chinese premier reportedly assured Obama that China will press ahead with currency reforms, thereby delaying the disagreement until their next meeting. In reality, Washington remains naive to expect any significant rise in the value of the yuan, and Beijing remains disingenuous in offering such a prospect in the first place.

Beijing does not doubt that a re-evaluation of the yuan upward is in China's long-term interest. As Chinese economists continually warn, their economy is way too dependent on exports and fixed investment and not enough on domestic consumption. The Chinese people consume around the same quantity of goods and services as France, a country with one-twentieth the population. Because China imports around half of its consumer products from overseas, making imports cheaper through major reform would be one way of boosting the purchasing power of its citizens.

Yet, since the **announcement** in June that China would eag its currency peg, the yuan has risen approximately 1.8 percent against the dollar and actually fell against a basket of major international currencies prior to this week [September 2010]. For Obama, the issue is ostensibly about creating export manufacturing jobs for Americans. But the president should be aware that Beijing is even more determined to ensure that its currency remains artificially low.

The first reason, though this is not a viewpoint widely held by the country's economists, is that a large number of Chinese Communist Party officials think that the United States is deliberately attempting to orchestrate a Chinese slowdown by pushing for the re-evaluation of the yuan. These officials point to the 1980s, when the U.S. Congress was making similar demands on Japan to revalue the yen upward. As the U.S. dollar fell from 240 yen to 160 yen over two years, Japanese growth subsequently slowed. Tokyo responded by boosting government spending and lowering interest rates, leading to the rise of a real estate bubble that eventually burst and is still haunting the Japanese economy today.

China now has its own real estate bubbles, the result of record government spending and bank lending in 2009. A recent study conducted by the People's Bank of China estimated that around a quarter of homes purchased in the first six months of 2010 in Beijing were bought for investment and speculation purposes. In "hot" regions such as Tongzhou district and Wangjing area, the figure is closer to 50 percent. Beijing is already committed to deflating these bubbles before they pop—meaning that its appetite for any further slowdown in exports is close to nil.

Second, like all governments, Beijing cares much more about maintaining jobs than it does about macroeconomic rebalancing. Although official unemployment rates are a healthy 4 to 5 percent, these figures measure less than one-tenth of the country's workforce. Local officials frequently admit that joblessness is probably more than double the official numbers released by their provinces. China lost an estimated 20 million to 40 million export-related jobs in the first few months of the global financial crisis, which explains why Beijing put an abrupt halt to the yuan's rise that occurred from 2005 to 2008.

China's export sector, moreover, is far less robust than it appears. Authorities conducted extensive "stress tests" on more than 1,000 export companies in the first quarter of this year [2010] to determine the effects of any significant yuan appreciation. The vast majority of firms were making do on profit margins of 2 to 4 percent. The **results revealed** that for every 1 percent rise in the yuan against the dollar, the profit margin of the labor-intensive exporters would decline by around 1 percent.

Finally, government policies enacted during the global financial crisis have worked to strengthen the state sector at the expense of the private sector. Between 80 and 90 percent of the 2008–2009 stimulus and bank loans were offered to state-controlled enterprises, according to official statistics compiled and analyzed by the *Australian Financial Review* in 2009. While the state sector grew from 2008 onward, the private sector has shrunk in both relative and absolute terms. This is important because private businesses, both in export and non-export sectors in China, are twice as efficient at job creation as the state-led sector, according to several Chinese Academy of Social Sciences studies that analyzed data from China's 12 largest provinces.

Given Chinese leaders' obsessive but understandable focus on employment, taking advantage of this greater efficiency would first require more emphasis on China's vibrant private enterprises to drive job creation, leading to a gradual loosening of the Communist Party's grip on economic power. Anyone want to guess whether Beijing is willing to take such a risk?

Aside from staying in power, employment is what matters to Chinese leaders. If they fail to create and sustain enough jobs, the party's hold on power is in danger, as is China's entire authoritarian model. Even before the financial crisis hit, in 2008, there were an estimated 124,000 instances of "mass unrest" in China. Much of this took place in rural areas and away from the seats of political power. Tens of millions of disgruntled workers in urban manufacturing centers would be much harder to contain.

Obama can lobby China on the yuan all he wants. But for the foreseeable future, Beijing will do nothing else but offer empty promises.

Critical Thinking

1. Why is the Chinese government reluctant to raise the value of the *yuan?*
2. How does the currency disagreement affect U.S.-China relations?

John Lee is director of foreign policy at the Center for Independent Studies in Sydney and a visiting fellow at the Hudson Institute in Washington. He is the author of Will China Fail?

China's Unbalanced Growth Has Served It Well

YUKON HUANG

Unbalanced growth in China inevitably prompts questions about the country's economic strength. Even premier Wen Jiaobao sees it as a problem, although he remains adamant that adjusting the exchange rate is not part of the solution; forcing Beijing to do so, he warned this week [October 2010], would be "a disaster for the world."

Investment as a share of gross domestic product is now well over 40 percent, while private consumption has fallen to 36 percent. Some observers believe that this imbalance can only lead to economic collapse; others see it as China's underlying source of power. The truth is in-between: while there is nothing fundamentally wrong with the country's economic strategy, China needs to move slowly towards more sustainable growth.

By definition, economic growth is driven by consumption, investment and net exports. No country has achieved rapid growth over extended periods with a consumption-driven approach. In China's case—contrary to popular belief—net exports have accounted for only 10–15 percent of growth over the past decade. Realistically, a country of China's size could not have sustained double-digit growth for three decades without an unusually high investment rate.

The double-digit growth of the past three decades could not have been sustained without a high investment rate

As noted by the World Bank's Growth Commission, the 13 countries that enjoyed growth rates exceeding 7 percent for several decades all had high investment rates and kept consumption down early in their development. However, against this group of mainly east-Asian countries, China stands out with its unusually high and steadily increasing rate of investment, and its success in avoiding any major economic slowdowns. Growth in GDP never fell below 8 percent over the past two decades and the investment rate increased steadily from about 35 percent to well over 40 percent of GDP. Stimulus programmes also helped China navigate unscathed through both the Asian and recent global financial crises.

While investment has been encouraged by subsidies and interest rate policies, this is also true for other countries. There is another explanation for China's high investment rates. As a former centrally planned economy, China continues to rely much more on the banking system and retained corporate earnings to finance investments, especially for infrastructure, that would normally be funded out of government budgets. Since all major banks are state-owned, any future loan write-offs are not private losses but absorbed by government and can thus be considered "quasi-fiscal deficits." The likelihood of a financial crisis emanating from these deficits depends on China's creditworthiness. A low debt-to-GDP ratio and high foreign reserves give it a comfortable cushion.

China's vulnerabilities stem more from whether or not such high levels of investment are being used efficiently. A worst-case scenario is the former Soviet Union, where high and largely wasteful expenditures led to economic collapse. But China's situation bears no resemblance. There is little evidence that China's investments are producing unusually low rates of return and the country is not operating in a closed environment. Chinese companies face competitive pressures—both domestically and globally—that limit inefficiencies.

Nor should China's low ratio of consumption to GDP be confused with anaemic increases in living standards. Private consumption and wages have actually been growing by an impressive 8–10 percent annually. An easy fix would be to mandate higher dividends out of the surging profits of state enterprises, which have been pushing up investment and channelling funds to consumption.

With both GDP and consumption increasing rapidly, why should China give up its unbalanced growth approach? The major concern is rather whether its high levels of investment will continue to generate adequate returns or are sustainable in a broader sense. While there have been misgivings about possible wasteful spending in the recent stimulus, it was designed with demand rather than efficiency in mind and any negative results are likely to be short-lived. The bigger fear is over the sustainability of an investment-driven industrialisation strategy increasingly challenged by environmental concerns and the social implications of widening income disparities.

Yukon Huang is a senior associate at the Carnegie Endowment and former country director for the World Bank in China.

Critical Thinking

1. After over 30 years of fast growth, where are China's vulnerabilities in terms of sustainable development?
2. Why has China's unbalanced growth served it well so far?

China Will not Be the World's Deputy Sheriff

DAVID PILLING

These days, people expect a lot from China. Beijing is expected to help Washington persuade (or force) North Korea and Iran to ditch their nuclear ambitions. It is expected to set the developing world's agenda on reducing carbon emissions. It is expected to keep buying US Treasuries, but not to create the requisite surpluses by selling Americans consumer goodies they can no longer afford. While it's at it, it is expected to bail out Greece. Oh, and it is expected to keep its own economy barrelling along at 10 percent a year. In short, it is expected to save the world.

The problem is China just does not see things that way. As Chinese officials may make clear in Davos, where such expectations are riding high, Beijing is not ready, or willing, to take up the leadership role being foisted upon it. Typical of what one hears in Beijing is the comment from Zhou Hong, director of the Institute of European Studies at the Chinese Academy of Social Sciences. "There will for a long time be a big gap between outside expectations and China's ability," she says. "China is big. But it is poor. Its preoccupation will still be internal."

That difference in perception has become a source of tetchiness, if not outright friction. David Shambaugh, a China specialist at George Washington university, says Barack Obama's administration put huge store in a joint document signed in November [2010]. That laid out the framework for a new era of shared responsibility in which the two will combine to tackle the world's biggest problems. But Prof Shambaugh, who detects a "hunkering down in Chinese diplomacy," says the plan was stillborn. "They've become very truculent, sometimes strident, sometimes arrogant, always difficult," he says of recent Chinese diplomacy.

Examples of China's truculence—as viewed from Washington—abound. Beijing played what many consider to have been a destructive role at the Copenhagen talks on climate change. It has shown no interest in supporting sanctions against Iran. Nor has it done as much as Washington would like to bring North Korea to heel. It has annoyed Japan by pressing ahead with exploitation of a disputed gas field in the East China Sea. Relations with India have also taken a turn for the worse over disputed territory. The list goes on.

Even the business community, which normally sticks up for Beijing, complains of a more hostile atmosphere. Google's threat to quit China has brought into the open previously muted complaints about non-tariff barriers and allegedly arbitrary regulation. Rio Tinto, the Anglo-Australian miner, is still smarting after last year's arrest of four of its employees, including Stern Hu, an Australian. "The world was collapsing and China did great. Who cares about foreigners?" is how one senior foreign business leader with 20 years of China experience characterises Beijing's new attitude.

Such talk of a hardening stance may be true on the margins. China could have concluded—and who can blame it?—that it has less to learn from the free-market model peddled at Davos than it once thought. It may also be more jumpy about internal stability because of recent outbreaks of violence in Tibet and Xinjiang and because it is beginning a tense period of transition to a new generation of political leaders in 2012.

Yet the basic problem is more fundamental. China has a different perception about how an emerging superpower should act. Beijing has no desire whatsoever to be the world's deputy sheriff.

China's mantra is "peaceful development." It even shuns the phrase "peaceful rise." Its priority is economic growth, both because that is required to recapture China's glory and because it is a vital ingredient of the Communist party's legitimacy. Beijing prefers to keep a low profile and get on with the hard slog of building an industrial economy. For that, it needs reasonably civil relations with an increasingly far-flung array of foreign countries, both those that supply its factories with fuel, minerals and components and those that buy its finished products.

Its much-trumpeted doctrine of non-intervention (said to have its roots in Confucian values of respecting others' opinions) suits that purpose well. China will continue to portray itself as a poor country with a less-than-decisive part in world affairs as long as it can. The role the US has mapped out for it looks dangerous. It involves picking fights, taking sides and—if recent history is any guide—even going to war.

Yet it is hard to see how China will be able to sustain its non-intervention doctrine as it grows richer and as its commercial and strategic interests become increasingly entangled in world affairs. There could eventually come a time when China begins to flex its muscles in the way we expect of a superpower, but on its own terms. Ms Zhou at the Chinese Academy of Social Sciences says of recent humiliations: "China was a loser in the last century or two. China was weak. China was occupied. China was attacked." China's first priority is to regain its strength. Then there may be some unfinished business.

Critical Thinking

1. What is the gap between outside expectations and China's ability and willingness?
2. What explains the difference in perception of recent Chinese diplomacy?
3. Why, according to the author, does China have no desire to be the world's deputy sheriff?

China Extends Trade with Iran

Rivals EU as biggest partner Beijing hesitant on Tehran sanctions.

NAJMEH BOZORGMEHR

China has overtaken the European Union to become Iran's largest trading partner, according to analysis of the commercial ties between the two countries.

The growing business links between Beijing and Tehran underline China's reluctance to agree to any further economic sanctions on Iran as western countries escalate their campaign to contain the country's nuclear ambitions.

The announcement by Mahmoud Ahmadi-Nejad, the Iranian president, that Iran will start enriching uranium to 20 percent purity—a step closer to the 90 percent required to build nuclear weapons—has given renewed impetus to western calls for the United Nations Security Council to impose more sanctions.

The Iranian atomic energy authority announced that further enrichment would begin today [February 2010].

The US yesterday sought to rally international support for sanctions against Iran in the wake of the announcement. Speaking in Paris, Robert Gates, US defence secretary, said that "the only path that is left" was to increase international pressure on Iran, but that it would require unity among the world's big powers to do so.

While Russia has softened its opposition to placing more pressure on the Iranian economy, China has not done the same.

Official figures say the EU remains Tehran's largest commercial partner, with trade totalling $35bn in 2008, compared with $29bn with China.

But this number disguises the fact that much of Iran's trade with the United Arab Emirates consists of goods channelled to or from China. Majid-Reza Hariri, deputy head of the Iran-China Chamber of Commerce, said that transhipments to China accounted for more than half of Tehran's $15bn (€10.9bn, £9.6bn) trade with the UAE.

When this is taken into account, China's trade with Iran totals at least $36.5bn, which could be more than with the entire EU bloc. No definite conclusion is possible because it is unclear how much of Iran's trade with Europe is channelled via the UAE.

Iran imports consumer goods and machinery from China and exports oil, gas, and petrochemicals.

Today, China depends on Iran for 11 percent of its energy needs, according to the chamber. In the past, China has allowed the passage of three UN resolutions imposing sanctions on Iran. But the country's ambassador emphasised the need for talks.

"Our approach is that dialogue and negotiations always produce better results," said Xie Xiaoyan, the Chinese ambassador to Tehran. "Sanctions will not produce the results set up [by the west], no matter how crippling."

But, some analysts believe this stance may change. Yin Gang, a Middle East expert at the Chinese Academy of Social Sciences, said: "China is extremely cautious in dealing with Iran, even over trade and energy. China would not keep a very close relationship with Iran because this could damage its relationships with lots of other countries."

If China were to prevent the Security Council from passing UN sanctions, the US and the EU would retain the option of imposing their own unilateral measures. The question is whether Iran's links with China would cushion the blow.

Already, US and EU energy companies have withheld investment in Iran's oil and gas industries. China has sought to fill the gap by signing agreements to develop oil and gas fields. But hardly any of these projects have gone on stream. A senior Iranian oil official has publicly complained about the poor quality of Chinese-made equipment.

Critical Thinking

1. How strong is the trade relationship between China and Iran?
2. How has China's trade with Iran affected its policy toward Iran and the Middle East in general?

Africa Builds as Beijing Scrambles to Invest

DAVID PILLING

A few years ago, Lukas Lundin, a mining executive, rode his motorbike 8,000 miles from Cairo to Cape Town. His journey, which took just five weeks, meandered through 10 countries, including Sudan, Ethiopia, Malawi, Zambia and Botswana. He was amazed to discover that 85 percent of the roads he travelled were tarred and of high quality. Many had been built by Chinese companies.

That was 2005. Since then, China's interest in Africa has intensified. In November 2006, Beijing hosted a lavish Sino-African summit at which it promised more than 40 of the continent's leaders a new era of co-operation. Giant elephants and giraffes appeared on hoardings across the capital to mark the occasion.

Beijing has offered more than long-necked symbolism. In 2006 alone, it signed trade deals with African countries worth $60bn. Investments, which often include a resources-for-infrastructure element, have poured in thick and fast. China's stock of foreign direct investment has shot well past $120bn (€81bn, £74bn). In 2006, Angola temporarily overtook Saudi Arabia as China's main supplier of oil, and Africa now accounts for nearly 30 percent of China's oil imports.

Nor is China's interest limited to oil and minerals. In 2007, Industrial and Commercial Bank of China, the biggest bank in the world by deposits, paid $5.6bn for a fifth of South Africa's Standard Bank. Only last month [November 2009], at yet another Sino-African jamboree, this one in Egypt, Beijing pledged $10bn of new low-cost loans to Africa. It also promised to eliminate tariffs on 60 percent of exports and to forgive the debt of several countries. Trade between Africa and China has already risen spectacularly: last year [2008], it jumped 45 percent to $107bn, a ten-fold increase over 2000.

Beijing's engagement with Africa has caused much hand-wringing. Western donors decry Beijing's supposedly scruples-free approach to investing in countries such as Sudan. In some African countries, too, China's growing shadow has provoked anger. Nigerian radicals likened an attempt by the China National Offshore Oil Corporation (CNOOC) to secure 6bn barrels of oil to being attacked by locusts.

Such objections are overdone. They are often disingenuous. China is no philanthropist, but its rise may still represent Africa's best hope of escaping poverty. In the eight years to 2007, before the financial crisis, African countries were growing, on average, by more than 4 percent a year, far higher than previously. That was thanks partly to better economic management, debt relief and increased capital flows (some from China), but also to the higher commodity prices driven by Chinese demand. Dambisa Moyo, the Zambian economist who riled western donors with her book *Dead Aid*, says:

Much of the criticism of China's influence rings hollow. As Chinese—and Japanese—officials point out, the west's record is less than exemplary. European contact with Africa can best be summed up as decades of naked rapaciousness followed by a spectacularly unsuccessful attempt to make amends. During the cold war western governments supported dictators and kleptomaniacs across the continent, from President Mobutu Sese Seko of what was then Zaire to Uganda's murderous British-trained Idi Amin. More recently, in the name of conditionality, benefactors have rammed frequently disastrous economic fads down the throats of hapless recipients. With donors like that, who needs enemies?

China's pragmatism may produce better results. First, an emphasis on infrastructure means that, even if deals are corroded by corruption, at least the recipient country ends up with a road, port or hospital. (OK, or perhaps a soccer stadium.) Much Asian growth, including that of China itself, was predicated on infrastructure. Officials in Tokyo often contrast Japan's own business-oriented approach to south-east Asia—where countries such as Thailand, Malaysia and Indonesia benefited greatly from Japanese trade and investment—with dubious development strategies pushed by the west in Africa.

Second, China's approach is built on trade. Ms Moyo argues that genuine business opportunity is more likely to catalyse development than government-to-government aid that is prone to being siphoned off. Robert Zoellick, president of the World Bank, told the FT *[Financial Times]* there was Chinese interest in helping to create low-cost manufacturing bases in Africa.

Third, and crucially, China is not alone in seeking opportunities on the continent. As well as the west, India, Brazil and Russia are also vying for business. That ought to give resource-rich African countries the ability to haggle for better terms, though of course there is no guarantee that increased funds will not simply line bigger pockets.

It would be wrong to be wide-eyed about China's investments. Some Chinese businesses are rightly condemned for lax safety standards and for shunning African labour. Critics are doubtless right that Chinese money has helped prop up unscrupulous regimes in Khartoum and Harare. Yet China is hardly alone in dealing with thieves and villains. Whatever its side-effects, a scramble to invest in Africa has got to be better than the European precedent; a scramble to carve it up.

Critical Thinking

1. How do we understand the statement that China's rise "may represent Africa's best hope of escaping poverty"?
2. According to the author, much of the criticism of China's influence in Africa ring hollow?
3. How may China's pragmatic approach in Africa produce better results?

A New China Requires a New US Strategy

"The worst thing Washington could do is to operate on autopilot, to assume that past strategies and policies (which have generally served the United States well) are ipso facto indefinitely useful."

DAVID SHAMBAUGH

The United States needs to revise its China strategy to deal with a complex new China. The China that is emerging today—domestically, regionally, and internationally—has to a large extent outgrown the old strategy.

The grand strategy toward the People's Republic that the United States, together with many of its European and Asian allies, has pursued over the past 30 years has not been a failure. Indeed, it was suitable to its era and proved generally successful. It contributed significantly to China's domestic reform and economic opening, to the integration of China into the international system, and to maintaining peace across the Taiwan Strait.

But now [2010] both China and the geostrategic landscape have changed, and the world is dealing with a different—and sometimes more difficult—People's Republic. This shift does not require a wholesale jettisoning of previous policies regarding Beijing, but it does mean that Washington needs to rethink and alter its macro-strategy.

The Old Strategy

No master document exists deep within the National Archives or the White House that reveals the US government's China strategy over the past three decades. And certainly, each US administration has reconsidered and revised the China policies of its predecessors after assuming office. Meanwhile, elite and public discourse has seen no shortage of debate on China policy. Yet, despite the alterations of successive administrations' policies and the cacophony of public discussion, US grand strategy toward China—since Richard Nixon opened talks with Beijing in the early 1970s—has shown remarkable continuity.

The strategy to date has rested on four pillars: *shaping, engagement, integration,* and *strategic hedging.*

The first of these reflects America's focus on shaping China's internal development. In 1978 China under the leadership of Deng Xiaoping launched "reform and opening." This policy was embraced by the Jimmy Carter administration, which began to devise a strategy intended to give the United States a role in the evolution of China's reform policies.

The "shaping strategy" was aimed largely at Chinese society instead of government. It was intended to build a wide range of educational, research, scientific, and nongovernmental exchanges—not only to facilitate contact between these sectors of Chinese and American society, but also to influence Chinese intellectuals and institutions in an open and liberal direction. The effort included significant philanthropic funding: The Ford Foundation, Asia Foundation, Luce Foundation, Rockefeller Foundation, and others began underwriting various programs in China aimed at improving social conditions and human rights, government capacities, university curricula, and intellectual knowledge, while opening up a public sphere for nongovernmental organizations (NGOs) to grow.

Funding also was provided for large numbers of Chinese students and researchers to go abroad for training. The assumption was that as students and intellectuals experienced Western societies and democracies firsthand, they would bring home to China not only professional expertise but also liberal values—which would, in turn, liberalize the academic world and possibly the political system.

This contemporary shaping strategy echoes a more longstanding "missionary complex" that Americans have held in relation to China since the late nineteenth century. From the late imperial period through the Nationalist era and into the People's Republic, America has exhibited a distinct paternalism toward China, and has sought to shape the nation's evolution. At times (the 1880s, 1920s–1940s), US largesse was readily accepted and Chinese intellectuals and government elites looked to the United States as a model. At other times (the early 1900s, 1919, 1949–72), America was seen in a more hostile light. No matter the fluctuations in Chinese responses to US paternalism, the American desire to help shape China politically, socially, religiously, economically, and strategically remained constant over time.

The second long-standing pillar has been bilateral "engagement." This term entered the American diplomatic lexicon during Bill Clinton's presidency, though it grew out of a strategy undertaken by the Carter and Ronald Reagan administrations. The idea was simple: The US government had to engage the Chinese government across the board in order to advance America's national interests and policy goals, to build institutionalized cooperation, and to ameliorate frictions in the relationship when they arose.

The Carter-Reagan strategy aimed to institutionalize the bilateral relationship by partnering counterpart ministries with each other and infusing them with positive, cooperative missions to replace the negative or hostile missions pursued by the bureaucracies during the cold war years of estrangement. This effort included, in addition to pairing the main ministries of the two national governments, pairing states with provinces and cities with cities in "sister" relationships.

The idea was to build a sound institutional infrastructure that would anchor the relationship, buffer it from inevitable disruptions, and establish bases of cooperation beyond strategic cooperation against the former Soviet Union. During the first Clinton administration, "engagement" meant restoring normal governmental interactions in the wake of the 1989 Tiananmen crackdown and the subsequent disruption of bilateral ties. Later, engagement became a euphemism for hard bargaining over contentious issues.

The third pillar of American strategy relates to the external manifestations of China's rise in world affairs and the uncertainties of Beijing's strategic intentions. To cope with such strategic uncertainty, for more than 30 years the United States has pursued a strategy of integrating China into the international institutional order. The premises of the integration strategy were threefold. First, as a large nation and a major rising power, China deserved a "seat at the table" in all global and regional institutions. Second, by becoming a "member of the club," China would (hopefully) absorb and begin to practice the operating norms and principles of the postwar international liberal order. Third, over time, it was expected that China would begin not only to obey the rules of the global liberal order, but also to make ever greater contributions to global governance.

The fourth pillar was fully established during the George W. Bush administration, though it had antecedents in the Reagan administration. This is the policy of "strategic hedging," sometimes referred to as the "Asia-first" China policy. Officials who promoted this strategy argued that the best way to deal with China was to bolster America's relations with allies and partners in Asia and thus to deal with Beijing from a position of consolidated diplomatic and military strength.

These officials rejected a "China first" Asia policy—the notion that, because China was the main regional power in Asia, a China-centric strategy had to be forged. Instead, they focused on building up military relationships (allied and non-allied) around China's periphery, so as to "hedge" strategically against the uncertainties surrounding China's regional aspirations and its rapidly modernizing military, while simultaneously deterring possible Chinese aggression.

The Scorecard

In retrospect, these four pillars have anchored American strategy toward China since the late 1970s. One could also add, as a kind of fifth pillar, the American version of the "one China policy" concerning Taiwan, though it is more a policy than a strategy. It has served well to maintain the peace across the Taiwan Strait for four decades. But now that the cross-strait dynamic has changed, so too must some elements of US policy toward Taiwan.

The relevant questions concerning the four pillars now are the extent to which they have been successful in achieving their intended aims; whether they need to be maintained, revised, or abandoned; whether they are appropriate to a changed China and changed regional circumstances; and whether a new set of strategies is called for. On balance, the four strategies have to be considered substantially successful, but not wholly so. It is also important to recognize that whatever positive developments have occurred were primarily the result of China's *own* domestic and foreign policy evolution: The West's impact has not been insignificant, but it has been supplemental and around the margins.

The shaping strategy has contributed to pluralizing Chinese society, intellectual life, and the public sphere, but it has only partially contributed to liberalizing Chinese politics. The vast majority of students sent abroad have not returned to China; nonetheless, Chinese intellectual life is far more open than it was 30 years ago, though the social sciences remain politically constricted.

Chinese politics is also considerably more pluralistic, predictable, and accountable than it was three decades ago, when the nation was emerging from the draconian Maoist era. Over time respect for human rights has been enhanced (albeit with continuing violations); there is greater political participation for citizens at all levels; direct multicandidate secret-ballot elections now occur in approximately 85 percent of village governments and in an increasing number of party committees; regular and more meritocratic procedures govern leadership promotion and retirement; laws and institutionalized procedures provide greater regularity and predictability in many policy spheres; and the party-state has shown growing transparency and greater responsiveness to society at all levels.

Yet, while exposure to the United States and the West has no doubt influenced China's political evolution, the American shaping strategy has only indirectly contributed to these positive developments. Indeed, the Chinese Communist Party (CCP) has carefully managed its own cautious political reforms. And as a political role model, the United States has lost the allure it enjoyed in China during the 1980s. In the 1990s a new distrust of America set in among the public that persists today. Curiosity and admiration have been replaced by derisiveness, cynicism, and considerable contempt for the United States.

In short, China is progressing politically—but more in spite of than because of the United States. Some private-sector actors, such as the Ford Foundation, have done important work in rule of law and civil society programs, but the US government has been a largely peripheral actor politically. By contrast, the European Union has been much more active across a wide range of areas. European governments, universities, and NGOs have trained large numbers of judges and lawyers as well as central and local officials. More than twice as many Chinese are studying in Europe (210,000) as in the United States (100,000) during the current academic year. Sanctions imposed by the US Congress in 1989, still on the books, have impeded many potentially constructive training initiatives and in-country assistance programs.

In the economic arena, the shaping strategy has produced more tangible results. Consistent efforts by various agencies of the US government, individual companies, and the American Chamber of Commerce in Beijing have all contributed a great deal to opening and marketizing the Chinese economy. Hard US bargaining over China's accession to the World Trade Organization (WTO) produced positive results. To be sure, China's economic reforms have developed because of China's own initiatives, but the United State and other Western and Asian nations—as well as international organizations like the World Bank, International Monetary Fund (IMF), and Asian Development Bank—have all been important stimulants to reform.

The engagement strategy has been more wholly successful. Over three decades the US and Chinese governments have come to know each other extremely well. Officials on both sides have long experience in dealing with each other personally and professionally, which contributes a certain degree of familiarity and trust to the relationship at the elite level. Government interactions are characterized by real professionalism, even on contentious issues.

For the most part, executive ministries and government agencies are now infused with positive and cooperative missions, guided by dozens of bilateral agreements in a wide variety of policy spheres. The joint statement issued during President Barack Obama's November 2009 visit to China is emblematic of such institutionalized cooperation. The only sectors that are not positively engaged are the militaries and intelligence agencies, though the United States and China do cooperate in law enforcement. State and provincial governments interact regularly as well.

The international integration strategy has also proved generally successful. China is certainly a member of "the club" today, taking its rightful place at the high table of major and rising powers (the Group of 20) as well as in virtually every international institution. It only remains outside of the Organization for Economic Cooperation and Development, the International Energy Agency, the Missile Technology Control Regime, and a handful of other international regimes.

In some institutions, such as the United Nations Security Council, China is proactive. Yet, in a wide range of other international organs, Beijing remains rather passive. China's own development has benefited greatly from its integration into the international institutional community of nations, and Beijing has generally acted as a "status quo" power—not seeking to disrupt or overturn the existing post–World War II order. But it is also evident that China is not wholly comfortable with, or supportive of, many of the norms and structures that undergird the Western liberal order. China is particularly interested in redistributing power and influence from North to South, and in this regard can be considered a "revisionist" actor in international affairs.

The efficacy of the fourth pillar—strategic hedging—is more uncertain. The US government, including the Obama administration, seems to believe it has deterred possible Chinese disruption to the American-dominant security system of alliances and partnerships in Asia, and Washington assumes that other Asian states share these goals. But that is not so certain. Some Asian governments do share them (Japan, Singapore, and possibly Vietnam and South Korea), but most are ambivalent at best.

For most Asian governments, the US effort effectively to encircle China with a ring of security relationships creates unease and places these governments in a difficult position vis-à-vis Beijing. While they do not wish the United States to withdraw from the region, neither are they comfortable with Washington's strategic posture against China. And China itself is clearly uncomfortable with it, pointing out that it is inconsistent with America's professed desire to build a global partnership. When combined with continuing US arms sales to Taiwan and assertive intelligence surveillance off China's coast, Beijing understandably sees hostile intent from Washington in general and the US military in particular.

Thus, on balance, the four pillars of American grand strategy toward China have had mixed effects. All strategies, in any case, require periodic revision to adjust to changing circumstances. What must a new US strategy toward China take into account? What is the nature of the "new" China today?

The New China

China today is highly complex. It contains multiple and often contradictory realities. A complex reality requires suitably complex policies, albeit built around some core strategies. And policy and strategy are not the same—the latter should guide the former. Without an underlying strategy, policies tend to become reactive, ad hoc, and fragmented. The strategies themselves must derive, in the first instance, from America's own national interests and the kind of world order it seeks—but, secondly, strategy should be based on a clear-eyed and hardheaded assessment of what kind of China is now emerging domestically and on the world stage.

A hardheaded analysis, however, does not necessarily lead to a hard-nosed policy. If the United States truly seeks to coexist peacefully and pursue cooperative policies with China, it must be respectful of China's own national interests. China has evolved, and American strategy must similarly evolve to account for new realities. There are in particular six dimensions of today's China that the United States needs to recognize.

First, China's economic growth is likely to continue apace indefinitely. This is not an economy that is going to go off the rails. Not only has China been the major economy that has best weathered the global financial crisis, but as a result it is better poised for more intensive growth in the years ahead. A $586 billion stimulus package put forward by the government has resulted in an orgy of infrastructure and transport construction, which, when complete, will serve to alleviate a number of previous bottlenecks.

Moreover, a relative decline in exports has provided the needed opportunity to "rebalance" the economy and increase domestic consumption as a driver of growth. The effort seems to be working, as citizens (particularly in cities) have gone on an unprecedented spending spree for property, consumer items, and travel. The government has also breathed new life into its "develop the west" (*xibu kaifa*) policy, channeling massive investment into China's central and western provinces and autonomous regions.

Some short-term economic concerns linger—an urban residential property bubble, creeping inflation, and still excessive personal savings—but the economy on balance remains robust. The National Bureau of Statistics predicts 9.1 percent growth for 2010, largely as a result of a surge in secondary industries (mining, manufacturing, and power). The government has also taken measures to cool off the red-hot property market and has tightened monetary lending nationwide. These short-term measures are likely to lead to continued strong growth over the medium and longer terms.

However, the real story of the Chinese economy—and a serious concern to the outside world—is the state's heavy "invisible" hand. State plans, state agencies, state investment, state banks, state trade protectionism, and state cadres still guide the national economy. China's discriminatory "indigenous innovation" policies, aimed at building "national champion" firms and allocating government procurement contracts to select national companies, are a particular concern today. And China has not complied with its WTO accession commitments in the areas of domestic distribution, financial services technology transfer, and intellectual property rights protection—all areas in which foreign firms suffer greatly.

Yet foreign corporations cannot ignore China's growing domestic economy; indeed, all major multinationals need to consider making China a "second home" in their global operations. Penetrating the China market, however, is now harder than ever. Recent surveys by the US and European Chambers of Commerce in Beijing all point to declining business confidence and substantial concerns over a bevy of barriers to investment and trade. While Google has had its highly publicized frustrations, recently General Electric's CEO Jeff Immelt accused China of being increasingly hostile to foreign multinationals: "I really worry about China," he said on July 1 [2010] in Rome. "I am not sure that in the end they want any of us to win, or any of us to be successful."

While foreign businesses cannot afford to pull out of China, some are beginning to do their own "rebalancing"—lowering their relative investment exposure and production dependency on China while diversifying into India, Vietnam, Indonesia, Mexico, and elsewhere in Latin America.

The Party Isn't Over

The second reality of today's China that America must take into account has to do with the country's political system: It will evolve, but slowly and in its own fashion, staunchly resistant to outside pressure.

Chinese politics, to be sure, is incrementally becoming more plural and liberal. Many domestic political reforms (including inside the CCP itself) are difficult to see from outside China, but the political system is getting more transparent, accountable, meritocratic, participatory, efficient, and responsive. Yet, despite (or perhaps owing to) these improvements, the CCP remains firmly in control, and it also enjoys widespread popular legitimacy.

The third reality is that the People's Liberation Army (PLA) is becoming a force to be reckoned with. Although the PLA remains a very long way from being able to project power globally (except with ballistic missiles and cyber capabilities, which it already possesses), it is rapidly emerging as a significant regional military power in the western Pacific, Northeast and Southeast Asia, increasingly the Indian Ocean, and space.

The fourth reality is that China has become truly a global actor—but is only a partial global power. Diplomatically and commercially, China is now present and active everywhere on the globe. As such, it is increasingly bumping up against the interests and equities of the only other truly global actor—the United States—as well as regional powers. Militarily, however, China's hard power is not being felt outside of Asia. And Beijing's soft power remains very soft, though the state is pumping huge resources into trying to improve China's global image. Outside of Africa, China's international image remains mixed at best and very negative in some regions (such as Europe).

China's global strategy is heavily influenced by its domestic development needs. Hence its preoccupation with acquiring steady and secure supplies of energy and raw materials. The result is a largely resource- and commercially-driven foreign policy. Beijing generally seeks to avoid foreign entanglements, while assertively protecting its four "core interests" (Taiwan, Tibet, Xinjiang, and the South China Sea) from foreign meddling. When it comes to contributing to global governance, China remains relatively low-key and cautious.

The fifth reality is attitudinal. China is demonstrating an odd combination of conflicting attitudes toward the world: confident abroad (sometimes overly so) but insecure at home; assertive but hesitant; occasionally arrogant but usually modest. China has a sense of entitlement growing out of historical victimization. It is risk-averse but increasingly engaged internationally. It exhibits a globalism mixed with parochialism, a cautious internationalism combined with strong nationalism. It is largely for national reasons that China is internationally involved. China tries to be a major global actor but is primarily a regional power. An often truculent and difficult negotiating partner, it is pragmatic when necessary. It is an increasingly modern and industrialized but still poor and developing country.

In short, China is a confused and conflicted rising power. We should expect these multiple international identities to play out simultaneously on the world stage.

The sixth reality is that cross-strait relations have now developed to such extent that the "Taiwan issue" has essentially been resolved. Game over. The ultimate form of union between the island and the mainland is still undetermined, but it will be more a question of form than substance. The essential interdependence that is being built will be enduring. A recently concluded landmark Economic Cooperation Framework Agreement will only further deepen Taiwan's dependence on the mainland economy. Although Taiwan's domestic identity will remain a wild card in the island's future, and geography will not change, the interdependence will only grow deeper at all levels, and the international community will continue to give Taiwan no other option but to accommodate itself to a "one China" formula.

A New Strategy

So, what do these new realities about China mean for the United States? They suggest seven needed revisions in US strategy and policy.

First, the United States would do well to discard its longstanding paternalistic attitude and missionary illusions about China. The "shaping strategy" is condescending and treats China as an object rather than a partner. The strategy may have been appropriate for a weak and developing China that was uncertain of its domestic and international future, but those days are over. China is now increasingly strong and confident. The United States should replace the shaping strategy with genuine mutual respect.

The US Congress in particular needs to reeducate itself about China and stop its ill-informed hectoring. Concomitantly, the military-industrial-intelligence complex needs to stop regarding China as a nascent threat and begin truly considering its potential as a genuine partner. Although there will always be limits to full partnership owing to differences in political systems, value systems, and worldviews, differences in these areas are only natural, and they need to be dealt with in a forthright way. A working partnership need not preclude serious discussion of US concerns about human, civil, religious, and political rights, as well as commercial concerns. China needs to understand that national values are national interests for the Untied States, Europe, and other democratic nations.

Americans also need to begin considering *how China will shape the United States* in the future, and what the United States can learn from China. In developing national plans for clean energy, scientific innovation, educational reform, hard infrastructure expansion, high-speed rail, urban public transport, eco-cities, and other sectors, Americans have much to learn from the Chinese.

Second, the United States must develop its own domestic strategy for dealing with China. It is imperative to reduce the US national debt, since it gives China enormous leverage—psychological and real—that constricts American independence of action. In addition, Americans should generally welcome Chinese investment into the United States, while carefully and clearly delimiting areas of genuine national security that should be off limits. The United States also needs to get serious about building, at all educational levels, national awareness of China and competence in the Chinese language.

Of equal importance is to strengthen the entire precollegiate educational system across America. Chinese secondary and university students attend school all but six weeks per year; average middle schoolers have four hours of homework per evening and face a never-ending battery of examinations. The real long-run competition with China is already taking place in the secondary school classroom, and China is winning.

Third, the US government needs to work out effective strategies and tactics to deal with China's state-led development juggernaut. This is best done in tandem with other nations and within international institutions, since most of the problems affecting US business affect all foreign companies and governments. Bargaining power is enhanced collectively, and China should be held to the established rules of the international system. Tough problems sometimes require tough responses, and the United States should not shy away from mixing economic sticks with carrots to achieve greater openness and competitiveness in the Chinese economy.

Toward Strategic Trust

Fourth, the United States needs to learn to live with—rather than trying to impede—China's military modernization. The policy of strategic hedging needs to be reexamined. That China is emerging as a significant regional military power and strategic actor in Asia is not in doubt. The United States cannot fundamentally affect this development, and can only marginally slow it by restricting contacts and transfers of technology and weapons. America's antiquated technology transfer policy in relation to China needs revision. It is discriminatory toward China and it hurts US exporters. The rules can be relaxed while still protecting key technologies with direct or indirect military applications.

Not only does the United States need to pursue robust exchanges with the PLA (which China always suspends following US arms sales to Taiwan), but both militaries need to move from *exchanges* to real *cooperation.* Exchanges of personnel are important for mutual learning and to reduce misperceptions, but they are insufficient for building a true partnership. The two armed services should first grasp the "low-hanging fruit" of joint search and rescue exercises, military medical research, and nontraditional security cooperation—and then move to *joint* maritime security patrols in the western Pacific, in tandem with Japan and other regional navies. The Pacific is no longer an American lake, and the US Navy needs to begin sharing the public good of providing regional maritime security. Other forms of military cooperation can follow.

At present, the US military establishment is disturbingly and increasingly oriented toward war-fighting scenarios against China (and vice versa). This is a dangerous example of a "security dilemma," in which each side's perceived defensive action is interpreted as offensive by the other, causing cyclical and inexorable reactions. This negative cycle needs to be broken before it escalates any further, and the negative security paradigm must be replaced by a positive one. The military dimension of US-China relations is currently far out of step with the rest of the relationship.

Fifth, the United States should immediately institute a moratorium on arms sales to Taiwan—preferably with a reciprocal move by the PLA to withdraw its forward-deployed missiles and attack fighters opposite the island. A "grand bargain" needs to be struck here. US arms sales to Taiwan make little sense anymore, given the dramatic improvement in cross-strait relations and the increasing imbalance between China's and Taiwan's relative military capabilities. Maintaining the so-called cross-strait "military balance" is an illusion given the PLA's increasing power. Moreover, on May 10, 2010, Taiwan's president, Ma Ying-jeou, shockingly asserted that "we will never ask the United States to fight for Taiwan." Freezing or halting US arms sales to Taiwan would address Beijing's long-standing "core" concern with the United States and would contribute to building strategic trust, while not jeopardizing American security interests.

The same applies to the US military's assertive intelligence activities along China's coastline within its 200-mile exclusive economic zone. The issue is not so much whether these daily probes are legal under the international law of the sea (a matter of dispute), but rather the hostile message that they convey to Beijing. Certainly the United States has other means of collecting intelligence to monitor the PLA. How would Americans like it if the PLA conducted regular air and naval surveillance dozens of miles off the California, Oregon, and Washington coasts?

The Multilateral Route

Sixth, China strategy needs, as always, to be embedded in a broader US regional strategy in Asia. There is wisdom in the traditional Reagan-Bush notion that it is best to deal with Beijing from a position of regional strength. But this is not the same as strategic hedging, which should no longer be the operative paradigm. Nor is the traditional US hub-and-spokes alliance-based strategy sufficient. A complex set of non-allied regional groupings (all involving China) is gradually supplanting the US alliance system, and the United States risks being left out of this emerging architecture.

The Obama administration early on expressed a commitment to this regional architecture, but to date there has been little follow-through. Secretary of State Hillary Clinton in 2009 proclaimed that "the United States is back" in Asia, but many Asians do not perceive it this way. The United States needs to really get in the game of regional institution building, and stop being so wedded to cold war alliance structures.

America is still prone to bilateralism while the region increasingly is moving in multilateral directions. And US diplomats maintain a very low public presence. The American embassy in Beijing, the world's second largest after Baghdad's, with 1,200 personnel, is inexplicably invisible in Chinese media and society. US diplomats need to emulate the proactive presence of their Chinese counterparts throughout Asia.

Finally, the United States needs to fashion an appropriate strategy for interacting with China on a global basis. These are still the early days of US-China global interaction, but this is also precisely the time to build in mechanisms of consultation concerning each side's interests and equities in literally every part of the world. Fortunately, the strategic track of the bilateral "Strategic and Economic Dialogue" now includes annual consultations between regional officials in the State Department and China's Foreign Ministry. This is a good beginning, but it is insufficiently institutionalized. What is needed is a truly comprehensive set of bilateral working groups—not dialogues—that operate 365 days a year across a broad range of global geographic and functional issues (such as China and Russia have).

The US and Chinese governments are always at pains to point out that they enjoy more than 60 bilateral dialogues, but many of these are episodic and unproductive. China has also mastered the art of using dialogues as diplomatic deflection devices, particularly in areas like human rights. A more deeply institutionalized relationship needs to be built to infuse both bureaucracies with cooperative, positive missions where possible, and to address differences directly where they arise.

America's global interaction with China need not become a geostrategic competition, as was the case during the cold war with the Soviet Union. China does not harbor the global strategic ambitions or practice the tactics of the former Soviet Union. But China's global footprint is only going to continue growing along all dimensions—diplomatic, commercial, cultural, energy-related, and military. To the extent that a nascent global US-China competition for influence does exist, the best tactic for the United States is to be more proactively engaged with precisely the nations and regions that China has targeted: Latin America, South Asia, Central Asia, and Africa. It is in these middle regions and with middle powers that the "battle for influence" between Washington and Beijing will be waged. Now is the time to begin to forge rules of engagement on a truly global level.

Thus far, China has expanded its global presence mostly in a narrowly self-interested way—pursuing energy resources, raw materials, and commercial opportunities. Its participation in global governance has been limited. Yet, in its contributions to UN peacekeeping operations, disaster relief, and aid programs in Africa and Southeast Asia, as well as its $50 billion contribution to the IMF in the wake of the global financial crisis, Beijing is beginning to do more. China should be encouraged to partner more with other nations to help address global problems.

Time for a "Reset"

In these seven areas, the United States can grasp an important opportunity to revise and reset its China strategy and policies. The worst thing Washington could do is to operate on autopilot, to assume that past strategies and policies (which have generally served the United States well) are ipso facto indefinitely useful.

A bottom-to-top review of China policy is sorely needed. Congress needs to hold hearings on China policy. Think tanks and universities should do what they do best: provide fresh ideas. Corporations need to think beyond their profit margins and get involved in the domestic discourse. The media should break out of their frequent stereotyping and demonizing of China, and begin to report the complexities of today's China.

If all actors do their part, a new vision and strategy for managing America's relations with China may be forged. If, on the other hand, the United States fails to grasp the magnitude and complexity of the challenge, it will find itself 20 years from now asking how it was that America "lost China."

Critical Thinking

1. What have been the four pillars of the US strategy toward China to date? How successful are they?
2. What are distinctive features of the "new China"?
3. What are the new US strategies proposed by David Shambaugh?
4. What do you think the United States can do to deal with a rising China?

David Shambaugh, a *Current History* contributing editor, is a professor at George Washington University and a nonresident senior fellow at the Brookings Institution. He has just completed a year as a Fulbright senior research scholar at the Institute of World Economics and Politics, a part of the Chinese Academy of Social Sciences.

China and Taiwan Sign Landmark Deal

ROBIN KWONG

China and Taiwan signed a landmark trade deal in June 2010 that marks the most dramatic improvement in cross-Strait relations in more than half a century.

The Economic Co-operation Framework Agreement is the centrepiece of Taiwanese president Ma Ying-jeou's effort to mend relations with China, which suffered under eight years of Chen Shui-bian, the former pro-independence Taiwanese president.

Taiwan hopes the deal will also smooth the path to sign free trade agreements with other countries in a bid to ensure that its export-oriented economy is not marginalised as trade deals flourish across Asia.

The agreement that negotiators signed in the Chinese city of Chongqing represents the first phase in trade liberalisation rather than a comprehensive free trade agreement. It also has China doing most of the economic opening in the initial round.

China will cut import tariffs across 539 products and services worth $13.84bn in trade. The cuts on Taiwan's side will only account for $3bn worth of goods. China also agreed not to ask for the opening of agriculture sectors or for Chinese labourers to be allowed to work in Taiwan.

Bonnie Tu, chief financial officer at Taiwan's Giant Manufacturing, the world's biggest bicycle maker by revenues, welcomed the agreement which she said would help her company better manage production. Giant currently produces its high-end bicycles in Taiwan with factories in China making more mid-ranged models.

"If there is no tax issue, we can really integrate our factories and shuffle [production] as we like," she said. "China's economy of scale for high-end bicycles could be really big."

China currently imposes a 17 percent tariff on bicycles and bicycle parts from Taiwan, which under the agreement would be cut to zero within three years. Ms Tu said the deal would likely allow Giant to export more high-end bicycles, such as carbon-fibre bikes, to China.

Kenichi Ohmae, the Japanese corporate strategist, said the agreement was a "very carefully crafted vitamin for Taiwan's continued success."

The semi-official Chung-hua Institute for Economic Research in Taipei estimates that the agreement could create 260,000 jobs and add 1.7 percent to Taiwan's economy.

Over the long term, economists Dan Rosen and Wang Zhi of the Washington-based Peterson Institute for International Economics think the deal could add a net 5.3 percent to Taiwan's economy by 2020.

"We can think of few (if any) other policy reforms available to Taipei that could deliver such gains," they said.

While the deal was hailed by business leaders and analysts, some Taiwanese have raised concerns about the political consequences of moving economically closer to China.

Tens of thousands of people braved pouring rain and took to the streets of Taipei to make a last effort to protest against the deal, which is expected to be approved by the ruling party-dominated legislature within this year [2010].

One banker in Taipei said China's main goal was to "draw Taiwan closer for eventual unification."

"It's easy to oppose moving closer to China now, but once people enjoy the economic benefits it would be very hard to reverse [the deal]," the banker said.

Chen Hsien-chiung, a businesswoman in her 40s, said she was concerned that the deal would produce an "unstable society" for Taiwan's children.

"Don't forget that they have not removed those missiles," she added, referring to the roughly 1,300 short-ranged missiles China has aimed at Taiwan.

Tsai Ing-wen, chairwoman of the opposition Democratic Progressive party, said that if the deal "was a vitamin, then it will only be eaten by big businesses and the rich-poor divide will be even greater in Taiwan."

However, as a sign of how popular opposition to the deal has waned in recent days, turnout at the weekend fell short of the 100,000 touted by the DPP. It was also far less than the 600,000 people who rallied in October 2008 to express disapproval of Mr Ma amid the financial crisis.

Critical Thinking

1. What are the features of this landmark deal between Taiwan and China?
2. How big is trade across the Taiwan Strait?
3. How will ECFA influence cross-strait relations?

Beijing and Taiwan Try Their Hand at Détente

Beijing and Taipei are on a conciliatory path these days. The Communist Party in China has desisted with its military threats and the small island has stopped pushing as vociferously for independence. Direct flights have also been re-established between the mainland and Taiwan. Both sides seem to be experiencing change through rapprochement.

SANDRA SCHULZ

Fan Guishan's voice is booming. The sky above Taipei may be overcast, the Chinese man bellows, but his heart is full of sunshine. Fan sees no reason to keep his voice down right now. He's here leading an important delegation, and he wants to speak in great sentences on this great day. "We will fulfill our long-held dream," he cries to the others assembled in the ballroom. "Chinese tourists will come to Taiwan!" Everyone applauds.

Thirty-nine representatives of Chinese tour operators have gathered here in a Taipei luxury hotel. They're at the vanguard of the new business of Chinese tourism in Taiwan—and they're here to see what the island has to offer. One highlight, for example, is "the best tea in the world," according to an official at the tourism office. Another is a jade cabbage from the Qing Dynasty housed at the Taipei National Palace Museum—an entire vegetable carved from a single piece of white-green jade that is as famous as the Mona Lisa. And then there's Sun Moon Lake. Even as far back as the 1980's the lake was listed by the Chinese Communist Party's *People's Daily* newspaper as one of China's 10 most beautiful places—even though it lies in Taiwan, the "renegade province," as Beijing officially calls this island.

Massive China and little Taiwan are beginning to bridge their differences again. On July 4, they reintroduced direct weekend flights from the mainland to the Taiwanese island, and a maximum of 3,000 Chinese visitors are allowed to enter Taiwan each day. Until this month, the only way for Chinese citizens to reach Taiwan was by first passing through Hong Kong, Macau or a third country. Now they can come directly to Taiwan and explore, draw some comparisons and tell friends and family back home what they've seen. They talk about the clean subways—and about the political talk shows on TV. Chinese tourists, according to the Taiwanese tourism official, like to stay at their hotels in the evening and watch freedom of the press in action. "All that," he says, "is our gentle power."

A newly gentle era seems to be arriving in this part of the world with the Olympic Games, which begin in Beijing on August 8.

Though the Communist Party in Beijing is still standing firm on its "One China" policy, it has stopped provoking Taiwan with military maneuvers. Taiwan has also suspended its demonstrations for independence and seems satisfied with the status quo. China and Taiwan are both counting on their people, on winning their hearts and minds. Who knows, say the Taiwanese, perhaps people on the mainland will soon start wondering why the people of Taiwan are allowed to elect their government, while they are not?

"Beijing has realized that Taiwan will continue to drift away if China doesn't establish contact," says Lin Chong-pin, President of the Institute for International Studies. After nearly 60 years of separation, a generation has grown up in Taiwan seeing the People's Republic of China as a foreign country. Beijing now wants to draw the 23 million Taiwanese closer to the "motherland." The time is ripe, too: New Taiwanese President Ma Yingjeou got elected partly on a platform of wanting to improve relations with China, thus breaking with the policy course of his political predecessor. The main hope is that improved ties will translate into benefits for Taiwan's economy.

"The whole world wants to do business with China," says Tsai Eng-meng. "But Taiwan, though geographically closest to China, has been the only country to be shy about doing it. That's the big joke." Tsai himself, a successful Taiwanese entrepreneur, certainly couldn't be accused shyness. "If there's a mouth and some money to go with it, there's a market," he says. And with 1.3 billion mouths in China, Tsai has succeeded in making himself a very rich man—complete with a private jet and a corporate headquarters in Shanghai. He's chairman of Want Want Group, a food brand that markets rice crackers and other snacks in China. Tsai has more than 40,000 employees, including 38,000 on the mainland. Almost all of his staff are Chinese.

Tsai opened his first factory on the mainland 16 years ago, and today he has 110. When he got started, he recalls, the Chinese welcomed Taiwanese investors with open arms.

The interconnections between Taiwan and China have only grown in the intervening years. Around 1 million Taiwanese business people now live on the mainland, and there are more than 250,000 marriages between Chinese and Taiwanese. Chinese women in Taiwan have never had it easy, though. In the past, these women were often suspected of being spies. Today people shout insults at women like Cui Yon-mei, such as: "You just want money."

Cui used to believe all Taiwanese were rich, since they were always so generous with the tips they gave out at Chinese hotels. But then she landed in Taipei and somehow the airport struck her as cheap, dimly lit and small—hardly comparable to the glitzy new terminals in Beijing or Guangzhou.

Fighting for Loyalty

The wives, the businesspeople, they're all seeing to it that Taiwan gradually moves closer to China—too close, some fear. Critics argue the new president is getting too chummy with the Chinese. In the past, Ma used the anniversary of the June 4, 1989, Tiananmen Square Massacre—when several hundred protesters were killed by the military in Beijing—as a time to direct accusations at the Communist leadership in Beijing. This year, though, he flattered and praised. China has made "certain progress," he says, thanks to the country's reforms. The Taiwanese opposition, however, argued it would have been better if Ma had reminded the Chinese about the unrest in Tibet.

And when the president agreed, out of respect, to let a negotiator from China call him simply "Mr. Ma," the Taiwanese political opposition immediately criticized him for smudging the nation's honor by allowing himself to be humbled in such a way.

Taiwan's honor seems to be in constant danger as the island fights for "international space," the term used to refer to the island's international activities, like diplomatic ties with foreign countries. The size of this space is a matter determined by Beijing. Just 23 countries have diplomatic relations with Taiwan—states like Tuvalu and the Vatican. This January, Malawi also decided to join Beijing's side, and before that Costa Rica, Chad and Senegal shifted allegiance. Memory of the $30 million lost in an attempt to win over Papua New Guinea—a middleman ultimately ran off with the money—still brings shame to the minds of the Taiwanese.

Taiwan has had the most success when it has been willing to be flexible about the name issue, making allowances for the "One China" policy. At the Olympic Games, Taiwan will compete as "Chinese Taipei," and within the World Trade Organization it is referred to as a "Separate Customs Territory." Those are the conditions under which Taiwan is allowed to participate.

An attraction that draws Taiwanese and Chinese both is the mausoleum of Chiang Kai-shek, Taiwan's national hero and a well-known figure in China as well. As the head of the Kuomintang (Nationalist Party) in China, Chiang agreed to an alliance with Mao's troops to fight against Japan during World War II. After subsequently fighting and losing to the Communists, he and his government fled to Taiwan in 1949. Mao's Communist Party gained control of mainland China, but Chiang Kai-shek ruled as dictator in Taiwan.

The path to the coffin of the "great leader," as he is called in the brochures for the tourists, passes through bamboo stands and palm trees. It leads past a small green lake, which the general loved because it reminded him of his home in mainland China. Guards stand at attention in the muggy heat, right hands to their rifles and left hands balled into fists, under attack by mosquitoes.

One Country, Two Systems?

Initially, Taiwan's military was none too happy about the prospect of mainland visitors coming to the general's provisional resting place. They feared the Chinese might not show enough respect—and that they might shuffle past the general's portrait in flipflops, or smoking cigarettes. But the Chinese visitors also bow to Chiang Kai-shek, just as a large sign nearby asks them to do. And when they come across the general's old soldiers, they're friendly and even exchange business cards.

Yang Jun-chi used to come here often, when his eyes were better. He's 83 now, a veteran, a small man in sweatpants and an undershirt. "I have nothing in Taiwan," he says. He lives in a small room with his songbird and his Thai wife, for whom he paid a marriage broker € 2,000 ($3,160). He left the cousin he loved behind on the mainland in 1949.

"Taiwan is a province of China," says Yang, and of course the politicians should be talking about reunification. It could be done the way it was in Hong Kong, where the "One Country, Two Systems" policy was established. Now in his twilight years, former soldier Yang, who once fought the Communists, has come to agree with the Communist Party.

In a couple months Yang will take one of the new direct flights to the mainland to attend a family celebration in his native village. While there, he'll also take a look at the urn he's ordered. Before he dies, Yang Jun-chi would like to finally leave Taiwan.

Critical Thinking

1. Why did tensions across the Taiwan Strait begin to ease in 2008?
2. Does China's "one country, two systems" policy enjoy widespread support in Taiwan? Why?

Who's Listening to Taiwan's People?

JULIAN BAUM

A decade ago, then Taipei Mayor Ma Ying-jeou led a small band of students in a candlelight memorial for the victims of the Tiananmen Square massacre. It was the 10th anniversary of the crackdown in Beijing, and tens of thousands were gathering in cities around the world.

There was little interest in the anniversary in Taipei though. Mr. Ma was accompanied by only one other official from his party, Shaw Yu-ming, vice chairman of the Kuomintang, along with a dozen or so members of the party's China Youth Corps. It was characteristic of Mr. Ma, with his anticommunist leanings and his concern for China, that he showed up that evening. In the years after 1989, he was one of the rare officials in the KMT to speak publicly against the brutality of the June 4 events and in support of Chinese democracy.

Since his inauguration as Taiwan's president, Mr. Ma's criticism of the Beijing government has gone mute. And unlike most other Taiwanese and even members of the KMT, he promotes a national identity under the label of "one China." The identity—which was dormant during his presidential election campaign, when he effusively praised the "Taiwan spirit" and submersed himself in "long stays" in the rural south of the island—has resurfaced in the past year. It has been an obvious factor in getting Beijing to sign on to agreements opening direct links across the Taiwan Strait.

Since becoming president, Mr. Ma has proclaimed that the people of Taiwan are members of the "Chinese nation," and has commended the virtues of the "Chinese race," most prominently in his inaugural address in May 2008. His language is often identical to that of Chinese President Hu Jintao, who has praised the "historic turning point" in cross-Strait relations after Mr. Ma came to office.

During the tenure of Taiwan-born Lee Teng-hui as president of the Republic of China in the 1990s, a majority of citizens for the first time began to publicly identify themselves as "Taiwanese." Polls since then have shown that most people on the island no longer conflate their ethnic and their political identities. Many Taiwanese say they see little to admire about the society across the Taiwan Strait except its economic vitality.

As Mr. Ma pursues an ambitious agenda that will require more accommodations with Beijing, he will need to deal with this popular affirmation of Taiwanese identity. "I can see how much the president wants to get our Chinese identity back, but I don't see anyone buying it," says a British-educated CEO of an import company in Taipei. "It was easier for Lee to let the genie of Taiwanese identity out of the bottle than it will be for Ma to put it back in."

With the easy ribbon cutting for a new era in cross-Strait relations behind him, Mr. Ma seeks a legacy that goes beyond opening direct links with China. But he faces many pitfalls that could prevent his re-election. He is struggling with dismal public approval ratings that have stayed below 30% for most of the past year.

Before even reaching the midpoint of his four-year term, Mr. Ma's government has already faced extreme difficulties. Not only has the administration had to confront the worst recession in Taiwan's history and record levels of unemployment, in August [2009] Typhoon Morakot, the worst natural disaster in a decade, left over 700 dead. Mass street protests have erupted over the erosion of democratic rights and civil liberties, and there has been ongoing international criticism over the politicization of criminal justice, with dozens of indictments and investigations involving the former government.

"Mr. Ma is losing time in the run-up to his re-election in 2012," says Liu Shihchung of the Brookings Institution in Washington, who served as a senior advisor to former President Chen Shui-bian. "He's in desperate need to come up with some concrete 'achievement' as a way to launch his re-election campaign beginning next year."

Mr. Liu has noted there is the prospect of a game-changing meeting in 2011 between Presidents Ma and Hu, coinciding with the 100th anniversary of the Qing dynasty's demise. In light of the popular disdain toward KMT leaders Lien Chan and James Soong for their meetings with senior Chinese leaders, this potential meeting would have considerable downside risk. But if carefully managed, Mr. Liu wrote in the Taipei Times, the symbolism could boost Mr. Ma's re-election and set the stage for political talks in a second presidential term.

Well before that could take place, Mr. Ma expects to sign a comprehensive economic framework agreement (ECFA) with China; talks between the two sides are expected to begin in December [2009]. As a tailored version of Hong Kong's Closer Economic Partnership Arrangement, which was signed with Beijing in 2003, it would put commercial and financial integration across the Taiwan Strait on an irreversible course and, among other things, widen access to China's domestic markets for Taiwanese manufacturers on the mainland.

Yet uncertainties about the content and unintended consequences of ECFA as well as suspicions that it would compromise Taiwan's autonomy may slow the process. Even members of the KMT say they are in the dark about the proposed agreement's provisions and its effects. The government's sales pitch on ECFA has been heavy-handed, which suggests to some critics they are promoting a done deal rather than preparing to negotiate an agreement.

A government-funded study by the Chunghwa Institution for Economic Research concluded that ECFA could add 1% to annual GDP. But the full findings by industrial sectors have not been released, and no independent economists have been as confident about its impact, especially since the exact provisions are unknown.

The uncertainty and general distrust have given rise to understandable paranoia. Former President Lee Teng-hui called the proposal Mr. Ma's "ultimate unification scheme." An editorial in the pro-independence Liberty Times described how many Taiwanese feel about an open-ended agreement with China: "The effect is like handing over your . . . id card, the contract for your house and a whole book of blank checks to a stranger. No one in their right mind would do such a thing."

Some businesses expect benefits from inbound Chinese investment, including Taiwanese construction companies, property developers and the island's overcrowded securities industry. But studies show these benefits could be heavily offset by industries that will see no upside and may be seriously disadvantaged from the "total normalization" of economic and financial relations that Mr. Ma advocates.

Others suspect that Mr. Ma has been disingenuous about an ECFA facilitating other free trade agreements, especially with Southeast Asia, where China and the Association of Southeast Asian Nations plan to launch the world's largest free trade zone next year [2010]. Mr. Ma recently told The Australian that he wanted to sign additional agreements, including with Australia. But there has been no official comment from Beijing, while one Chinese economist has commented that Beijing would be absolutely opposed. If so, it would abort the KMT's strongest talking point: that the pact is "absolutely necessary" for Taiwan's greater integration into global commerce.

Other analysts see the main beneficiaries of ECFA as Taiwanese companies operating in China who want broader access to Chinese domestic markets. "What's good for Acer or Honhai is not necessarily good for Taiwan," says a Taipei securities analyst, naming two electronics manufacturers. "These companies don't even provide new jobs in Taiwan anymore. They pay almost no taxes and don't repatriate profits. They are 'Taiwanese' in name only."

A survey conducted by Global Views, a Taipei magazine, showed that most Taiwanese still prefer the so-called political status quo-Taiwan as a de facto independent state which is clearly not a part of the People's Republic of China, with informal relations with most countries. Surprisingly, only a fraction of "mainlanders"—families that arrived in Taiwan with Chiang Kai-shek in the late 1940s and who are presumed to be the core constituency of the KMT's unification doctrine—supported reunification with the P.R.C.; in the survey a mere 23% supported unification now or in the future, while 56% rejected it.

The Global Views survey shows how divided the island continues to be on fundamental issues of identity and nationhood. The divisions pose political risks for the government's aggressive agenda to move the cross-Strait relationship beyond the obvious steps taken so far.

In his National Day address in October [2009], Mr. Ma acknowledged the public's fears about sacrificing sovereignty and national security. He seemed ready to listen more. "We are willing to engage in dialogue via all sorts of channels, including the Legislature and political parties, in order to forge a public consensus on this government's mainland China policy," he said.

Yet there is no indication that the president is ready to step back and either consider a less compromising strategy or abandon his "one China" framework. Meanwhile, Beijing is pressuring him to do more, and Taiwan's election calendar has its own momentum. Thus, the government's attitude appears to be full speed ahead. If this means that Taiwan is beginning to look reminiscent of Hong Kong's "one country, two systems" model, then the voters could choose differently in 2012.

Critical Thinking

1. How has Ma Ying-jeou changed his predecessor's policy toward China?
2. Is Ma's "one China" framework the same as that of Beijing?

Mr. Baum, a former Taiwan correspondent for the *Far Eastern Economic Review,* is based in Richmond, Virginia.

Article 29

Behind the Dalai Lama's Taiwan Visit

JULIAN BAUM

The irony of the Dalai Lama's house call on Taiwan in early September was in plain view. No one, even in the Byzantine world of Taiwan-China relations, is more adept at keeping politics always in mind but subtly out of sight than the Tibetan Nobel laureate.

With advice from his "good friend" President Ma Ying-jeou to limit himself to religious and humanitarian activities, the Tibetan leader promised strictly to fulfill his "moral responsibility" as a Buddhist monk with "no political agenda."

In visits to those suffering from the island's deadliest typhoon, he would "share some of their sorrow, some of their sadness," he told CNN. As he went about his allotted task during five media-saturated days—his first visit to Taiwan since 2001 everyone else was free to do and say what they wanted. Among those least constrained were Dalai Lama's hosts, a clutch of local government chiefs from Taiwan's deep south—historically a stronghold of the pro-independence party—who jointly invited him to visit the disaster-stricken area and comfort survivors of the storm. After Typhoon Morakot passed through on Aug. 7–8, the need for consolation was real as the suffering was severe, especially in mountainous areas where the storm dropped up to 9 feet of rain in less than two days. Hundreds died; many were buried in mudslides. Three weeks after the deluge, even Taoists and Christians were moved to see the world's most famous monk show up on their doorstep with words of comfort and goodwill.

For Taiwan's separatist opposition, including Chen Chu, feisty mayor of Kaohsiung, the visit was a political coup. They have been watching Mr. Ma's nonstop courting of Beijing for the past year with growing anxiety. Many were asking whether their government would ever stand up to China again. The president was clearly not ready to do so last December [2008] when he announced the time was not right for a visit from China's most famous "splittist." It might disrupt the reconciliation and deal making across the Taiwan Strait that are the signature achievements of Mr. Ma's government.

In the wake of Morakot, the circumstances had changed, at least long enough for the invitation to go out while the government struggled to get on top of its worst domestic crisis so far. In this moment of weakness, one Taiwanese paper said, Mr. Ma showed rare political courage by allowing the Tibetan back on Taiwan soil. Though newspapers have issued various opinions on the event, the controversy has confirmed the discomforting depth of Mr. Ma's commitment to preserving a working relationship with Beijing. "They are certainly taking Beijing's attitude into priority consideration," said Hsiao Bi-khim of the Democratic Progressive Party, the main pro-independence group.

It's clear that the Tibetan leader's visit would not have happened when it did if August had not been such an excruciating month. For Mr. Ma, August [2009] exposed an aloof and seemingly uncaring side of his leadership. Nearly everyone saw television clips of mainly aboriginal villagers in remote mountain areas pleading with the once popular president for more help. As an emotionless leader struggled to respond to the outpouring of anger and grief, television cameras recorded many distressed moments of a government caught off-guard. Some critics say much of the blame rests squarely on the president's shoulders for failing to mobilize resources by declaring a state of emergency which could have sped up rescue efforts by several days. Another dysfunctional moment came at a press conference in which Mr. Ma and his senior ministers were subject to tough questioning. Perhaps attempting to sooth public anger, the president announced that the military's main mission in the future would be disaster prevention and rescue. The enemy was not Beijing, but "nature," he said, discounting China's 1,500 short-range missiles targeting the island. Defense Minister Chen Chao-min gently corrected the record, saying that the military's responsibilities included defense as well as disaster relief. But his resignation was accepted the next day.

The failure to keep humanitarian activities free of politics extends mostly to the uncertain relationship with China. When the invitation went out to the exiled Tibetan government in late August, the National Security Council met for many hours to consider whether to allow the visit and how to minimize the inevitable tongue lashing. The ruling Kuomintang dispatched an emissary to explain the circumstances, though Chinese officials had already protested that they were "resolutely opposed" to the visit. "Obviously this is not for the sake of disaster relief," a spokesman for China's Taiwan Affairs Office said. "It's an attempt to sabotage the hard-earned good situation in cross-Strait relations." Beijing took care to single out Taiwan's pro-independence opposition for the exploiting the crisis "to plot the Dalai Lama's visit" while shielding their negotiating partner. There was no mention of Mr. Ma's role in approving the arrangements.

"Such behavior of Chinese officials runs contrary to the expectations of the Taiwanese people who anticipate the spiritual compassion and peace that the Dalai Lama represents," the leading opposition party said. On the first full day of the Dalai Lama's itinerary, the Kyodo News Agency said the Taiwan government had agreed with Beijing to limit his schedule, starting first with the abrupt cancellation of an arrival press conference, then with the relocation of a lecture on Buddhism from a 15,000-seat sports stadium to a hotel conference room, and announcements from senior Kuomintang officials they would not see him.

It's not as if Taipei was not attuned to Beijing's concerns from the start. In the immediate aftermath of the disaster, opposition leaders accused the Ma government of delaying the acceptance of international aid for several days as they first consulted with Beijing. Three days after the storm, the foreign ministry instructed Taiwan's foreign missions to turn down international offers, except for cash donations. Those instructions were rescinded the next day and help eventually arrived from more than 70 countries. The ministry's flip-flop led to speculation that senior officials were worried about Beijing objecting to aid from the United States and Japan in the form of military personnel arriving on the island. But the general demeanor of the government was enough cause for complaint. "Looking back at the first week after Morakot, even if the Ma administration has not been colluding with the Chinese government, its arrogance and public detachment, combined with its eagerness to push cross-Strait relations, could still send the wrong message . . . and place Taiwan at risk," commentator Hoong Ting wrote in the pro-independence Taipei Times.

After the Nantou earthquake in 1999, the worst natural disaster in Taiwan's history, Beijing asserted that it alone could authorize international assistance for Taiwan. Yet most governments dealt directly with Taipei and sensibly ignored the interference. In the event, Beijing offered only $100,000 in cash and a lesser amount in material assistance. This time Beijing was more circumspect and more generous. Almost certainly with Beijing's permission, a United Nations team from the Office of Coordination of Humanitarian Affairs flew in to assess emergency relief operations in the disaster zone, a rare contact with any U.N. agency since China pushed Taiwan out of the UN in 1971. Beijing also engaged more directly, including donations of 1,000 prefabricated houses and cash of $122 million, according to the People's Daily.

For the Dalai Lama, Taiwan is a unique place, not just one more place from which to plot the destruction of the People's Republic of China, as Beijing caricatures his intentions. For hundreds of Tibetan monks who have taken up residence on the island in recent years, Taiwan looks like the antithesis of the People's Republic of China.

Earlier this year in a talk on the 50th anniversary of his escape from Lhasa and exile to India, the Dalai Lama described the repressive and violent campaigns against Tibetans under Chinese rule. "These thrust Tibetans into such depths of suffering and hardship that they literally experienced hell on earth," he said in unusually bitter comments.

Taiwan, on the other hand, is an ethnic Chinese society that is open, democratic and practices freedom of religion while it respects Tibetan culture and religious beliefs. If not exactly a celestial kingdom with its crass consumerism and divisive identity politics, there is a broad tolerance of different religious faiths and diverse lifestyles. In some respects, Taiwan exemplifies what Tibetans would like China to be. Taiwanese and Tibetans also know from experience what it's like to struggle against Han Chinese chauvinism, though the Tibetans have suffered much more from the encounter. These mutual interests and common challenges have drawn them closer in the past decade. During the Dalai Lama's first visit to Taiwan in 1997, one of his brothers expressed delight with the crowds that greeted their entourage. It was the first time the Tibetan leader had visited a predominantly Chinese society since his exile from Tibet in 1959. The outpouring of respect and admiration was a welcome antidote to the hostility and rejection heaped upon him by Beijing over many years.

The Dalai Lama has now reciprocated this support. Besides the many messages of encouragement to disaster victims as he made his rounds, he frequently affirmed the value of the democracy. He urged the Taiwanese to safeguard their democratic way of life. "The most important [thing] is democracy. That you must protect!" he said on one occasion. "You enjoy democracy, you must preserve it," he said elsewhere. Whatever their politics, Taiwanese could not have heard more timely advice.

When the Dalai Lama first visited Taiwan in 1997, the island was nearing the high tide of its democratic transformation. Lee Teng-hui had been chosen as the first popularly elected president one year earlier. There was still much discussion about the need for deeper political reform and a new, more appropriate constitution than the one imported from China a half century earlier. Even the KMT agreed that the remedy for the island's problems was more democracy, not less. The prospects for life under democratic government were bright.

Now many observers worry whether a functioning democracy is part of Taiwan's future. There has been noticeable backsliding on civil liberties and democratic rights in the 15 months since the KMT returned to power. Many citizens are uncomfortable with the revival of an unpopular Chinese nationalist ideology while they despair over a dysfunctional legislature with its supermajority of KMT seats. There is talk of "creeping unification." Even outside the separatist movement, many Taiwanese question the government's heavy focus on China-centric economic and social policies and the secretive party-to-party negotiations with Beijing that are setting the course for an uncertain future.

Morakot's devastation may have briefly interrupted these ominous trends. Popular discontent over managing the rescue and recovery operations along with the worst recession in Taiwan's history have forced the president to pay closer attention to domestic matters and be more solicitous of public opinion. In a poll taken in August [2009], fewer than 20% of Taiwanese adults approved of his performance in office.

No one really expects the popular backlash over the typhoon to bring down Mr. Ma's government or divert relations with China. A few officials have resigned, including the highly regarded Andrew Hsia, a vice-minister of foreign affairs. That, as well as the "apology tours" by Mr. Ma and senior officials across the island, may help to mitigate public outrage. Ultimately, Taiwan's voters should judge whether the Ma administration has acted responsibly on their behalf. The island holds local elections later this year [2009], and there will be contests for big city mayors in early 2010.

Whether voters believe that the government's deference to China will be a slippery slope, leading to the avalanche of mistaken priorities that Beijing is counting on, will depend partly on the economic payoffs. So far, those pay-offs have been invisible as the economy is expected to shrink by 4% to 5% this year, the sharpest decline in more than 60 years. Unemployment is at record levels. With the KMT's ambitious program of economic and financial integration with China just getting underway, Mr. Ma is asking for more time, even as he denies concerns about compromising Taiwan's sovereignty. Thanks in small part to a visiting Buddhist monk, voters now may be less willing to give him the benefit of the doubt.

Critical Thinking

1. Why did the Dalai Lama visit Taiwan in 2009?
2. Do Tibet and Taiwan present the same or different challenges to China?

Mr. Baum, a former Taiwan correspondent for the *Far Eastern Economic Review,* is based in Richmond, Virginia.

Taiwan Caters to China's Giant Fish Appetite

ROBIN KWONG

Choosing your own dinner from a large fish tank is a central attraction of Hong Kong's famous seafood restaurants.

As the rest of China grows increasingly enamoured with Cantonese cuisine, a Taiwanese company is hoping to cash in, with a boost from a trade deal between Taiwan and China.

The most valuable fish from Long Diann Marine Bio Technology's aquaculture farms in the Taiwanese seaside town of Fangliao is unlikely to be found in any restaurant tank. The giant grouper, whose steamed flesh is the most prized among Hong Kong gastronomes, can reach 2.5 metres long and weigh 600 kilos.

Tai Kun-tsai, the owner of Long Diann, was the first in the world to reliably breed the huge fish for farming. Mr Tai, who keeps two 150kg pet grouper in a tank in his living room, has since emerged as Taiwan's biggest grouper farmer.

Partly because of Mr Tai's technology, Taiwan has emerged as the world's only significant exporter of farmed giant grouper. Other countries such as Malaysia, Indonesia and Thailand ship wild giant grouper caught from the sea.

Almost all of Mr Tai's T$200m ($6.3m) sales last year were to Hong Kong, the world's biggest market for grouper. Over the past year, however, Mr Tai has seen demand surge from mainland Chinese cities such as Shanghai and Beijing. Prices for farmed giant grouper have almost doubled from T$280 a kilo a year ago to T$480 ($15).

"The surge in demand and the rise in prices for grouper over the past year was the biggest I have seen since I began this business more than 20 years ago," Mr Tai says.

China's growing appetite stands to make Mr Tai and Fangliao's other grouper farmers significant beneficiaries of the Economic Co-operation Framework Agreement, the first formal trade deal between Taiwan and China. China considers Taiwan part of its territory even though the island has been ruled as a de facto independent state since 1949.

Besides eliminating the current 13 percent Chinese tariff on live fish, the deal, which came into effect this month, will see Beijing remove tariffs on more than 500 products from Taiwan to the benefit of the island's bicycle, petrochemicals and machinery industries.

Breeding breakthrough

Giant grouper are particularly hard to breed because they are protogynous hermaphrodite, meaning that all grouper begin life as female but some later change sex to male. This, however, happens rarely and, in the wild, only after the grouper is at least seven years old.

Farming giant grouper was therefore economically unviable until Tai Kun-tsai, Taiwan's foremost grouper farmer, developed a way to artificially force the fish to change genders.

Mr Tai first began experimenting with forcing a gender change through giving the fish hormonal medicine in 1990. "I bought 50 giant groupers at a cost of T$5m ($157,000), and we tried giving them the medicine mixed with their food. But this was difficult to control, and overnight all 50 fish died," he recalls.

He succeeded in 1995 after spending more than T$70m on the project. By operating on the fish and injecting the medicine into their bodies, he can ensure an 80 percent success rate in forcing gender change. The technique has cut the breeding cycle of the giant groupers to one year.

Mr Tai did not rest on that accomplishment, however. "Of the 108 types of edible grouper in the world, only eight can be farmed now and we developed the breeding technique for six of them," he said. Mr Tai is working on finding a way to breed leopard coral grouper on land in Taiwan.

"The future is in fish farming because catching fish in the wild has become unsustainable," he said.

Taiwan grouper production has already risen from T$2bn in 2007 to T$3bn last year, according to Mr Tai, and huge capacity expansions are now being planned to tap growing Chinese demand. Over the next three years, Mr Tai plans to invest T$300m to T$500m to triple the size of his fish farm, betting he can quintuple annual revenue to T$1bn.

"The growth of the mainland Chinese market has really just begun," Mr Tai said. "We expect the greater China market to grow by 20 percent every year."

Mr Tai's efforts are backed by the government, which is promoting grouper as a new national export.

James Sha, director-general of Taiwan's Fisheries Agency, says grouper avoids direct competition with China, whose coastal areas become too cold for the species during the winter.

"There are very few grouper farmers in China," he said. "They mostly focus on freshwater fish."

Mr Sha says the trade deal opens the door for grouper farmers, typically small family businesses, to develop into bigger companies by entering new markets such as supplying frozen fish to inland Chinese provinces.

"We want to help make sure the farmers have alternative markets in case prices crash," Mr Sha says.

Critical Thinking

1. To what extent can grouper farmers in Taiwan benefit from the Economic Co-operation Framework Agreement?
2. What advantages does Taiwan enjoy vis-à-vis other grouper exporters when catering to China's growing market?

Wen Hints at Scrapping Taiwan-Facing Missiles

ROBIN KWONG

Wen Jiabao, the Chinese premier, has for the first time raised the possibility of removing some of the thousand-plus missiles China has deployed facing Taiwan.

Mr Wen's comment in September 2010 to overseas Chinese media while on a visit to the US makes clear Beijing's wish to engage in political and military discussions with the island it regards as a runaway province. His conciliatory tone stands in contrast to China's increasingly assertive stance towards other neighbours.

Mr Wen earlier threatened retaliation against Japan over Tokyo's arrest and detention of a Chinese fishing boat captain in contested waters. Beijing has also expressed annoyance about joint US-South Korean naval exercises and moves by the US and south-east Asian nations to promote regional discussion of overlapping claims in the South China Sea

Cold war rivals Taiwan and China have been separately governed for the last century except for a brief period after the second world war. While relations between the two sides have improved dramatically over the past two years, cross-Strait discussions have been limited to economic issues. Those links were cemented by a broad trade pact that took effect this month [September 2010].

Taiwan officials have long called for the redeployment of Chinese missiles pointed at the island, viewing them as a stark reminder that Beijing has not ruled out the use of force to achieve unification. Taiwanese and US defence officials have expressed alarm in recent months that China has appeared to be increasing the number and sophistication of its missiles despite the recent détente.

When asked about the possibility of removing the missiles, Mr Wen said the trade pact with Taiwan provides a foundation for talks on other issues including politics and military confidence building. " I believe that [removal] will eventually happen," he said.

In Taipei, Mr Wen's comments were cautiously welcomed. Wu Den-yih, Taiwan premier, said that Taiwan hoped China would remove the missiles "as quickly as possible," rather than "eventually." Tsai Ing-wen, opposition leader, said Mr Wen's comments were "so vague that it may be considered a meaningless answer."

"You also have to ask, how quickly and easily can [China] redeploy the removed missiles?" she said. "If the answer is very easily, then what [Mr Wen] said was just lip service."

Critical Thinking

1. How many Chinese missiles are aimed at Taiwan?
2. Why does China deploy missiles near Taiwan?
3. Why would China consider withdrawing those missiles?

Hong Kong Closes in on Financial Top Spot

Asian centres rise up leading cities' index London and New York losing edge.

Brooke Masters

London and New York are still the world's leading cities for banking and other financial services, but Hong Kong is breathing down their necks, according to the latest Global Financial Centres Index.

The twice-yearly ranking of 75 world cities by Z/Yen Group think-tank is based both on surveys of industry professionals and objective factors such as office rental rates, airport satisfaction and transport. In previous versions, London has generally led the pack, followed closely by New York, with other cities lagging well behind.

But GFCI 8, which will be released on Monday [September 2010], found that London and New York are statistically indistinguishable, with 772 and 770 points, respectively, out of a possible 1,000, and Hong Kong is just 10 points behind, at 760. Hong Kong was 81 points behind the other two as recently as March 2009.

"London and New York are going down a little bit. Their competitiveness isn't quite as good as it once was, while Hong Kong is steaming ahead," said Mark Yeandle, the Z/Yen associate director who developed and runs the survey.

He predicts that Singapore, currently 32 points behind Hong Kong, could join the top ranks soon. Between them, the top four centres account for 70 per cent of all equity trading.

Shanghai also posted significant gains, rising five places to sixth, which puts it in the top 10 for the first time since the survey began in 2005. The mainland Chinese centre showed marked improvements in its ratings for both the "people" and "business environment" categories, rising four places, to sixth and seventh respectively. Overall, there are now four Asian cities in the top 10.

"It is no surprise that centres in Asia are closing rapidly. This is a long-standing trend that reflects the importance of this rapidly growing marketplace," said Stuart Fraser, policy committee chairman of the City of London, which both promotes the area and provides some essential services.

"The level of ambition demonstrated by financial centres in Asia—many of them shiny and new with state-of-the-art infrastructure and seemingly endless amounts of office space—makes it all the more important that London continues to work to retain its international competitiveness," he added.

The City of London used to sponsor the GFCI but has recently been replaced by the Qatar Financial Centre Authority. Qatar ranked 34th overall but came in seventh in the new category of "wealth management and private banking."

London continues to lead for asset management and professional services, while New York is top in the categories of government and regulatory and banking, while Hong Kong is number one for insurance.

Mr Yeandle said the ratings reflect growing concerns about the business and tax environment in London, as well as rising enthusiasm for Asia and a concerted effort by Chinese authorities to reach out to foreigners and improve their standing, particularly on regulation issues. The UK imposed a one-off tax on bonuses last year and plans to tax bank balance sheets.

The survey also found that offshore centres, including the Channel Islands, the Caymans and Gibraltar, continue to face headwinds. Virtually all of them fell in the rankings, continuing a trend that has lasted more than a year. Tax crackdowns appear to be at the root of the shift.

Critical Thinking

1. Which cities in Asia could join the top ranks for international banking and financial services?
2. What explains the better rankings of Asian cities now?

Glossary of Terms and Abbreviations

Ancestor Worship Ancient religious practices still followed in Taiwan, Hong Kong, and the People's Republic of China. Ancestor worship is based on the belief that the living can communicate with the dead and that the dead spirits to whom sacrifices are ritually made can bring about a better life for the living.

Brain Drain A migration of professional people (such as scientists, professors, and physicians) from one country to another, usually in search of higher salaries or better living conditions.

Buddhism A religion of East and Central Asia founded on the teachings of Siddhartha Gautama (the Buddha). Its followers believe that suffering is inherent in life and that one can be liberated from it by mental and moral self-purification.

Capitalist A person who has capital invested in business, or someone who favors an economic system characterized by private or corporate ownership of capital goods.

Chinese Communist Party (CCP) Founded in 1921 by a small Marxist study group, its members initially worked with the Kuomintang (KMT) under Chiang Kai-shek to unify China and, later, to fight off Japanese invaders. Despite Chiang's repeated efforts to destroy the CCP, it eventually ousted the KMT and took control of the Chinese mainland in 1949.

Cold War A conflict between the communist and anti-communist (democratic-capitalists) blocs, without direct military conflict.

Communism In theory, a system in which most goods are collectively owned and equally distributed. In practice, a system of governance in which a single authoritarian party controls the political, legal, educational, and economic systems in an effort to establish a more egalitarian society.

Confucianism Often referred to as a religion, actually a system of ethics for governing human relationships and for ruling. It was established during the fifth century B.C. by the Chinese philosopher Confucius.

Cultural Revolution Formally, the Great Proletarian Cultural Revolution. In an attempt to rid China of its repressive bureaucracy and to restore a revolutionary spirit to the Chinese people, Mao Zedong (Tse-tung) called on the youth of China to "challenge authority" and "make revolution" by rooting out the "reactionary" elements in Chinese society. The Cultural Revolution lasted from 1966 until 1969, but the term is often used to refer to the 10 year period from 1966 to 1976. It seriously undermined the Chinese people's faith in the Chinese Communist Party's ability to rule and led to major setbacks in the economy.

De-Maoification The rooting-out of the philosophies and programs of Mao Zedong in Chinese society.

Democratic Centralism The participation of the people in discussions of policy at lower levels. Their ideas are to be passed up to the central leadership; but once the central leadership makes a decision, it is to be implemented by the people.

ECFA, Economic Cooperation Framework Agreement A special free trade agreement between Taiwan and mainland China signed in 2010.

ExCo The Executive Council of Hong Kong, consisting of top civil servants and civilian appointees chosen to represent the community. Except in times of emergency, the governor must consult with the ExCo before initiating any program.

Feudal In Chinese Communist parlance, a patriarchal bureaucratic system in which bureaucrats administer policy on the basis of personal relationships.

Four Cardinal Principles The Chinese Communists' term for their commitment to socialism, the leadership of the Chinese Communist Party, the dictatorship of the proletariat, and the ideologies of Karl Marx, Vladimir Lenin, and Mao Zedong.

Four Modernizations A program of reforms begun in 1978 in China that sought to modernize agriculture, industry, science and technology, and defense by the year 2000.

Gang of Four The label applied to the four "radicals" or "leftists" who dominated first the cultural and then the political events during the Cultural Revolution. The four members of the Gang were Jiang Qing, Mao's wife; Zhang Chunqiao, former deputy secretary of the Shanghai municipal committee and head of its propaganda department; Yao Wenyuan, former editor-in-chief of the *Shanghai Liberation Daily;* and Wang Hongwen, a worker in a textile factory in Shanghai.

Great Leap Forward Mao Zedong's alternative to the Soviet model of development, this was a plan calling for the establishment of communes and for an increase in industrial production in both the cities and the communes. The increased production was to come largely from greater human effort rather than from more investment or improved technology. This policy, begun in 1958, was abandoned by 1959.

Great Proletarian Cultural Revolution *See* Cultural Revolution.

Gross Domestic Product (GDP) A measure of the total flow of goods and services produced by the economy of a country over a certain period of time, normally a year. GDP equals gross national product (GNP) minus the income of the country's residents earned on investments abroad.

Guerrilla A member of a small force of "irregular" soldiers. Generally, guerrilla forces are used against numerically and technologically superior enemies in jungles or mountainous terrain.

Han Of "pure" Chinese extraction. Refers to the dominant ethnic group in the P.R.C.

Ideograph A character of Chinese writing. Originally, each ideograph represented a picture and/or a sound of a word.

Islam The religious faith founded by Muhammad in the sixth and seventh centuries A.D. Its followers believe that Allah is the sole deity and that Muhammad is his prophet.

Kuomintang (KMT) The Chinese Nationalist Party, founded by Sun Yat-Sen in 1912. *See also* Nationalists.

LegCo Hong Kong's Legislative Council, which reviews policies proposed by the governor and formulates legislation.

Long March The 1934–1935 retreat of the Chinese Communist Party, in which hundreds of thousands died while journeying to the plains of Yan'an in northern China in order to escape annihilation by the Kuomintang.

Glossary of Terms and Abbreviations

Mainlanders Those Chinese in Taiwan who emigrated from the Chinese mainland during the flight of the Nationalist Party in 1949.

Mandarin A northern Chinese dialect chosen by the Chinese Communist Party to be the official language of China.

Mao Thought In the post-1949 period, originally described as "the thoughts of Mao Zedong." Mao's "thoughts" were considered important because he took the theory of Marxism-Leninism and applied it to the concrete conditions existing in China. But since Mao's death in 1976 and the subsequent reevaluation of his policies, Mao Thought is no longer conceived of as the thoughts of Mao alone but as the "collective wisdom" of the party leadership.

May Fourth Period A period of intellectual ferment in China, which officially began on May 4, 1919, and concerned the Versailles Peace Conference. On that day, the Chinese protested what was considered an unfair secret settlement regarding German-held territory in China. The result was what was termed a "new cultural movement," which lasted into the mid-1920s.

Nationalists The Kuomintang (KMT). The ruling party of the Republic of China, but its army was defeated by 1949. Was the only political party in Taiwan until the 1990s.

Newly Industrialized Country (NIC) A term used to refer to those developing countries that have enjoyed rapid economic growth. Most commonly applied to the East Asian economies of South Korea, Taiwan, Hong Kong, and Singapore.

Offshore Islands The small islands in the Formosa Strait that are just a few miles off the Chinese mainland but are controlled by Taiwan, nearly 90 miles away.

Opium A bitter, addictive drug made from the dried juice of the opium poppy.

Opium War The 1839–1842 conflict between Britain and China, sparked by the British import of opium into China. After the British victory, Europeans were allowed into China and trading posts were established on the mainland. The Treaty of Nanking, which ended the Opium War, also gave Britain its first control over part of Hong Kong.

People's Procuracy The investigative branch of China's legal system. It determines whether an accused person is guilty and should be brought to trial.

People's Republic of China (P.R.C.) Established in 1949 by the Chinese Communists under the leadership of Mao Zedong after defeating Chiang Kai-shek and his Nationalist supporters.

Pinyin A newer system of spelling Chinese words and names, using the Latin alphabet of 26 letters, created by the Chinese Communist leadership.

Proletariat The industrial working class, which for Marx was the political force that would overthrow capitalism and lead the way in the building of socialism.

Republic of China (R.O.C.) The government established as a result of the 1911 Revolution. It was ousted by the Chinese Communist Party in 1949, when its leaders fled to Taiwan.

Second Convention of Peking The 1898 agreement leasing the New Territories of Hong Kong to the British until 1997.

Severe Acute Respiratory Syndrome (SARS) A grave respiratory illness that emerged in 2003 as an epidemic in Hong Kong and part of mainland China.

Shanghai Communique A joint statement of the Chinese and American viewpoints on a range of issues in which each has an interest. It was signed during U.S. President Richard Nixon's historic visit to China in 1971.

Socialism A transitional period between the fall of capitalism and the establishment of "true" communism. Socialism is characterized by the public ownership of the major means of production. Some private economic activity and private property are still allowed, but increased attention is given to a more equal distribution of wealth and income.

Special Administrative Region (SAR) A political subdivision of the People's Republic of China that is used to describe Hong Kong's status following Chinese sovereignty in 1997. The SAR has much greater political, economic, and cultural autonomy from the central government in Beijing than do the provinces of the P.R.C.

Special Economic Zone (SEZ) An area within China that has been allowed a great deal of freedom to experiment with different economic policies, especially efforts to attract foreign investment. Shenzhen, near Hong Kong, is the largest of China's Special Economic Zones.

Taiwan Relations Act (TRA) U.S. domestic law passed by Congress in 1979 to regulate unofficial relations between the U.S. and Taiwan.

Taiwanese Independence Movement An organization of native Taiwanese who wanted to declare Taiwan an independent state. Had to organize outside of Taiwan, as its leaders were persecuted in Taiwan by the KMT. Only with the recognition of the legitimacy of competing political parties in the 1990s could they adopt the goal of an independent Taiwan.

Taoism A Chinese mystical philosophy founded in the sixth century B.C. Its followers renounce the secular world and lead lives characterized by unassertiveness and simplicity.

United Nations (UN) An international organization established on June 26, 1945, through official approval of the charter by delegates of 50 nations at a conference in San Francisco. The charter went into effect on October 24, 1945.

Yuan Literally, "branch"; the different departments of the government of Taiwan, including the Executive, Legislative, Judicial, Control, and Examination Yuans.

Bibliography

SOURCES FOR STATISTICAL REPORTS

U.S. State Department *Background Notes* (2008)
C.I.A. *World Factbook* (2008)
World Bank *World Development Reports* (2008)
UN *Population and Vital Statistics Reports* (2008)
World Statistics in Brief (2008)
The Statesman's Yearbook (2008)
Population Reference Bureau *World Population Data Sheet* (2008)
The World Almanac (2008)
The Economist Intelligence Unit (2008)

PEOPLE'S REPUBLIC OF CHINA

Periodicals and Newspapers

The following periodicals and newspapers are excellent sources for coverage of Chinese affairs:
Asiaweek
Asian Survey
Asia Times Online
Beijing Review
China Business Review
China Daily
The China Journal
The China Quarterly
The Economist
Far Eastern Economic Review
Journal of Asian Studies
Journal of Chinese Political Science
Journal of Contemporary China
Modern China
Pacific Affairs
People's Daily
South China Morning Post

GENERAL AND BIOGRAPHIES

Jasper Becker, *The Chinese* (New York: Free Press, 2000).
Insightful portraits of peasants, entrepreneurs, corrupt businessmen and party members, smugglers, and ethnic minorities by a resident journalist. Reveals much about the effect of the government's policies on the lives of ordinary people.

Ma Bo, *Blood Red Sunset* (New York: Viking, 1995).
Perhaps the most compelling autobiographical account by a Red Guard during the Cultural Revolution. Responding to Mao Zedong's call to youth to "make revolution," the author captures the intense emotions of exhilaration, fear, despair, and loneliness. Takes place in the wilds of Inner Mongolia.

Jung Chang, *Wild Swans: Three Daughters of China* (New York: Simon and Schuster, 1992).
A superb autobiographical/biographical account that illuminates what China was like for one family for three generations.

Kwang-chih Chang, *The Archeology of China*, 4th ed. (New Haven, CT: Yale University Press, 1986).

_____*Shang Civilization* (New Haven, CT: Yale University Press, 1980).Two works by an eminent archaeologist on the origins of Chinese civilization.

Nien Cheng, *Life and Death in Shanghai* (New York: Grove Press, 1987).
A gripping autobiographical account of a woman persecuted during the Cultural Revolution because of her earlier connections with a Western company, her elitist attitudes, and her luxurious lifestyle in a period when the Chinese people thought the rich had been dispossessed.

B. Michael Frolic, *Mao's People: Sixteen Portraits of Life in Revolutionary China* (Cambridge, MA: Harvard University Press, 1980).
A must read. Through composite biographies of 16 different types of people in China, the author offers a humorous but penetrating view of "unofficial" Chinese society and politics. Biographical sketches reflect political life during the Maoist era, but the book has enduring value for understanding China.

Rob Gifford, *China Road: A Journey into the future of a rising power* (New York: Random House 2008).
Fascinating stories about China and Chinese people by a journalist as he traveled from Shanghai to Kazakhstan.

Peter Hessler, *River Town: Two Years on the Yangtze* (New York: HarperCollins, 2001).
Insights into Chinese culture by a Peace Corps volunteer who lived in a Yangtze River city from 1996 to 1998. The author gains considerable insights into the life of Fuling, a city that partly flooded when the Three Gorges Dam was completed.

Yarong Jiang and David Ashley, *Mao's Children and the New China* (New York: Routledge, 2000).
More than 20 ex-Red Guards who participated in the Cultural Revolution were interviewed in Shanghai in the mid-1990s. They reminisce about their lives then, revealing much about life in Shanghai during a critical period in China's political history.

Zhisui Li, *The Private Life of Chairman Mao* (New York: Random House, 1994).
A credible biography of the Chinese Communist Party's leader Mao Zedong, written by his physician, from the mid-1950s to his death in 1976. Fascinating details about Mao's daily life and his relationship to those around him.

Heng Liang and Judith Shapiro, *Son of the Revolution* (New York: Vintage, 1984).
A gripping first-person account of the Cultural Revolution by a Red Guard. Offers insights into the madness that gripped China during the period from 1966 to 1976.

Anchee Min, *Becoming Madame Mao* (Boston: Houghton Mifflin, 2000).
This novel vividly portrays Mao's wife, Jiang Qing, tracing her life from early childhood through her failed career as an actress, her courtship with Mao Zedong in the caves of Yenan, and her ultimate demise as a member of the notorious Gang of Four. A real page-turner.

Chihua Wen, *The Red Mirror: Children of China's Cultural Revolution* (Boulder, CO: Westview Press, 1995).
A former editor and reporter presents the heartrending stories of a dozen individuals who were children when the Cultural Revolution started. It shows how rapidly changing policies of the period shattered the lives of its participants and left them cynical adults 20 years later.

James and Ann Tyson, *Chinese Awakenings: Life Stories From the Unofficial China* (Boulder, CO: Westview Press, 1995).
Lively verbal portraits of Chinese people from diverse backgrounds (for example, "Muddy Legs: The Peasant Migrant"; "Turning Iron to Gold: The Entrepreneur"; "Bad Element: The Shanghai Cosmopolite").

HISTORY, LANGUAGE, AND PHILOSOPHY

Johan Bjorksten, *Learn to Write Chinese Characters* (New Haven, CT: Yale University Press, 1994).
A delightful introductory book about writing Chinese characters, with many anecdotes about calligraphy.

William Theodore De Bary, ed., *Sources of Chinese Tradition,* Vols. I and II (New York: Columbia University Press, 1960).
A compilation of the major writings (translated) of key Chinese figures from Confucius through Mao Zedong. Gives readers an excellent understanding of intellectual roots of development of Chinese history.

William Theodore De Bary and Weiming Tu, eds. *Confucianism and Human Rights* (New York: Columbia University Press, 1998).
Articles debate whether the writings of Confucius and Mencius (a Confucian scholar) are relevant to today's human rights doctrine (as defined by the United Nations).

John DeFrancis, *Visible Speech: The Diverse Oneness of Writing Systems* (Honolulu, HI: University of Hawaii Press, 1989).
Discusses the evolution of the Chinese written language and compares it with other languages that use "visible" speech.

Patricia Buckley Ebrey, *The Cambridge Illustrated History of China* (New York: Cambridge University Press, 1996).
A beautifully illustrated book on Chinese history from the Neolithic Period to the People's Republic of China. Includes photos of artifacts (such as bronze vessels) and art (from Buddhist art to modern Chinese paintings), which enrich the historical presentation.

John King Fairbank and Merle Goldman, *China: A New History,* 2nd Enlarged Edition (Cambridge, MA: Harvard University Press, 2006).
Examines forces in China's history that define it as a coherent culture from its earliest recorded history to the present. Examines why the ancient and sophisticated China had fallen behind other areas by the nineteenth century. The Chinese Communist Revolution and its aftermath are reviewed.

William Hinton, *Fanshen: A Documentary of Revolution in a Chinese Village* (New York: Random House, 1968).
A gripping story based on the author's eyewitness account of the process of land reform carried out by the CCP in the north China village of Long Bow, 1947 to 1949.

Edgar Snow, *Red Star Over China* (New York: Grove Press, 1973).
This classic, which first appeared in 1938, is a journalist's account of the months he spent with the Communists' Red Army in Yan'an in 1936, in the midst of the Chinese Civil War. It is a thrilling story about the Chinese Revolution in action and includes Mao's own story (as told to Snow) of his early life and his decision to become a Communist.

Jonathan D. Spence, *The Search for Modern China, 2nd edition* (New York: W. W. Norton & Co., 2000).
A lively and comprehensive history of China from the seventeenth century through the 1990s. Looks at the cyclical patterns of collapse and regeneration, revolution and consolidation, and growth and decay.

Song Mei Lee-Wong, *Politeness and Face in Chinese Culture* (New York: Peter Lang, 2000).
Part of a series on cross-cultural communication, this book discusses how politeness is portrayed in speech and how it relates to a central concept in Chinese culture: "face" and "losing face."

POLITICS, ECONOMICS, SOCIETY, AND CULTURE

Julia F. Andrews, *Painters and Politics in the People's Republic of China, 1949–1979* (Berkeley, CA: University of California Press, 1994).
A fascinating presentation of the relationship between politics and art from the beginning of the Communist period until the eve of major liberalization in 1979.

Jasper Becker, *Dragon Rising: An Inside Look At China Today* (National Geographic, 2007).
Weaves analysis with anecdotes to address today's pressing uncertainties: How will China cope with pollution, unemployment, and demand for energy? What form will its government take? Can Shanghai's success with urban capitalism be replicated elsewhere?

N. Susan D. Blum and Lionel. M. Jensen, eds., *China Off Center: Mapping the Margins of the Central Kingdom* (Honolulu, HI: University of Hawaii Press, 2002).
Arguing that there are many "Chinas," these articles offer new insights into the complexity and diversity of China. Interpretative essays on topics such as linguistic diversity, regionalism, homosexuality, gender and work, popular music, magic and science. Ethnographic reports on minorities.

Susan Brownell and Jeffrey Wasserstrom, eds., *Chinese Femininities and Chinese Masculinities: A Reader* (Berkeley, CA: University of California Press, 2002).
A reader that investigates various issues through the lens of feminist and gender theory.

Thomas Buoye, Kirk Denton, Bruce Dickson, Barry Naughton, and Martin K. Whyte, *China: Adapting the Past, Confronting the Future* (Ann Arbor, MI: The University of Michigan Center for Chinese Studies, 2002).
Articles on China's geography and pre-1949 history, including environmental history, Confucianism, and the Boxer Uprising. It also examines the last few decades, including homosexuality, the Internet, and culture; several short stories.

Allen Carlson, Mary E. Gallagher, Kenneth Lieberthal, and Melanie Manion, eds., *Contemporary Chinese Politics: New Sources, Methods, and Field Strategies* (Cambridge University Press, 2010).
Considers how new and diverse sources and methods are changing the study of Chinese politics. Contributors spanning three generations in China studies place their distinct qualitative and quantitative methodological approaches in the framework of the discipline and point to challenges or opportunities (or both) of adapting new sources and methods to the study of contemporary China.

Guidi Chen and Chuntao Wu, *Will the Boat Sink the Water? The Life of China's Peasants* (New York: Public Affairs, Perseus Books, 2006).
An award-winning book of reportage. While most of the world focuses on China's rapid economic growth, these reporters present stories about the poor peasants in China's vast countryside. Theme challenges mainstream view that China's peasantry was

the primary beneficiary of the Chinese Communist revolution, and argues that even under a Chinese-style market economy, the peasantry continues to suffer.

Deirdre Chetham, *Before the Deluge: The Vanishing World of the Yangtze's Three Gorges* (New York: Palgrave MacMillan, 2002).
A portrait of life along the Yangtze River just before it was flooded to fill up the Three Gorges Dam. Examines the policies that led to the dam, the criticisms of was, and the hopes and fears of what this dam might generate other than electricity.

Paul Close, David Askew and Xu Xin, *The Beijing Olympiad: The Political Economy of a Sporting Mega-Event* (New York: Routledge), 2007.
Looks at the motivations for Beijing to host the Olympics in 2008, and the opportunities and dangers for China embedded in this event. Chapters on the relationship of the individual, nationalism, and capitalism to the Olympics, as well as on how the Olympics can serve as a "coming out party" for an ambitious state.

Elisabeth Croll, *China's New Consumers: Social Development and Domestic Demand* (New York: Routledge), 2006.
Examines the expansion of the domestic market for Chinese-made goods, but challenges the conventional wisdom that there is an insatiable demand for goods among Chinese consumers. Looks at new expectations and social aspirations because of the consumer revolution, the livelihoods and lifestyles of various categories of Chinese consumers, and the government's policy of encouraging internal consumption. Includes chapters on consumption patterns of the rural poor, children, youth, and the elderly.

Deborah S. Davis, ed., *The Consumer Revolution in Urban China* (Berkeley, CA: University of California, 2000).
Articles cover the impact of China's consumer revolution on urban housing, purchases of toys, clothes, and leisure activities for children, and bridal consumerism.

Michael S. Duke, ed., *World of Modern Chinese Fiction: Short Stories & Novellas From the People's Republic, Taiwan & Hong Kong* (Armonk, NY: M. E. Sharpe, Inc., 1991).
A collection of short stories written by Chinese authors from China, Taiwan, and Hong Kong during the 1980s. The 25 stories are grouped by subject matter and narrative style.

Elizabeth Economy, *The River Runs Black: The Environmental Challenge to China's Future* (Ithaca: Cornell University Press), 2004.
The central government's inability to cope with the growing environmental crisis has led to serious social, economic, and health issues, as well as a steadily rising involvement of citizens in non-governmental organizations. Such civic participation may lead to greater democratization and the development of civil society.

Barbara Entwisle and Gail E. Henderson, eds., *Re-drawing Boundaries: Work, Households, and Gender in China* (Berkeley, CA: University of California Press, 2000).
Looks at how gender inequality affects types of work, wages, and economic success. Examines issues of work and gender in China's cities and countryside and among the "floating" population.

Joseph Fewsmith, ed., *China Today, China Tomorrow: Domestic Politics, Economy, and Society* (Rowman & Littlefield Publishers, 2010).
Taking the thirtieth anniversary of China's adoption of reform and opening as an occasion to reflect on the course of development over the past three decades, the contributors consider where the country may be going in the future.

Merle Goldman and Elizabeth. J. Perry, eds., *Changing Meanings of Citizenship in Modern China* (Cambridge, MA: Harvard University Press, 2002).
Studies of citizenship in China over the last century. Focuses on the debate over the relationship of the individual to the state, the nation, the community, and culture.

Peter Gries and Stanley Rosen, eds., *Chinese Politics: State, Society and the Market* (Routledge, 2010).
Despite the continuing economic successes there has been increasing social protests over corruption, land seizures, environmental concerns, and homeowner movements. Such political contestation presents an opportunity to explore the changes occurring in China today—what are the goals of political contestation, how are CCP leaders legitimizing their rule, who arc the specific actors involved in contesting state legitimacy today, and what are the implications of changing state-society relations for the future viability of the People's Republic?

Ellen Hertz, *The Trading Crowd: An Ethnography of the Shanghai Stock Market* (Cambridge, England: University of Cambridge Press, 1998).
An anthropologist examines the explosion of "stock fever" since the stock market opened in Shanghai in 1992. Looks at the dominant role of the state in controlling the market, resulting in a stock market quite different from those in the West.

Yasheng Huang, *Capitalism with Chinese Characteristics: Entrepreneurship and the State* (Cambridge University Press, 2008).
The book presents a story of two Chinas—an entrepreneurial rural China and a state-controlled urban China. In the 1980s, rural China gained the upper hand. In the 1990s, urban China triumphed. While GDP grew quickly in both decades, the welfare implications of growth differed substantially. The single biggest obstacle to sustainable growth and financial stability in China today is its poor political governance.

Alan Hunter and Kim-kwong Chan, *Protestantism in Contemporary China* (New York: Cambridge University Press, 1993).
Examines historical and political conditions that have affected the development of Protestantism in China.

William R. Jankowiak, *Sex, Death, and Hierarchy in a Chinese City* (New York: Columbia University Press, 1993).
Written by an anthropologist with a discerning eye, this is one of the most fascinating accounts of daily life in China. Particularly strong on rituals of death, romantic life, and the on-site mediation of disputes by strangers (e.g., with bicycle accidents).

Maria Jaschok and Suzanne Miers, eds., *Women and Chinese Patriarchy: Submission, Servitude and Escape* (New York: Zen Books, 1994).
Examines Chinese women's roles, the sale of children, prostitution, Chinese patriarchy, Christianity, and feminism, as well as social remedies and avenues of escape for women. Based on interviews with Chinese women who grew up in China, Hong Kong, Singapore, and San Francisco.

Yarong Jiang and David Ashley, *Mao's children and the New China* (New York: Routledge, 2000).
More than 20 ex-Red Guards who participated in the Cultural Revolution were interviewed in Shanghai in the mid-1990s. They reminisce about their lives then, revealing much about life in Shanghai during the critical period in China's political history.

Ian Johnson, *Wild Grass: Three Stories of Change in Modern China* (New York: Pantheon Books, 2004).
The author portrays three ordinary citizens who, by testing the limits of reform, may cause China to become a more open country.

Lane Kelley and Yadong Luo, *China 2000: Emerging Business Issues* (Thousand Oaks, CA: Sage Publications, 1998).
Looks to the emerging business issues for Chinese domestic firms and foreign firms.

Conghua Li, *China: The Consumer Revolution* (New York: Wiley, 1998).
An impressive account of China's rapidly growing consumer society. Looks at the forces that are shaping consumption, China's cultural attitudes toward consumerism, consumer preferences of various age groups, and the rapid polarization of consumer purchasing power.

Jianhong Liu, Lening Zhang, and Steven F. Messner, eds., *Crime and Social Control in a Changing China* (Westport, CT: Greenwood Press, 2001).
Focuses on crime in the context of a rapidly modernizing China. Shows the deeply rooted cultural context for Chinese attitudes toward crime, criminals, and penology that might well interfere with reform.

Stanley B. Lubman, *Bird in a Cage: Legal Reform After Mao* (Stanford, CA: Stanford University Press, 1999).
Traces the victories and frustrations of legal reform since 1979, but is based on a thorough examination of the pre-reform judicial system.

Michael B. McElroy, Christopher P. Nielsen, and Peter Lydon, eds., *Energizing China: Reconciling Environmental Protection and Economic Growth* (Cambridge, MA: Harvard University Press, 1998).
Research reports address the dilemmas, successes, and problems in China's efforts to reconcile environmental protection with economic development. Addresses issues such as energy and emissions, the environment and public health, the domestic context for making policy on energy, and the international dimensions of China's environmental policy.

Joanna McMillan, *Sex, Science and Morality in China* (New York: Routledge), 2006.
Looks at the supposed "opening up" of the sexual world in China and discovers a world still defined by a deep conservatism, a propensity to judge sexual practices based on old-style morality, as well as intolerance of difference. Describes such topics as the coverage of sexual anatomy and sexual function by marriage manuals, transsexuals, homosexuals, masturbation, Viagra, sexual dysfunction, sex shops, prostitution, and many other related topics as presented by China's sexologists and the media.

Gina Marchetti, *From Tian'an Men to Times Square: Transnational China and the Chinese Disaspora on Global Screens, 1989–1997* (Philadelphia: Temple University Press), 2006.
The portrayal of China in the media and in film since the crackdown on demonstrators in 1989. Interviews with Chinese and non-Chinese film makers provide basis for analysis of how global capitalism and other political and social forces have affected the aesthetics of film and presentation of China and the Chinese throughout the world's Chinese communities.

Katherine Morton, *International Aid and China's Environment: Taming the Yellow Dragon* (New York: Routledge), 2005.
Case studies on the three major donor approaches to giving environmental aid to China: helping to build the environmental infrastructure (Japan), introducing market measures and incentives (World Bank), and increasing stakeholder participation in order to improve decision-making (UNDP). Contrasts these with the Chinese emphasis on regulatory control. Examines impact of these three approaches to strengthening sustainable environmental capacity in China.

Andrew J. Nathan and Perry Link, eds., and Liang Zhang, compiler, *The Tiananmen Papers: The Chinese Leadership's Decision to Use Force Against Their Own People—In Their Own Words* (New York: Public Affairs, 2001).
Widely believed to be authentic documents that reveal what was said among China's top leaders behind closed doors during the Tiananmen crisis in 1989. These leaked documents lay out the thinking of China's leaders about the students and workers occupying Tiananmen Square for almost six weeks—and how they eventually decided to use force.

Kevin J. O'Brien and Lianjiang Li, *Rightful Resistance in Rural China* (New York: Cambridge University Press, 2006).
Examines question of how weak, unorganized groups go about resisting state control and articulating their demands. The focus is on resistance in rural China that relies on the use of officially-sanctioned policies, values, laws, and rhetoric to challenge political and economic elites who have abused their power, not implemented policies, or failed to live up to their professed ideals. Looks at how rightful resisters gain legitimacy by using approved channels and not resorting to illegal or criminal activity.

Suzanne Ogden, *Inklings of Democracy in China* (Cambridge, MA: Harvard University Asia Center and Harvard University Press, 2002).
Asks whether liberal democracy is possible or even appropriate in China, given its history, culture, and institutions. Looks at a broad array of indicators. Argues for fair and consistent standards for evaluating freedom and democracy in China and for comparing it with other states.

Suzanne Ogden, Kathleen Hartford, Lawrence Sullivan, and David Zweig, eds., *China's Search for Democracy: The Student and Mass Movement of 1989* (Armonk, NY: M. E. Sharpe, 1992).
A collection of wall posters, handbills, and speeches of the prodemocracy movement of 1989. These documents capture the passionate feelings of the student, intellectual, and worker participants.

Elizabeth J. Perry and Mark Selden, eds., *Chinese Society: Change, Conflict and Resistance,* 2nd edition (New York: Routledge, 2003).
A collection of articles on the resistance generated by economic reforms since 1979. Topics include suicide as resistance, resistance to the one-child campaign, and religious and ethnic resistance.

Paul G. Pickowicz and Yingjin Zhang, eds., *From Underground to Independent: Alternative Film Culture in Contemporary China* (Lanham, Md.: Rowman & Littlefield), 2006.
Examines the evolution beginning in the early 1990s from underground to quasi-independent film making in China: film making that is independent from the state-controlled system of film production, distribution, and showing, and which depends on private (including foreign) funding. Includes articles on diverse topics, such as independently-made documentaries, and film clubs in Beijing.

James Seymour and Richard Anderson, *New Ghosts, Old Ghosts: Prisons and Labor Reform Camps in China* (Armonk, NY: M. E. Sharpe, 1998).
A look inside labor camps in China's northwestern provinces, including details about prison conditions and management, the nature of the prison population, excesses perpetrated in prisons, and fate of released prisoners.

David Shambaugh and Richard H. Yang, *China's Military in Transition* (Oxford: Clarendon Press, 1997).
Collection of articles on China's military covers such topics as party–military relations, troop reduction, the financing of defense, military doctrine, training, and nuclear force modernization.

Susan L. Shirk, *China, Fragile Super Power* (New York: Oxford University Press, 2007).
China portrayed as very pragmatic in both its foreign and domestic policies because of a leadership motivated by fear of its own citizens. Chinese Communist Party leaders seen as insecure and afraid of losing power in spite of economic success and development.

Stockholm Environment Institute and United Nations Development Program (UNDP) China, *China Human Development Report 2002: Making Green Development a Choice* (New York: Oxford University Press, 2002).
Examines the key issues for sustainable development in China. Also looks at the government's response and the creation of environmental associations to address the issues.

Kellee S. Tsai, *Capitalism Without Democracy: The Private Sector in Contemporary China* (Cornell University Press, 2007).
Once illegal, the private business sector now comprises over 30 million businesses employing more than 200 million people and accounting for half of China's GDP. Yet the triumph of capitalism has not led to substantial democratic reforms. Chinese entrepreneurs are not agitating for democracy. Most are working to stay in business, while others are saving for their one child's education or planning to leave the country. Many are Communist Party members.

United Nations Development Program, *China: Human Development Report* (New York: UNDP China Country Office, Annual report).
Provides measurements of the effect of China's economic development on human capabilities to lead a decent life. Areas examined include health care, education, housing, treatment and status of women, and the environment.

Jianying Zha, *China Pop: How Soap Operas, Tabloids, and Bestsellers Are Transforming a Culture* (New York: W. W. Norton, 1995).
Examines the impact of television, film, weekend tabloids, and best-selling novels on today's culture. Some of the material is based on remarkably revealing interviews with China's leading film directors, singers, novelists, artists, and cultural moguls.

Yuezhi Zhao, *Media, Market, and Democracy in China: Between the Party Line and the Bottom Line* (Urbana, IL: University of Illinois Press, 1998).
Raises the basic question of whether the expected value of a "free press" will be realized in China if the party-controlled press is replaced by private entrepreneurs, and a state-managed press is required to make a profit.

Zhiqun Zhu, ed., *The People's Republic of China Today: Internal and External Challenges* (World Scientific Publishing, 2010).
This book provides the most up-to-date, comprehensive, and authoritative assessments of the PRC's political, economic, social, ethnic, energy, security, military, diplomatic and other developments and challenges today. It focuses on the efforts needed by China to grow in a sustainable manner and to become a respected global power.

TIBET AND MINORITY POLICIES

Robert Barnett, *Lhasa: Streets with Memories* (New York: Columbia University Press, 2006).
Examines the interplay of forces from Tibetan history and culture, Chinese control, and modernization to create the streets and life of Tibet's capital today.

Melvyn C. Goldstein, *The Snow Lion and the Dragon: China, Tibet, and the Dalai Lama* (Berkeley, CA: University of California Press, 1997).
The best book on issues surrounding a "free" Tibet and the role of the Dalai Lama. Objective presentation of both Tibetan and Chinese viewpoints.

Melvyn C. Goldstein and Matthew T. Kapstein, eds., *Buddhism in Contemporary Tibet: Religious Revival and Cultural Identity* (Hong Kong: Hong Kong University Press, 1997).
An excellent, nonpolemical collection of articles by cultural anthropologists on Buddhism in Tibet today. Studies of revival of monastic life and new Buddhist practices in the last 20 years are included.

Hette Halskov Hansen, *Lessons in Being Chinese: Minority Education and Ethnic Identity in Southwest China* (Seattle, WA: University of Washington Press, 1999).
Examines Chinese efforts to achieve cultural and political integration through education of a minority population in Chinese cultural values and communist ideology.

Donald S. Lopez, *Prisoners of Shangri-la: Tibetan Buddhism and the West* (Chicago, IL: University of Chicago Press, 1998).
Explodes myths about Tibetan Buddhism created by the West. Shows how these myths have led to distortions that do not serve well the cause of greater autonomy for Tibet.

Orville Schell, *Virtual Tibet: Search for Shangri-la from the Himalayas to Hollywood* (New York: Henry Holt and Co., 2000).
Examines the journals of those hoping to find a spiritual kingdom in Tibet. Notes the perilous journeys undertaken for the last 200 years in pursuit of this quest, and the disappointment of almost all in what they found.

FOREIGN POLICY: CHINA AND THE INTERNATIONAL SYSTEM

Gerald Chan, *China's Compliance in Global Affairs: Trade, Arms Control, Environmental Protection, Human Rights* (Hackensack, NJ: World Scientific Publishing Co., 2006).
Assesses China's compliance with international rules and norms in the four areas of the title. Asks whether China has acted "responsibly" from the perspective of US-China relations. First looks at how China sees its "responsibility" to the world community, given its own history, culture, ethics, and level of economic development. Case studies of compliance with WTO rules, with the norms and regulations of the arms control and disarmament regime, with newly formed international environmental norms, and with Western notions of human rights, which are at odds with China's primary definition of human rights.

Jian Chen and Shujie Yao, eds. *Globalization, Competition and Growth in China* (New York: Routledge, 2006).
Chapters look at reforms in the financial sector, foreign direct investment, globalization, and China's strategies for development. Considerable technical analysis as well.

Yong Deng, *China's Struggle for Status* (Cambridge University Press, 2008).
Examines how the once beleaguered country has adapted to, and proactively realigned, the international hierarchy, great-power politics, and its regional and global environment in order to carve out an international path within the globalized world.

Sujian Guo and Jean-Marc F. Blanchard, eds., *"Harmonious World" and China's New Foreign Policy.* (Lexington Books, a Division of Rowman & Littlefield, 2008).
The concept of the harmonious world has become the basis for the new principles and goals of Chinese foreign policy under the fourth generation CCP leadership. The question remains, however, about the exact meanings of these principles and slogans, and their implications for Chinese foreign policy.

Yufan Hao, C.X. George Wei, and Lowell Dittmer, eds., *Challenges to Chinese Foreign Policy: Diplomacy, Globalization, and the Next World Power* (The University Press of Kentucky, 2009).

Defines the "peacefully rising" position currently articulated by Beijing and its implications for international peace and security. Also describes the world's evolving perceptions of Chinese foreign relations, as well as Beijing's diplomatic strategy toward the United States, Europe, Japan, Russia, and other Asian nations.

Alastair Iain Johnston and Robert S. Ross, eds., *Engaging China: The Management of an Emerging Power* (New York: Routledge, 1999). A collection of articles on how various governments, including Korea, Singapore, Indonesia, Japan, Taiwan, the United States, and Malaysia, have tried to "engage" an increasingly powerful China.

Samuel S. Kim, ed., *China and the World, Chinese Foreign Policy Faces the New Millennium (4th edition)* (Boulder, CO: Westview Press, 2001).
Examines theory and practice of Chinese foreign policy with the United States, Russia, Japan, Europe, and the developing world as China enters the new millennium. Looks at such issues as the use of force, China's growing interdependence with other countries, human rights, the environment, and China's relationship with multilateral economic institutions.

Joshua Kurlantzick, *Charm Offensive: How China's Soft Power Is Transforming the World* (New Haven, CT, Yale University Press 2008). Looks at how China is using soft power to appeal to its neighbors and to distant countries alike.

David M. Lampton, *The Three Faces of Chinese Power: Might, Money, and Minds* (University of California Press, 2008).
Investigates the military, economic, and intellectual dimensions of China's growing influence. Provides a fresh perspective from which to assess China—how its strengths are changing, where vulnerabilities and uncertainties lie, and how the rest of the world, not least the United States, should view it.

Richard Madsen, *China and the American Dream: A Moral Inquiry* (Berkeley, CA: University of California Press, 1995).
Looks at the emotional and unpredictable relationship that the United States has had with China from the nineteenth century to the present.

James Mann, *About Face: A History of America's Curious Relationship With China, From Nixon to Clinton* (New York: Alfred A. Knopf, 1999).
A journalist's account of the history of U.S.–China relations from Nixon to Clinton. Through examination of newly uncovered government documents and interviews, gives account of development of the relationship, with all its problems and promises.

Ramon H. Myers, Michel C. Oksenberg, and David Sham-baugh, eds., *Making China Policy: Lessons From the Bush and Clinton Administrations* (New York: Rowman and Littlefield, 2002).
Examines the policy of the United States toward China during the George Bush and Bill Clinton administrations (1989–2000). Includes an account of China's perception and response to America's China policies.

Robert G. Sutter, *Chinese Foreign Relations: Power and Policy since the Cold War,* 2nd ed. *(Asia in World Politics)* (Rowman & Littlefield Publishers, 2009).
A comprehensive and thoroughly updated introduction to Chinese foreign relations. It discusses the opportunities and limits China faces as it seeks increased international influence.

Michael D. Swaine and Zhang Tuosheng, eds., *Managing Sino-American Crises: Case Studies and Analysis* (Washington D.C.: Carnegie Endowment for International Peace, 2006).
Looks at the pattern of management during crises between the U.S. and China. Case studies of wars in Korea, Vietnam, and conflicts over Taiwan, as well as incidents such as the U.S. bombing of the Chinese Embassy in Belgrade and the U.S.–China aircraft collision.

Ian Taylor, *China and Africa: Engagement and Compromise* (New York: Routledge, 2006).
China's policies toward African states reflect its stated foreign policy imperative of opposing the spread of "hegemonism" while trying to find a place for the expansion of its own economic interests. Need for China to fulfill aspirations as a great power complicated by an international system that has been hostile to its ambitions, in part by shutting China out from access to natural resources in many places. Chapters on the history of China's relationship with Africa, and its present relationship with specific African countries.

David Zweig, *Internationalizing China: Domestic Interests and Global Linkages* (Ithaca, NY: Cornell University Press, 2002).
Case studies on issues that connect domestic interests to China's foreign policy and international linkages, and the diminished role of bureaucrats in regulating the internationalization of China.

HONG KONG

Periodicals and Newspapers

Asia Times Online
Hong Kong Commercial Daily
Hong Kong News Online
The Standard
South China Morning Post

POLITICS, ECONOMICS, SOCIETY, AND CULTURE

Robert Ash, Peter Ferdinand, Brian Hook, and Robin Porter, *Hong Kong in Transition: One Country, Two Systems* (New York: Routledge Curzon, 2003).
Investigates changes since the 1997 handover in Hong Kong's business environment, including the role of public opinion and government intervention, and the evolving political culture.

Ming K. Chan and Alvin Y. So, eds., *Crisis and Transformation in China's Hong Kong* (Armonk, NY: M. E. Sharpe, 2002).
Examines political and social changes in Hong Kong since it was returned to China's sovereignty in 1997.

Robert Cottrell, *The End of Hong Kong: The Secret Diplomacy of Imperial Retreat* (London: John Murray, 1993).
Exposes the secret diplomacy that led to the signing of the "Joint Declaration on Question of Hong Kong" in 1984, the agreement that ended 150 years of British colonial rule over Hong Kong. Thesis is that Britain was reluctant to introduce democracy into Hong Kong before this point because it thought it would ruin Hong Kong's economy and lead to social and political instability.

Michael J. Enright, Edith E. Scott, and David Dodwell, *The Hong Kong Advantage* (Oxford: Oxford University Press, 1997).
Examines the special relationship between the growth of Hong Kong's and mainland China's economies, such topics as the role of the overseas Chinese community in Hong Kong and the competition Hong Kong faces from Taipei, Singapore, Seoul, and Sydney as well as from such up-and-coming Chinese cities as Shanghai.

Bruce Kam-kwan Kwong, *Patron-Client Politics and Elections in Hong Kong* (Routledge 2010).
Studies whether patron-client relations are critical to the electoral victory of candidates; how the political elites cultivate support from clients in order to obtain more votes during local elections; and tests the extent to which patron-client relations are crucial in order for candidates to obtain more ballots during elections.

Wai-man Lam, *Understanding the Political Culture of Hong Kong: The Paradox of Activism and Depoliticization* (Armonk, N.Y.: M. E. Sharpe), 2004.
Through case studies of protest, Lam challenges the view of a politically apathetic Hong Kong populace. Looks at role of

ideology, nationalism, gender, civil rights, and economic justice as motivating political participation in Hong Kong.

C. K. Lau, *Hong Kong's Colonial Legacy: A Hong Kong Chinese's View of the British Heritage* (Hong Kong: Chinese University Press, 1997).
Engaging overview of the British roots of today's Hong Kong. Special attention is given to such problems as the "identity" of Hong Kong people as British or Chinese, the problems in speaking English, English common law in a Chinese setting, and the strictly controlled but rowdy Hong Kong "free press."

Jan Morris, *Hong Kong: Epilogue to an Empire* (New York: Vintage, 1997).
Witty and detailed first-hand portrait of Hong Kong by one of its long-term residents. Gives the reader the sense of actually being on the scene in a vibrant Hong Kong.

Christopher Patten, *East and West: China, Power, and the Future of Asia* (New York: Random House), 1998.
The controversial last governor of Hong Kong gives a lively insider's view of the British colony in the last 5 years before it was returned to China's sovereignty. Focuses on China's refusal to radically change Hong Kong's political processes on the eve of the British exit. Argues against the idea that "Asian values" are opposed to democratic governance, and suggests that "Western values" have already been realized in Hong Kong.

Mark Roberti, *The Fall of Hong Kong: China's Triumph and Britain's Betrayal* (New York: John Wiley & Sons, Inc., 1994).
A fast-paced, drama-filled account of the decisions Britain and China made about Hong Kong's fate beginning in the early 1980s. Based on interviews with 150 key players in the secret negotiations between China and Great Britain.

Ming Sing, *Hong Kong's Tortuous Democratization: A Comparative Analysis* (New York: Routledge Curzon), 2004.
An examination of the governance in Hong Kong since the 1940s, and the constraints to democratization. Looks beyond the limits imposed by Beijing to other forces, including lack of public support and weak pro-democracy forces, to explain why democracy has not yet emerged.

Ming Sing, *Politics and Government in Hong Kong: Crisis under Chinese sovereignty* (Routledge, 2008).
Examines the government of Hong Kong since 1997, focusing in particular on the anti-government mass protests and mobilizations since 2003. Argues that Hong Kong has been poorly governed since transferring to Chinese rule, and that public frustration with governmental performance, including anti-subversion laws and slow democratization, has resulted in the regular and massive protests, which have been rare in Hong Kong's past political development.

Alvin Y. So, *Hong Kong's Embattled Democracy: A Societal Analysis* (Baltimore: Johns Hopkins University Press, 1999).
Traces Hong Kong's development of democracy.

Steven Tsang, *A Modern History of Hong Kong* (New York: I.B. Tauris), 2004.
History of British colonial rule from before the Opium Wars. Examines problems in creating the rule of law and an independent judiciary in Hong Kong, and the impact of trade with China on Hong Kong's society and economy.

Frank Welsh, *A Borrowed Place: The History of Hong Kong* (New York: Kodansha International, 1996).
Best book on Hong Kong's history from the time of the British East India Company in the eighteenth century through the Opium Wars of the nineteenth century to the present.

TAIWAN

Periodicals and Newspapers

China Post
Taipei Journal
Taipei Review
Taipei Times

POLITICS, ECONOMICS, SOCIETY, AND CULTURE

Bonnie Adrian, *Framing the Bride: Globalizing Beauty and Romance in Taiwan's Bridal Industry* (Berkeley, CA: University of California Press), 2003.
A fascinating ethnographic study of Taipei's bridal photography as a narrative on contemporary marriages, intergenerational tensions, how the local culture industry and brides use global images of romance and beauty, and the enduring importance of family and gender.

Muthiah Alagappa, ed., *Taiwan's Presidential Politics: Democratization and Cross-Strait Relations in the Twenty-first Century* (Armonk, NY: M. E. Sharpe, 2001).
Focuses on Taiwan's presidential elections in March 2000 and the impact of those elections one year later on the democratic transition from a one-party-dominant system to a multiparty system. Also examines the degree to which Taiwan under the leadership of Chen Shui-bian was able to consolidate democracy.

Robert Ash and J. Megan Greene, eds., *Taiwan in the 21st Century: Aspects and Limitations of a Development Model* (New York: Routledge, 2007).
Examines what is unique, or at least special to Taiwan's economic and political development that makes taking Taiwan as a model for China or other Asian countries questionable. Also notes those aspects of Taiwan's development model that might be replicable.

Christian Aspalter, *Understanding Modern Taiwan: Essays in Economics, Politics, and Social Policy* (Burlington, VT: Ashgate, 2001).
A collection of articles on Taiwan's "economic miracle" and such topics as Taiwan's "identity," democratization, policies on building nuclear-power plants and the growing antinuclear movement, labor and social-welfare policies, and the role of political parties in developing a welfare state.

Melissa J. Brown, *Is Taiwan Chinese? The Impact of Culture, Power, and Migration on Changing Identities* (Berkeley: University of California Press, 2004).
Author explores the meaning of identity in Taiwan. From 1945–1991, Taiwan's government claimed that Taiwanese were ethnically and nationally Chinese. Since 1991, the government has, in a political effort to claim national and cultural distance from the mainland, moved to a position asserting that their identity has been shaped by a mix of aboriginal ancestry and culture, Japanese cultural influence, and Han Chinese cultural influence and ancestry. Examines cultural markers of identity, such as folk religion, footbinding and ancestor worship as well as how identities change.

Wei-Chin Lee, *Taiwan's Politics in the 21st Century: Changes and Challenges* (World Scientific Publishing, 2010).
Highlights Taiwan's ongoing efforts to mediate between competing political actors, a means to ensure domestic stability and national security without severely affecting its continuous economic growth and sovereign status in the international society.

Bibliography

Richard Madsen, *Democracy's Dharma: Religious Renaissance and Political Development in Taiwan* (University of California Press, 2007).

Explores the remarkable religious renaissance that has reformed, revitalized, and renewed the practices of Buddhism and Daoism in Taiwan. Connects these noteworthy developments to Taiwan's transition to democracy and the burgeoning needs of its new middle classes.

Fen-ling Chen, *Working Women and State Policies in Taiwan: A Study in Political Economy* (New York: Palgrave, 2000).

A study of the impact of social welfare and state policies on the relationships between men and women since 1960. "Gender ideology" has changed and, with it, women's views of the workplace and their role in society. Examines related issues of childcare, wages, the women's movement, and women in policy-making system.

Richard C. Bush, *At Cross Purposes: U.S.-Taiwan Relations* (Armonk, N.Y.: M. E. Sharpe, 2004).

The former head of the American Institute in Taiwan (1997–2002) examines why President Roosevelt decided that Taiwan ought to be returned to China after World War II, the U.S. position on the Kuomintang's repressive government rule, the nature of the U.S. "2-China" policy from 1950 to 1972, and the basis for U.S. military and political relations with Taiwan.

Michael S. Chase, *Taiwan's Security Policy: External Threats and Domestic Politics* (Lynne Rienner Publishers, 2008).

Offers a comprehensive analysis of what has caused Taiwan's underactive response to the accelerating pace of China's military upgrading. Focuses on three key factors—Taiwan's security ties with the US, its perceptions of China's capabilities and intentions, and Taiwanese domestic politics.

Ko-lin Chin, *Heijin: Organized Crime, Business, and Politics in Taiwan* (Armonk, N.Y.: M. E. Sharpe), 2003.

An examination of the connection between Taiwan's underworld (*hei*—black) and business/money (*jin*—gold) to politics that has accompanied Taiwan's efforts to democratize since emerging from martial law after 1987. Looks at ways in which black-gold politics have undercut democratization through vote buying, political violence, bid rigging, insider trading, and violence.

Bernard D. Cole, *Taiwan's Security: History and Prospects* (New York: Routledge, 2006).

An objective, well-written, and interesting account of Taiwan's complex security issues by a faculty member at the National War College. Full of valuable insights concerning the strategic issues of Taiwan's defense that only someone with military training and academic research capabilities can offer. The perfect starting point for understanding the strategic standoff that continues in the triangular relationship among the U.S., the P.R.C., and Taiwan.

John F. Copper, *Taiwan: Nation-state or Province?* (Westview, 2009).

Examines Taiwan's geography, history, society, culture, economy, political system, and foreign and security policies in the context of Taiwan's uncertain political status, whether a sovereign nation or a province of the People's Republic of China.

Bruce J. Dickson and Chien-min Chao, eds., *Assessing the Lee Teng-hui Legacy in Taiwan's Politics: Democratic Consolidation and External Relations* (Armonk, NY: M. E. Sharpe, 2002).

Focuses on the impact of Lee Teng-hui presidency (1996–2000) on democratic consolidation, the role (and demise) of the Nationalist Party and the rise of the Democratic Progressive Party, and the economy. Also examines President Lee's impact on security issues.

Dafydd Fell, *Government and Politics in Taiwan* (Routledge, 2011).

Introduces the reader to the big questions that concern change and continuity in how politics operates and how Taiwan is governed. Taking a critical approach, it provides students with the essential background to the history and development of the political system as well as an explanation of the key structures, processes, and institutions that have shaped Taiwan over the last few decades.

J. Megan Greene and Robert Ash, eds., *Taiwan in the 21st Century: Aspects and Limitations of a Development Model* (Routledge, 2011).

Analyzes Taiwan's economic and political achievements, and asks whether it is possible to identify through the experience of a single nation—Taiwan— the makings of a replicable model.

Dennis V. Hickey, *Foreign Policy Making in Taiwan: From Principle to Pragmatism* (Routledge, 2006).

Examines the future direction of Taiwan's foreign policy, focusing on the internal and external forces that shape its foreign policy decisions today. Suggests four levels of analysis—the international system, governmental structure, societal forces, and individual factors—when seeking to understand Taipei's foreign policy behavior.

A-Chin Hsiau, *Contemporary Taiwanese Cultural Nationalism* (New York: Routledge, 2000).

Traces the development of Taiwanese cultural nationalism. Includes the impact of Japanese colonialism, post–World War II literary development, and the spawning of a national literature and national culture.

Chen Jie, *Foreign Policy of the New Taiwan: Pragmatic Diplomacy in Southeast Asia* (Northampton, MA: Edward Elgar, 2002).

Outstanding book on Taiwan's foreign policy (1949–2000). Shows patterns in Taiwan's diplomacy and provides basis for theories and insights about Taiwan's policies, frustrations, sensitivities, and motivations in international affairs. Also covers Taiwan's policy toward the millions of "overseas Chinese."

David K. Jordan, *Gods, Ghosts, and Ancestors: The Folk Religion of a Taiwanese Village* (Berkeley, CA: University of California Press, 1972).

A fascinating analysis of folk religion in Taiwan by an anthropologist, based on field study. Essential work for understanding how folk religion affects the everyday life of people in Taiwan.

Robert M. Marsh, *The Great Transformation: Social Change in Taipei, Taiwan, Since the 1960s* (Armonk, NY: M. E. Sharpe, 1996).

An investigation of how Taiwan's society has changed since the 1960s when its economic transformation began.

Shelley Rigger, *Taiwan's Rising Rationalism: Generations, Politics, and "Taiwanese Nationalism,"* Policy Studies 26. (Washington D.C.: East-West Center Washington, 2006).

Challenges conventional assumptions that identifying as a Taiwanese equates to a pro-independence stance or to opposition to improved ties with the China Mainland. Looks at generational differences in attitudes among Taiwanese.

Denny Roy, *Taiwan: A Political History* (Ithaca, NY: Cornell University Press, 2003).

A comprehensive narrative of the island's history from the first Chinese settlements to the Chen Shui-bian presidency.

Murray A. Rubinstein, ed., *Taiwan: A New History (expanded edition)* (Armonk, NY: M. E. Sharpe, 2007).

A collection of articles on a wide range of topics, from aborigines and the historical development of Taiwan during the Ming Dynasty, to topics in Taiwan's more recent history. These include such topics as Taiwanese new literature, identity and social change in Taiwanese religion, socioeconomic modernization, and aboriginal self-government.

Scott Simon, *Sweet and Sour: Life-Worlds of Taipei Women Entrepreneurs* (Lanham, Maryland: Rowman & Littlefield Publishers), 2003.
Examines the contradictions and tensions that characterize the lives of Taiwan's female entrepreneurs, who spear-headed Taiwan's economic "miracle." Presents portraits of these women, including street vendors, a hairdresser, a café owner, a fashion designer, and more. Sheds light on urban life and on impact of patriarchal culture on male-female relations.

John Q. Tian, *Government, Business, and the Politics of Interdependence and Conflict across the Taiwan Strait* (New York: Palgrave MacMillan, 2006).
Examines the complexities of the Taiwan-China mainland relationship generated by the many situations in which both sides must compromise in order to advance their interests. Looks at the specifics of cross-strait trade and investment, how industrial organization and the financial system affect economic interactions, and how local governments in Mainland China attract Taiwanese investors.

Alan Wachman, *Why Taiwan? Geostrategic Rationales for China's Territorial Integrity* (Stanford University Press, 2007).
Traces the evolution, explains the appeal, and suggests implications of the strategic calculations that pervade PRC strategic considerations of Taiwan.

Index

A

Abortion, "one-child policy" and, 13, 14, 15
Aging population, 14–15
Agriculture
  communes and, 20
  economic liberalization and, 25
  in Manchu Dynasty, 6
  programs for, 23–24
  Special Economic Zones and, 26
  in Taiwan, 82
Air pollution, 28, 81
Air travel, to Taiwan, 86
All-China Lawyers Association, 30
American Institute, in Taiwan, 75, 88
Animism, 18
Anti-Rightest Campaign, 12
ARATS. *See* Association for Relations Across the Taiwan Strait (ARATS)
Arms sales, to Taiwan, 88–89
Article 23, of Basic Law, 63–64
Asian financial crisis, 58
Association for Relations Across the Taiwan Strait (ARATS), 90
Authoritarianism, 32
Authority, Cultural Revolution and, 9
Avian flu, 65

B

Basic Law, 54, 62, 63–64
BBC. *See* British Broadcasting Corporation (BBC)
Belgrade embassy bombing, 20, 40
Bible, 19
Bird flu, 65
Birth-control campaign, 13
"Black" societies, 19
Blair, Tony, 54
British Broadcasting Corporation (BBC), 36
Buddhism
  in PRC, 18
  in Tibet, 17
Bureaucracy, in imperial China, 6
Burma, 52
Bush, George W., 42, 45, 88

C

Cars
  increasing number of, 27
  in use, in PRC, 5
Carter, Jimmy, 88
Caterpillar cultivation, in Tibet, 16
Catholicism, 19, 79
CCP. *See* Chinese Communist Party (CCP)
CCTV-9, 38
Chen Shui-bian, 42, 81, 86
Chen Yunlin, 90
Chiang Ching-kuo, 74, 84
Chiang Kai-shek, 7
  in Japanese invasion of Manchuria, 8
  Kuomintang and, 75
  in Taiwanese history, 73, 74, 75
  in World War II, 52
Chiang Pin-kung, 90
Children, "one-child policy" and, 13
China. *See also* People's Republic of China (PRC); Republic of China (ROC)
  bureaucracy in, 6
  Confucius in history of, 5
  contraction and expansion of, 6
  as empire, 5–6
  gross domestic product of, historic, 5
  as historical technological leader, 5
  in Manchu Dynasty, 6
  Republican, 6–9
Chinese Association for Human Rights, 85
Chinese Communist Party (CCP)
  agricultural programs and, 24
  Buddhism and, 18
  civilian organizations and, 34
  collectivization and, 24
  Confucianism and, 18
  Cultural Revolution and, 11
  in establishment of PRC, 9–10
  ethical basis of law and, 28
  Falun Gong and, 19
  founding of, 7
  future of, 46
  Gang of Four and, 10
  Hong Kong and, 61–62
  ideology of, 20–21
  Kuomintang and, 7, 8–9
  Mao in, 7, 8
  McCarthyism and, 74
  organization of, 31
  in political system, 30–32
  repression by, 13
  rise of, 5
  Soviet Union and, 7, 9, 10
  state and, 30–32
  State Council in, 30–32
  student movement and, 12
  Taiwanese constitution and, 84
Christianity
  in PRC, 19
  in Taiwan, 79
Civilian organizations, Communist Party and, 33
Civil law, 30
Civil rights
  in Hong Kong, 66–67
  in Taiwan, 85
Civil service examinations, 6
Clinton, Bill, 44
Coal, energy from, 27, 28
Cold War, 42, 74
Collectivization, 24
Comintern (Communist International), 7
Command economy, 23
Commercial law, 30
Communes, 20
Communication
  in Hong Kong, 51
  in PRC, 4–5
  in Taiwan, 73
Communist Party. *See* Chinese Communist Party (CCP)
Computers, Chinese characters and, 21
Condyceps sinensis, 17
Confiscation, of land, 25
Confucius and Confucianism
  in Chinese history, 5
  civil service examinations and, 6
  history of, 6
  overview of, 18
  in Taiwan, 79
  women and, 80
Constitution, of Taiwan, 83–84
Corruption
  defining, 32
  market policy and, 27
  "one-child" policy and, 14
  in political system, 32–33
  reform and, 32
Crime
  in Hong Kong, 60–61
  in PRC, 25–26
Criminal law, 29–30
Cultural Revolution
  authority and, 9
  education and, 9
  Gang of Four and, 10
  Hong Kong and, 52
  Mao and, 8, 11
  objectives of, 11
  political education and, 22
  Red Guards in, 9
  start of, 11
Culture
  authoritarianism and, 32
  Christianity and, 19
  foreign, 40
  nationalism and, 40
  symbolism in, 12
  of Taiwan, 76–77
Curse of the Golden Flower (film), 38
Czechoslovakia, 42

D

DAB. *See* Democratic Alliance for the Betterment of Hong Kong (DAB)
Dalai Lama, 16
Decentralization, 25
Defense, in legal system, 30
De-Maoification, 77
Democracy
  business and, 35–36
  forces for, 35–36
  globalization and, 36

in Hong Kong, 54
human rights and, 38–39
in ROC history, 32
socialist, 32–33, 46
Democratic Alliance for the Betterment of Hong Kong (DAB), 64–65
Democratic Party, in Hong Kong, 64
Democratic Progressive Party (DPP), 77–78, 85–87
Deng Xiaoping
death of, 46
Gang of Four and, 10
legacy of, 12
minorities and, 16
political education and, 22
"rehabilitation" of, 11
Dialects, 21
Diplomatic relations
of Hong Kong, 60
between US and PRC, 44
Dollar, Hong Kong, exchange rate of, 51
Domestic Violence Prevention Law, 85
DPP. *See* Democratic Progressive Party (DPP)
Drug trade, in Hong Kong, 61

E

East India Company, 7
Economic Cooperation Framework Agreement or ECFA, 82
Economic reform, in PRC, 25
Economic system
in Hong Kong, 57–58
in PRC, 23–28
in Taiwan, 80–83
Education
Cultural Revolution and, 9
democratization and, 35–36
in Hong Kong, 55–56
minorities and, 16
political, 22
in PRC, 21–22
study abroad and, 22–23
in Taiwan, 79–80
Egalitarianism, 11
Elections
in Hong Kong, 64–65
in Taiwan, 76
Embassy bombings, 20, 40
Empire, Chinese, 5–6
Energy
pollution and, 28
in Taiwan, 78
Environmental protection, 28, 77–78
Ethics, law and, 28
Exchange rate
of Hong Kong dollar, 51
of Taiwan dollar, 73
of yuan, 5
Executive Council, 61
Export commodities
of Hong Kong, 51
of PRC, 5
in Taiwan, 73

F

Falun Gong, 18–19
Family, "one-child policy" and, 13
Female babies, "one-child policy" and, 14
Fertility rate
in Hong Kong, 51
in Taiwan, 72
Festivals, in Taiwan, 78–79
Film industry, 38
Folk religions, 18
Foreign investment, in PRC, 41
Freedom of press, in Hong Kong, 67
Free enterprise, 25
Free trade zones, 82
Fungus, cultivation in Tibet, 17

G

Gang of Four, 8, 9, 10, 24
Garbage production, in Taiwan, 77, 81
GDP. *See* Gross domestic product (GDP)
Gender Equality Labor Law, 85
Gender ratio
in PRC, 4, 14
in Taiwan, 72
Geography
of Hong Kong, 50
of PRC, 4
of Taiwan, 72
"Ghost Month," 79
Globalization, democracy and, 36
Gorbachev, Mikhail, 12, 42
"Great Leap Forward," 11, 12, 20
Gross domestic product (GDP)
historic, of China, 5
in Hong Kong, 51
of PRC, 5
of Taiwan, 4, 80
Growth rate, population
in Hong Kong, 50
in PRC, 14
in Taiwan, 72
Guangzhou, 7
Guilt, presumption of, 29
Gulf War, 42

H

Han Chinese population, 16
Health
pollution and, 28
in Taiwan, 79, 80
Historical authoritarianism, 32
HIV/AIDS, 46
Hong Kong
Article 23 of Basic Law and, 63–64
Asian financial crisis and, 58
Basic Law and, 54, 62, 63–64
civil rights in, 66–67
communication in, 51
cooperation with, and PRC, 61
crime in, 60–61
Cultural Revolution and, 52
Democratic Alliance for Betterment of, 64–65
Democratic Party in, 64
diplomatic relations of, 60
economy in, 57–58
education in, 55–56
elections in, 64–65
emigration from, to UK, 60
export commodities of, 51
fertility rate in, 51
as financial center, 58–59
future of, 67–68
GDP of, 51
growth rate in, 50
history of, 51–54
immigrant population in, 54–55
import commodities of, 51
Internet use in, 51
Joint Declaration on, 53
Joint Liaison Group on, 53–54
judiciary in, 66
Korean War and, 59
Kowloon Peninsula and, 52
language in, 55–56
legal system in, 66
living conditions in, 56–57
media in, 66–67
military spending in, 5
negotiations over status of, 53–54
New Territories of, 52
people of, 54–61
physicians in, 51
political parties in, 64–65
political structure in, 61–67
population of, 50
poverty in, 56, 57
refugees in, 55
religion in, 66–67
SARS and, 58, 60
security in, 66
sensitivity of economy in, 60
Shenzhen and, 59, 60
social clubs in, 56
society of, 54–61
as Special Administrative Region, 53
Special Economic Zones and, 59–60
statistics on, 50–51
terrorism and, 63
as trade center, 58–59
transportation in, 51
unemployment in, 51
Vietnam War and, 55
wealth in, 56–57
in WTO, 54
Hong Kong Alliance in Support of the Patriotic Movement in China, 66
Hong Kong dollar, exchange rate of, 51
Hong Kong Transition Project, 62
Hu Jintao, 86
Human rights, 38–39
US relationship and, 44–46
Hunger strike, in student movement, 12
Hu Yaobang, 12, 13

I

Immigrant population, in Hong Kong, 54–55
Import commodities
of Hong Kong, 51
of PRC, 5
in Taiwan, 73
India, opium from, 7
Infant mortality
in Hong Kong, 51
in PRC, 4
Inflation
in PRC, 5
in Taiwan, 73

Inner Mongolia, 17
Instability, 25–26
Interest groups, in Taiwan, 87
Internet, 38
Internet use
in Hong Kong, 51
in PRC, 4
in Taiwan, 73
Intrauterine devices (IUDs), 13
Islam, 17–18
Isolation, 39
IUDs. *See* Intrauterine devices (IUDs)

J
Japan
invasion of Manchuria by, 7–8
Taiwan and, 73
Jiang Qing, 10
Jiang Zemin, 12, 30, 42, 67
Joint Declaration on the Question of Hong Kong, 53
Joint Liaison Group, 53–54
Judiciary, in Hong Kong, 66
Jurisdiction, legal, in Hong Kong, 66

K
Kazakhstan, 17, 42
Keyboards, Chinese language and, 21
Khruschev, Nikita, 11, 41
Kissinger, Henry, 74
KMT. *See* Kuomintang (KMT)
Korean War, 43, 59, 74
Kosovo War, 20
Kowloon Peninsula, 52
Kuomintang (KMT)
Chiang Kai-shek and, 75
Communist Party and, 7, 8–9
martial law in Taiwan and, 84–85
ousting of, 8–9
"peace offensive" and, 89–90
as political party, 86
Taiwan and, 73–74, 75
Taiwanese constitution and, 83–84
Taiwanese reform and, 85
"Taiwanization" and, 75
United States-PRC relationship and, 43
Kyrgyzstan, 17

L
Labor Pension Retirement Act, 80
Labor Standards Law, 80
Land disputes, 25
Language
computers and, 21
dialects, 21
education, 16
history of, 21
in Hong Kong, 50, 55–56
pinyin, 19, 21
spoken, 21
written, 19, 21
Law
Basic Law, in Hong Kong, 54, 62, 63–64
civil, 30
commercial, 30
criminal, 29–30
ethics and, 28
lawyers and, 30
martial, in Taiwan, 84–85
politics and, 28–29
presumption of guilt in, 29
Lawyers, 30
Lee Teng-hui, 75–76, 85
Legal jurisdiction, in Hong Kong, 66
Legal system
in Hong Kong, 66
in PRC, 28–30
Legislative Council, 61
Legislature, in Hong Kong, 54
Leninism, 11, 19–21
Lhasa, 17
Liberalization, economic, 25
Lien Chan, 86
Life expectancy
in Hong Kong, 50
in PRC, 4
in Taiwan, 72
Lin Zexu, 51
Literacy
in Hong Kong, 51
increasing, 21
in PRC, 4
Long March, 7
Luxury cars, in Hong Kong, 57

M
MacArthur, Douglas, 43
Macau, 53
Magazines, 37
Malay, 52
Male:female ratio, in population, 14
Manchu Dynasty, 6
Manchuria, invasion of, 7–8
Manchus, 6, 73
Mandarin language, 16
Manufacturing
command economy and, 23
in Hong Kong, 59
Mao Zedong, 7, 8, 10, 11, 19–21, 74
Martial law, in Taiwan, 84–85
Marxism, 7, 11, 19–21. *See also* Chinese Communist Party (CCP)
Matsu, 43, 44, 76
Ma Ying-jeou, 44, 77
McCarthyism, 74
McDonald's, 56
Media
economic reforms and, 38
exposure of corruption by, 32
globalization and, 36
in Hong Kong, 66–67
in Taiwan, 87–88
Medicine, in Taiwan, 79
Military spending
in Hong Kong, 51
in PRC, 5
Minorities, 16–18
Missile defense, 42
Missile tests, Taiwan and, 44, 85
Monasteries, in Tibet, 17
Mongol Dynasty, 6
Mongolia, Inner, 17
Motor vehicles
in Hong Kong, 57
increasing number of, 27
in use, in PRC, 5
Muslim minorities, 17–18

N
National Cable Company, 38
Nationalism, in PRC, 40
Nationalist Army, Red Army *vs.*, 8
Nationalist Party. *See* Kuomintang (KMT)
National minorities, 16
National People's Congress (NPC), Hong Kong and, 54, 66
Natural resources, 82
Neolithic Period, 5
New Culture Movement, 7
New Party, 86
Newspapers, 4, 37
New Territories, of Hong Kong, 52
NGOs. *See* Nongovernmental organizations (NGOs)
Nixon, Richard, 74, 88
Nongovernmental organizations (NGOs), 33
North Korea, 43
NPC. *See* National People's Congress (NPC)
Nuclear power, 78
Nuclear Test Ban Treaty, 41

O
Offshore Islands, 11, 76
Olympics, 38, 40, 43
"One-child policy," 13–15
"Open door" policy, 22
Openness, 41
Opium, in history of Hong Kong, 51
Opium War, 6, 7, 51
Organized crime, in Hong Kong, 61
Orphanages, 14

P
"Pan-blue alliance," 87
"Pan-green alliance," 87
Participation, political, in PRC, 33
Patten, Christopher, 60, 61
"Peace offensive," 89–91
Peasants
agricultural programs and, 23–24
collectivization and, 24
communes and, 20
inequality of, 27
People First Party, 86
"People's congresses," 30
People's Liberation Army (PLA)
in Hong Kong, 66
in PRC, 42–43
People's Republic of China (PRC)
communication in, 4–5
crime in, 25–26
economic reform in, 25
economic system in, 23–28
education in, 21–22
establishment of, 9–10
export commodities of, 5
future of, 46
GDP of, 5
geography of, 4

import commodities of, 5
infant mortality in, 4
inflation in, 5
instability in, 25–26
international recognition of, Taiwan and, 75–76
international relations of, 39–41
Internet use in, 4
investment in, from Taiwan, 81
legal system in, 28
life expectancy in, 4
literacy rate in, 4
military spending in, 5
motor vehicles in use in, 5
in negotiations over Hong Kong, 53
"one-child policy in, 13–15
physicians in, 4
political system in, 30–32
population of, 4, 13
protests in, 12–13
reform in, 12–13
Soviet Union and, 10–11, 41–43
statistics on, 4–5
telephones in, 4
Tibet and, 16–17
transportation in, 5
unemployment in, 5
United States, relationship with, 43–45
women in, 15–16
in WTO, 25
People's war, 42–43
"Permanent normal trading relations" (PNTR), with US, 44–45
Physicians
in Hong Kong, 51
in PRC, 4
Pinyin, 19, 21
PLA. *See* People's Liberation Army (PLA)
Political education, 22
Political participation, in PRC, 33
Political parties
in Hong Kong, 64–65
in Taiwan, 86–87
Political system
in PRC, 30–32
in Taiwan, 83–88
Politics
in Hong Kong, 61–67
law and, 28–29
Pollution, 28, 81
Population
aging, 14–15
control, 13–14
gender ratio in, 14
growth, 14
of Hong Kong, 50
immigrant, in Hong Kong, 54–55
of PRC, 4, 13
of Taiwan, 72
Poverty
in Hong Kong, 56, 57
in Tibet, 17
PRC. *See* People's Republic of China (PRC)
Presbyterian Church, 79
Press, in Hong Kong, 66–67
Prices, determination of, 26
Privatization, in Taiwan, 81
Procuracy, 29–30
Profit, 26
Property disputes, 25
Protests, 12–13
Public security, in Hong Kong, 66

Q

Qin Dynasty, 5
Qing Dynasty, 6
Qin Shi Huang Di, emperor, 5
Quality control, 23
Quemoy, 43, 44, 76

R

Reagan, Ronald, 88
Red Army, Nationalist Army *vs.*, 8
Red Guards, 9
"Red Sun, The" (song), 8
Refugees, in Hong Kong, 55
Religion. *See also* Buddhism; Christianity; Confucius and Confucianism
in Hong Kong, 66–67
in PRC, 18–19
in Taiwan, 78–79
Repression, 13
Republic of China (ROC). *See also* Taiwan
democracy in history of, 32
founding of, 6–7
Reunification, Taiwan, 91–92
Revolution (1911), 6
River Elegy (documentary), 36
ROC. *See* Republic of China (ROC)
Roman Catholicism, 19, 79
Romanization, of written language, 19, 21

S

SAR. *See* Special Administrative Region (SAR)
SARS. *See* Severe acute respiratory syndrome (SARS)
Security, in Hong Kong, 66
SEF. *See* Strait Exchange Foundation (SEF)
Self-sufficiency, 24
Severe acute respiratory syndrome (SARS), 58, 60
SEZs. *See* Special Economic Zones (SEZs)
Shang Dynasty, 5, 21
Shenzhen, 26, 59, 60
Sino-Japanese War, 73
Small business, in Taiwan, 81
Social clubs, in Hong Kong, 56
Socialist democracy, 32–33, 46
Social security, in Taiwan, 80
South Korea, 43
Soviet Union
breakup of, 42
Chinese Communist Party and, 7, 9, 10
as model, 10
relationship of PRC with, 40–42, 41–43
souring relations with, 11
Special Administrative Region (SAR), 53
Special Committee on Colonialism, 53
Special Economic Zones (SEZs), 59–60
Hong Kong and, 59
overview of, 26
Spelling, pinyin and, 19
Spratlys, 42
State Council, 30–32
Stock markets, 37, 82
Strait Exchange Foundation (SEF), 90
Student movement, 12, 13
Students, democratization and, 35–36
Study abroad
by PRC students, 22–23
by Taiwanese students, 79–80
Sun Yat-sen, 6, 7, 83
Symbolism, in Chinese culture, 12

T

Taipei Economic and Cultural Office, 88
Taiwan
agriculture in, 82
air travel to, 86
arms sales to, 88–89
Chiang Kai-shek and, 73, 74, 75
Christianity in, 79
civil rights iun, 85
communication in, 73
Confucianism in, 79
constitution of, 83–84
culture of, 76–77
Democratic Progressive Party in, 77–78, 85–87
diplomatic relations with, 75
economy of, 80–83
education in, 79–80
elections in, 76
energy in, 78
environment and, 77–78
evolving status of, 74–76
export commodities in, 73
free trade zones and, 82
garbage production in, 77, 81
GDP of, 4, 80
geography of, 72
health care in, 80
history of, 73–74
import commodities in, 73
income in, 80
inflation in, 73
interest groups in, 87
international recognition of, 75–76
international recognition of PRC and, 75–76
Internet use in, 73
investment in PRC from, 81
Japan and, 73
Korean War and, 43
Kuomintang and, 73–74
Lee Teng-hui and, 75–76
life expectancy in, 72
martial law in, 84–85
Matsu and, 76
media in, 87–88
missile tests and, 44, 85
as model for economic development, 82–83

Taiwan—*(Cont)*
natural resources in, 82
nuclear power and, 78
Offshore Islands and, 76
"peace initiatives" for, 76
"peace offensive" and, 89–91
political parties in, 86–87
political reform in, 85–86
political system of, 83–88
pollution in, 81
population of, 72
privatization in, 81
Quemoy and, 76
religion in, 78–79
representation of, 75
reunification prospects, 91–92
Sino-Japanese War and, 73
small business in, 81
social security in, 80
society of, 76–77
transportation in, 73, 86
unemployment in, 73
United States and, 74, 87, 88–89
US-PRC relationship and, 44
women in, 80
World War II and, 73
Taiwan dollar, exchange rate of, 73
Taiwan Independence Party, 86
Taiwan Relations Act, 88
Taiwan Security Enhancement Act, 88
Taiwan Stock Exchange (TSEC), 82
Tajikstan, 17
Taoism, 18
Technology, China as historic leader in, 5
Telephones
in PRC, 4
Television, culture and, 40
Television ownership, 36–37
Terrorism
Hong Kong and, 63
minorities and, 16
Muslim minority and, 17–18
US relations and, 45
Thatcher, Margaret, 53
Theater missile defense (TMD), 42
Tiananmen Square, 12, 18, 19, 36
Tibet, 16–17
TMD. *See* Theater missile defense (TMD)
Tomb-Sweeping Festival, 78
Tourism
in Hong Kong, 65
in Tibet, 17
Transportation
in Hong Kong, 51
in PRC, 5
in Taiwan, 73, 86
Treaty of Nanjing, 6, 7, 52
Trials, in Chinese legal system, 29
Truman, Harry, 74
Tsang, Donald, 58, 62
TSEC. *See* Taiwan Stock Exchange (TSEC)
Tung Chee-hwa, 57, 58, 61

U
Unemployment
in Hong Kong, 51
in PRC, 5
in Taiwan, 73
United Kingdom
emigration to, from Hong Kong, 60
in history of Hong Kong, 51–52
in Joint Liaison Group, 53–54
Kowloon Peninsula and, 52
negotiations with, over Hong Kong, 53–54
in Opium War, 6, 7, 51
United States
arms sales to Taiwan, 88–89
Belgrade embassy bombing by, 20, 40
diplomatic relations with, 44
human rights and PRC relationship with, 44–45
permanent normal trading relations with, 44–45
relationship with PRC, 43–45
Taiwan and, 44, 74, 87, 88–89
Uyghurs, 17, 18

V
Vietnam War, 41–42, 43, 55
VOA. *See* Voice of America (VOA)
Voice of America (VOA), 36

W
Wade-Gilles system, 19
Wages, "one-child policy" and, 13
Water pollution, 81
Wealth, in Hong Kong, 56–57
Wen Jiabao, Tung and, 62
WHO. *See* World Health Organization (WHO)
Women
Confucianism and, 80
in PRC, 15–16
in Taiwan, 80
World Expo, 46
World Health Organization (WHO), 80
World Trade Organization (WTO)
Hong Kong in, 54
PRC membership in, 25, 40, 41, 43
World War II
Hong Kong and, 52
Japanese invasion of Manchuria in, 7–8
Taiwan and, 73
Written language, 19, 21

X
Xenophobia, 40
Xinjiang, 16
Xinjiang Autonomous Region, 17

Y
Yao Wenyuan, 10
Yuan, exchange rate of, 5
Yuan Dynasty, 6
Yuan Shikai, 7, 83

Z
Zhang Chunqiao, 10
Zhao Ziyang, 13